Expert Systems and Geographical Information Systems for Impact Assessment

Expert Systems and Geographical Information Systems for Impact Assessment

Agustin Rodriguez-Bachiller
with John Glasson
Oxford Brookes University, UK

Taylor & Francis
Taylor & Francis Group
LONDON AND NEW YORK

First published 2004
by Taylor & Francis
11 New Fetter Lane, London EC4P 4EE

Simultaneously published in the USA and Canada
by Taylor & Francis Inc,
29 West 35th Street, New York, NY 10001

Taylor & Francis is an imprint of the Taylor & Francis Group

Typeset in Sabon by
Integra Software Services Pvt. Ltd, Pondicherry, India
Printed and bound in Great Britain by
MPG Books Ltd, Bodmin, Cornwall

British Library Cataloguing in Publication Data
A catalogue record for this book is available
from the British Library

Library of Congress Cataloging in Publication Data
Rodriguez-Bachiller, Agustin, 1942–
Expert systems and geographical information systems for impact assessment/Agustin Rodriguez-Bachiller with John Glasson.
p. cm.
Includes bibliographical references and index.
1. Geographical information systems. 2. Expert systems (Computer science) I. Glasson, John, 1946– II. Title;
G70. 212 .R64 2003–03–04
910′ .285′633—dc21

2003002535

ISBN 0–415–30725–2 (pbk)
ISBN 0–415–30724–4 (hbk)

Contents

Acknowledgements

Grateful acknowledgement is owed to various groups of persons who helped with some of the preparatory work leading to this book. These include, for Part I, the experts in the Regional Research Laboratories who kindly agreed to be interviewed (in person or by telephone):

- Peter Brown (Liverpool University)
- Mike Coombes (University of Newcastle)
- Derek Diamond (London School of Economics)
- Peter Fisher (Leicester University)
- Richard Healey (Edinburgh University)
- Graeme Herbert (University College, London)
- Stan Openshaw (University of Leeds)
- David Walker (Loughborough University)
- Chris Webster (University of Wales in Cardiff)
- Craig Whitehead (London School of Economics)

For Part II, many thanks are also given to those experts consulted on various aspects of Impact Assessment, most working at the time in Environment Resources Management Ltd (ERM) at its branches in Oxford or London (although some of these professionals have now moved to other jobs or locations, they are listed here by their position at the time [1994]), and one from the Impact Assessment Unit (IAU) at Oxford Brookes University:

- Dave Ackroyd, ERM (Oxford)
- Roger Barrowcliffe, ERM (Oxford)
- Nicola Beaumont, ERM (Oxford)
- Sue Clarke, ERM (Oxford)
- Stuart Dryden, ERM (Oxford)
- Gev Edulgee, ERM (Oxford, Deputy Manager)
- Chris Ferrari, ERM (London)
- Nick Giesler, ERM (London)
- Karen Raymond, ERM (Oxford, Manager)

- John Simonson, ERM Enviroclean (Oxford)
- Joe Weston, IAU

Also for Part II, this acknowledgement includes a group of graduates from the Master Course in Environmental Assessment and Management at Oxford Brookes University who helped with the amalgamation of material for the discussion of different types of Impact Assessment:

- Mathew Anderson
- Andrew Bloore
- Duma Langton
- Owain Prosser
- Julia Reynolds
- Joanna C. Thompson

Finally, many thanks to Rob Woodward, from the School of Planning at Oxford Brookes University, for the prompt and competent preparation of the figures.

Part I

GIS and expert systems for impact assessment

This book started as a research project[1] to investigate the potential of integrating Expert Systems (ES) and Geographical Information Systems (GIS) to help with the process of Impact Assessment (IA). This emergent idea was based on the perception of the potential of these two technologies to complement each other and help with impact assessment, a task that is growing rapidly in magnitude and scope all over the world. Part I discusses these three fields, their methodology and their combined use as recorded in the literature. In Part II we discuss the potential – and limitations – of these two computer technologies for *specific* parts of IA, as if replicating in the discussion what could be the first stage in the design of computer systems to automatise these tasks.

1 Funded by PCFC from 1991 and directed by Agustin Rodriguez-Bachiller and John Glasson.

1 The potential of expert systems and GIS for impact assessment

1.1 INTRODUCTION

Impact assessment is increasingly becoming – mostly by statutory obligation but also for reasons of good practice – part and parcel of more and more development proposals in the United Kingdom and in Europe. For instance, while the Department of the Environment (DoE) in Britain was expecting about 50 Environmental Statements each year when this new practice was introduced in 1988, the annual number soon exceeded 300. As the practice of IA developed, it became more standardised and *good practice* started to be defined. In the early years – late 1980s – a proportion of Environmental Statements in the UK still showed relatively low level of sophistication and technical know-how, but the quality soon started to improve (Lee and Colley, 1992; DoE, 1996; Glasson *et al.*, 1997), largely due to the establishment and diffusion of expertise, even though the overall quality is still far from what would be desirable. And it is here that the idea of expert systems becomes suggestive.

The idea of expert systems – computer programs crystallising the way experts solve certain problems – has shown considerable appeal in many quarters. Even though their application in other areas of spatial decision-making – like town planning – has been rather limited (Rodriguez-Bachiller, 1991) and never fully matured after an initial burst of enthusiasm, a similar appeal seems to be spreading into IA and related areas as it did in town planning ten years earlier (see Rodriguez-Bachiller, 2000b).

Geographical information systems are visually dazzling systems becoming increasingly widespread in local and central government agencies as well as in private companies, but it is sometimes not very clear in many such organisations how to make pay off the huge investment which GIS represent. Early surveys indicate that mapping – the production of maps – tends to be initially the most important task for which these expensive systems are used (Rodriguez-Bachiller and Smith, 1995). Only as confidence grows are more ambitious jobs envisaged for these systems, which have significant potential for impact assessment (see also Rodriguez-Bachiller, 2000a; Rodriguez-Bachiller and Wood, 2001).

The proposition behind the work presented here is that these three areas of IA, ES and GIS are potentially complementary and that there would be mutual benefits if they could be brought together. This first chapter outlines their potential role, prior to a fuller discussion in subsequent chapters.

1.2 EXPERT SYSTEMS: WHAT ABOUT SPACE?

Although a more extensive discussion of expert systems will be presented in the next chapter, a brief introduction is appropriate here. Expert Systems are computer programs that try to encapsulate the way experts solve particular problems. Such systems are designed by crystallising the expert's problem-solving logic in a "knowledge base" that a non-expert user can then apply to similar problems with data related to those problems and their context. An expert system can be seen as a synthesis of problem-specific expert knowledge and case-specific data.

Expert systems first came onto the scene in America in the 1960s and 1970s, as a way forward for the field of Artificial Intelligence after its relative disappointment with "general" problem-solving approaches. This new approach also coincided with trends to develop new, more interactive and personalised approaches to computer use in their full potential. Jackson (1990) argues that Artificial Intelligence had gone, until the mid-1970s, through a "romantic" period characterised by the emphasis on "understanding" the various intelligent functions performed automatically by humans (vision, language, problem-solving). It was partly as a result of the disappointments of that approach that what Jackson calls the "modern" period started, and with it the development of expert systems, less interested in understanding than in building *systems that would get the same results as experts*. In this context, the power of a problem solver was thought to lie in relevant subject-specific knowledge. It is this *shift from understanding to knowledge* that characterises this movement and, with it, the shift to relatively narrow, domain-specific problem-solving strategies (Hayes-Roth *et al.*, 1983a).

Although in the early days many of these systems were often suggested as capable of simulating human intelligence, this proved to be more difficult than at first thought. Today, a safer assumption underpinning expert systems work is that, while to "crack" the really difficult problems requires the best of human intelligence beyond the capabilities of the computer, *after* the solution to a problem has been found and articulated into a body of expertise, expert systems can be used to transfer such expertise to non-experts. This view translates into the more modest – but all the more achievable – expectation that ES can help solve *those problems that are routine for the expert but too difficult for the non-expert.*

Following from this lowering of expectations, when textbooks and manuals on expert systems started to appear – like the early one by Waterman

(1986) – the range of problems to which ES could be realistically expected to be applied with some degree of success had been considerably narrowed down, and it is instructive in this respect to remind ourselves of the main "rules of thumb" suggested by Waterman to identify the kind of problem and circumstances for which the use of expert systems is considered to be practicable:

- The problem should be not too large or complicated, it should be the kind that would take an expert only a few hours to solve (hours, rather than days).
- There should be established procedures to solve the problem; there should be some degree of consensus among experts on how the problem should be solved.
- The sources of the expertise to solve the problem (in the form of experts and/or written documentation) should exist and be accessible.
- The solution to the problem should not be based on so-called "common sense", considered to be too broad and diffuse to be encoded in all its ramifications.

In addition to this, a good reason for using ES is found in the need to replicate expert problem-solving expertise in situations where it is *scarce* for a variety of reasons: because experts are themselves becoming scarce (through retirement or because they are needed simultaneously in many locations), because their expertise is needed in hostile environments (Waterman, 1986), or simply because experts find themselves overloaded with too much work and unable to dedicate sufficient time to each problem. In this context, expert systems can be used to liberate experts from work which is relatively routine (for them), but which prevents them from dedicating sufficient time to more difficult problems. The idea is that overworked experts can off-load their expertise to non-experts via these systems and free up time to concentrate their efforts on the most difficult problems. This aspect of expert systems as instruments of *technology transfer* (from top to bottom or from one organisation to another) adds another more political dimension to their appeal.

Although classic reference books on the subject like Hayes-Roth *et al.* (1983b) list many different types of expert systems according to the different areas of their application, practically all expert systems can be classified in one of four categories:

- *diagnostic/advice* systems to give advice or help with interpretation;
- *control* systems in real time, helping operate mechanisms or instruments (like traffic lights);
- *planning/design* systems that suggest how to do something (a "plan");
- *teaching/training* systems.

Most of the now classic pioneering prototypes that started the interest in expert systems were developed in the 1970s – with one exception from the 1960s – in American universities, and it is instructive to note that most of them were in the first category (diagnostic/advice), with a substantial proportion of them in medical fields. This dominance of diagnostic systems has continued since.

With the advent of more and more powerful and individualised computers (both workstations and PCs) the growth in expert systems in the 1980s was considerable, mostly in technological fields, while areas more concerned with social and spatial issues seemed to lag behind in their enthusiasm for these new systems. In town planning, the development of expert systems seems to have followed a typical sequence of *stages* (Rodriguez-Bachiller, 1991) which is useful to consider here, given that there are signs that developments in fields like IA seem to follow similar patterns:

- First, *eye-opener* articles appear in subject-specific journals calling people's attention to the potential of expert systems for that field.
- In a second *exploratory* stage, differences seem to appear between the nature of the exploratory work in America and Europe: while European research turns to *soul-searching* (discussing feasibility problems with the new technology and identifying unresolved problems), American work seems to plunge directly into application work, with the production of *prototypes*, often associated with doctoral work at universities. Sooner or later, European research also follows into this level of application.
- In the next stage, *full systems* are developed, even if these are few and far between.
- In what can be seen as a last stage in this process, expert systems start being seen as "aids" in the context of more general systems that take advantage of their capacity to incorporate logical reasoning to the solution of a problem, and they tend to appear *embedded* in other technologies, sometimes as intelligent interfaces with the user, sometimes as interfaces between different "modules" in larger decision-support systems.

What is interesting here is the parallel with IA, as ES started attracting fresh interest in the early 1990s following a similar process, and we can now see the first stages of the same cycle sketched above beginning to develop. Articles highlighting the potential of ES for IA started to appear early in the Environmental literature (Schibuola and Byer, 1991; Geraghty, 1992). The first prototypes combining ES and EIA – leaving GIS aside for the moment – also started to emerge (Edwards-Jones and Gough, 1994; Radwan and Bishr, 1994), and we shall see in Chapter 5 how this field has flourished (see also Rodriguez-Bachiller, 2000b).

This fresh interest in ES may be interpreted in rather mechanistic style as a new field like IA following in the steps of older fields like town planning – similarly concerned with the quality of the environment – developing similar expectations from similar technologies, and in that respect maybe also doomed to be a non-starter in the same way. Another possible interpretation is that IA is (or has been until now) a much more *technical* activity than town planning ever was (where the technocratic approach advocated in the 1960s never really caught on), concerned with a much narrower range of problems – specific impacts derived from specific projects – more likely to be the object of technical analysis and forecasting than of political policy-making and evaluation.

One of the limitations that ES showed in trying to deal with town planning problems lay in the difficulty that traditional expert system tools have had from the start in dealing directly (i.e. automatically) with *spatial information.* Some rare early experiments with this problem apply to a very local scale, dealing with building shapes (Makhchouni, 1987) or are confined to the micro-scale of building technology (Sharpe *et al.*, 1988), and all involved considerable programming "from scratch". It is in this respect that other off-the-shelf technologies like GIS might prove productively complementary to expert systems.

1.3 GEOGRAPHICAL INFORMATION SYSTEMS: MORE THAN DISPLAY TOOLS?

As opposed to expert systems – discussed in detail in Chapter 2 – we are not going to discuss GIS in detail beyond this introductory chapter, and interested readers are directed to the very good and accessible literature available. In the GIS field we have the good fortune of having two benchmark publications (Maguire *et al.*, 1991; Longley *et al.*, 1999)[2] which summarise most of the research and development issues up to the 1990s and contain a collection of expert accounts which can be used as perfectly adequate secondary sources when discussing research or history issues in this field. Also, Longley *et al.* (2001) contains an excellent overview of the whole field at a more accessible level.

Computerised databases and "relational" databases (several databases related by common fields) are becoming quite familiar. GIS take the idea of relational databases one step further by making it possible to include spatial positioning as one of the relations in the database, and it is this aspect of GIS that best describes them. Despite the considerable variety of definitions suggested in the literature (Maguire, 1991), GIS can be most simply seen as *spatially referenced databases.* But what has made these systems so popular and appealing is the fact that the spatial referencing of

2 Although Longley *et al.* (1999) is presented as a "second edition" of Maguire *et al.* (1991), it is an entirely new publication, with different authors and chapters; so the two should really be taken *together* as a quite complete and excellent source on GIS.

information can be organised into maps, and automated mapping technology can be used to perform the normal operations of database management (subset extraction, intersection, appending, etc.) *in map form*. It is the manipulation and display of maps with relative speed and ease that is the trademark of GIS, and it is probably fair to say that it is this graphic efficiency that has contributed decisively to their general success. A crucial issue for the development of this efficiency has been finding efficient ways of holding spatial data in computerised form or, in other words, how maps are represented in a computer, and two basic models of map representation have been developed in the history of GIS:

(a) The *raster* model is cell-based where the mapped area is divided up into cells (equal or unequal in size) covering its whole extension, and where the attributes of the different map features (areas, lines, points) are simply stored as values for each and everyone of those cells. This model can be quite economical in storage space and is simple, requiring relatively unsophisticated software, and for these reasons the first few generations of mapping systems all tended to use it, and the importance of raster systems research cannot be overestimated in the history of GIS (Foresman, 1998). Raster-based systems tend to be cheaper, but this approach has the drawback that the *accuracy* of its maps will be determined by the size of the cells used (the smaller the cells the more accurate the map will be). To obtain a faithful representation of maps the number of cells may have to grow considerably, reducing partially the initial advantages of economy and size.

(b) The *vector* model, on the other hand, separates maps from their attributes (the information related to them) into different filing systems. The features on a map (points and lines) are identified reasonably accurately by their co-ordinates, and their relationships (for example, the fact that lines form the boundaries of areas) are defined by their "topology", while the attributes of all these features (points, lines, areas) are stored in separate but related tables. From a technical point of view, GIS are particular types of relational databases that combine attribute files and map files so that (i) attribute databases can be used to identify maps of areas with certain characteristics and (ii) maps can be used to find database information related to certain locations. The accuracy of these systems does not depend any more – as it does in raster-based systems – on the resolution used (the size of the smallest unit) but on the accuracy of the source from which the computer maps were first derived or digitised. Despite vector systems being more demanding on the computer technology (and therefore more expensive) their much improved accuracy is leading to their growing domination of the GIS market. However, raster-based systems still retain advantages for certain types of application – for instance when dealing with satellite data – and it is increasingly common to find vector systems which can also transform their own maps into cell-based representations – and *vice versa* – when needed.

The development of GIS has been much more gradual than that of expert systems (full prototypes of which were developed right from the start), probably due to the fact that, for GIS to be practical, computer technology had to take a quantum leap forward – from raster to vector – to handle maps and the large databases that go with them. This leap took decades of arduous work to perfect the development in all the directions in which it was needed:

- *Hardware* to handle maps had to be developed, both to encode them at the input stage, and to display and print them at the output stage. On the *input* side, the digitiser – which proved to be one of the cornerstones of GIS development – was invented in the UK by Ray Boyle and David Bickmore in the late 1950s, and Ivan Southerland invented the sketchpad at MIT in the early 1960s. *Output* devices suitable for mapping had started to be developed by the US military in the 1950s, and by some public and private companies (like the US oil industry, also some gas and public-service companies) in the 1960s, while universities – who couldn't afford the expensive equipment – were concentrating on software development for the line printer until the 1970s.
- It is argued that the development of map-handling *software* can be traced back to when Howard Fisher moved from Northwestern University to chair the newly created Harvard Computer Graphics laboratory in 1964, bringing with him his recently created thematic mapping package for the line printer (SYMAP), which he would develop fully at Harvard. While this is true of cell-based mapping – most systems in the 1950s and 1960s belonged to this type – interactive screen display of map data was being developed at the same time for the US military. Computer Aided Drafting was being developed at MIT, and Jack Dangermond – a former researcher at the Harvard Graphics laboratory – produced in the early 1970s the first effective vector polygon overlay system, which would later become Arc-Info.
- Also crucial was the development of software capable of handling large spatially referenced *databases* and their relationships with the mapping side of these systems. The pioneering development of some such large systems was in itself a crucial step in this process. These included the Canada Geographic Information System started in 1966 under the initiative of Roger Tomlinson, the software developments to handle such spatial data, like the MIADS system developed by the US Forest Service at Berkeley from the early 1960s to store and retrieve attributes of a given map cell and perform simple overlay functions with them, and the new methods for encoding census data for the production of maps developed at the US Census Bureau from 1967.

Good accounts on the history of GIS can be found in Antenucci *et al.* (1991) and also Coppock and Rhind (1991), and the latter authors argue that four distinct *stages* can be identified in the history of GIS, at least in the US and the UK:

1 The first stage – from the 1950s to the mid-1970s – is characterised by the pioneering work briefly mentioned above, research and developmental work by *individuals* – just a few names like those mentioned above – working on relatively isolated developments, breaking new ground in the different directions required by the new technology.

2 The second stage – from 1973 to the early 1980s – sees the development of formal experiments and government-funded research, characterised by *agencies and organisations* taking over GIS development. The New York Department of Natural Resources developed, from 1973, the first State-wide inventory system of land uses, the first of many States in the US to develop systems concerned with their natural resources and with environmental issues. The US Geological Survey developed, from 1973, the Geographical Information Retrieval and Analysis System (GIRAS) to handle information on land use and land cover from maps derived from aerial photography. Jack Dangermond had started ESRI (Environmental Systems Research Institute) in 1969 as a non-profit organisation and, with the development in the 1970s of what would become Arc-Info, ESRI turned into a commercial enterprise with increasing environmental interests. At the same time, Jim Meadlock (who had developed for NASA the first stand-alone graphics system) had the idea of producing turn-key mapping systems for local government – which he implemented for the first time in Nashville in 1973 – and he would later go on to found INTERGRAPH. This is a period that Coppock and Rhind characterise as one of "lateral diffusion" (still restricted mostly to within the US) rather than innovation, with the characteristic that it all tended to happen (whether in the private or the public sector) outside the political process, with no government policy guidance.

3 The third stage – from 1982 – can be characterised as the *commercial* phase, still with us, and characterised by the *supply-led* diffusion of the technology *outside* the US. GIS is becoming a worldwide growth industry, nearing a turnover of $2 billion per year (Antenucci, 1992), with the appearance on the market of hundreds of commercial systems, more and more of them being applicable on smaller machines at lower and lower prices. Even if the market leaders are still the large organisations (like ESRI and INTERGRAPH) which grew out of the previous stage, smaller and more flexible systems like SPANS (small, for PC computers) or Map-Info (with a modular structure that makes its purchase much easier for smaller organisations) – to mention but a few – are increasing their market presence.

4 Coppock and Rhind see a new stage developing in which commercial interests are gradually being replaced by *user dominance*, although an alternative interpretation is simply that – rather than commercial interests being displaced – increasing competition among GIS manufacturers is letting the needs of the users dictate more and more what the industry produces, in what could be seen as a transition to a more *demand-led* industry. Also, this new stage can be seen as characterised by the transition from the use of individual data on isolated machines, to dealing with distributed databases accessed through computer networks, with increased availability of data (and software) through networks in all kinds of organisations and at all levels, including the World Wide Web.

Although British research was at the very heart of GIS developments, it is probably fair to say that, after the first "pioneering" stage mentioned, the second stage in GIS development and diffusion of use has been largely dominated by developments in the US. Subsequent growth of GIS outside the United States can be seen as a process of *diffusion* of the technology from America to other countries – the UK included – despite the continuation of GIS work at academic British institutions like the Royal College of Art (where David Bickmore had founded the Experimental Cartography Unit in 1967) and later at Reading University and its Unit for Thematic Information Systems since 1975, as well as those resulting from the Regional Research Laboratories in the 1980s (see Chapter 5).

Apart from isolated developments in the early 1980s – like the SOLAPS system developed in-house in South Oxfordshire (Leary, 1989) – Coppock and Rhind underline the importance in the UK of three *official surveys* that mark the evolution of GIS: (i) the Ordnance Survey Review Committee (1978) looking at the prospect of changing to digital mapping; (ii) the Report of the House of Lords' Select Committee on Science and Technology (1984) investigating digital mapping and remote sensing, which recommended a new enquiry; (iii) the Committee of Enquiry into the Handling of Geographical Data, set up in 1987 and chaired by Chorley, which launched the ESRC-funded Regional Research Laboratories (RRL) programme (Masser, 1990a) from February 1987 to October 1988 and then to December 1991, with funding of over 2 million pounds. This programme tried to diffuse the new GIS technology into the public arena by establishing 8 laboratories spread over 12 universities: Belfast, Cardiff, Edinburgh, Lancaster, Leicester, Liverpool, London-Birkbeck, London School of Economics, Loughborough, Manchester, Newcastle and Ulster. Research and application projects were to be undertaken *in-house*, with the double objective of spreading the technology to the academic and private sectors and, in so doing, making the laboratories self-financing beyond the initial period supported by the ESRC grant by increasingly attracting private investment. While the first objective was fully achieved – as a result, a new generation of GIS experts was created in the academic sector, and numerous GIS courses started to appear, diffusing their expertise to others – the private sector never became sufficiently involved in these developments to support them after the period of the ESRC experiment.

In the US there was a parallel experience of the National Centre for Geographic Information and Analysis funded with a comparable budget by the National Science Foundation. Concentrated in only three centres for the whole country (Santa Barbara, Buffalo and Maine) and financing research projects done both inside and outside those centres, it had mainly theoretical aims (Openshaw *et al.*, 1987; Openshaw, 1990).

1.4 GIS PROBLEMS AND POTENTIAL

It is also productive to look at the development of GIS in terms of typical problems and *bottlenecks* that have marked the different stages of its progress, problems which tend to move from one country to another as the technology becomes diffused:

1. First there is (was) what could be called the *research bottleneck*, mainly manifest in the UK and mostly in the US, where much of the fundamental research was carried out during the 1950s and 1960s, taking decades to solve specific problems of mapping and database work, as mentioned above.
2. Next, the *expertise bottleneck* – the lack of sufficient numbers of competent professionals to use and apply GIS – became apparent especially outside the US, when the new technology was being diffused to other countries before they had developed educational and training programs to handle it. In the UK this was evident in the 1980s and it is this bottleneck that the ESRC's RRL Initiative sought to eliminate. It is now appearing in other countries (including developing countries) as the wave of GIS diffusion spreads more widely.
3. Finally, the *data bottleneck*: beyond the classic problems or data error well identified in the literature (Chrisman, 1991; Fisher, 1991), this bottleneck refers more to problems of data quality discussed early by De Jong (1989) and – most importantly – data availability and cost, especially when GIS is "exported" to developing countries (Masser, 1990b; Nutter *et al.*, 1996; Warner *et al.*, 1997).

These general bottlenecks can be seen as the main general obstacles to the adoption of GIS in countries other than the US, but they work in combination with other factors specific to each particular case. *Organisational resistance* has been widely suggested (Campbell and Masser, 1994) as responsible for the relatively lower take-up of GIS in Europe than in America, and the magnitude of the *financial cost* (and risk) involved in the implementation of such systems may be another factor growing in importance even after the initial resistance has been overcome (Rodriguez-Bachiller and Smith, 1995).

In terms of what GIS can *do*, it can be said that these systems are still to some extent prisoners of their cartographic background, so that part of their functionality (Maguire and Dangermond, 1991) was initially directed towards solving *cartographic* problems related to the language of visualisation, three-dimensional map displays and, later, the introduction of multi-media and "hyper-media" with sound and images. Also, because the development of GIS has been largely *supply-led*, spearheaded by private software companies who have until recently concentrated on improving the "graphical" side of GIS – their mapping accuracy, speed and capacity – their analytical side has been somewhat neglected, at least initially. As Openshaw (1991) pointed out ironically some years ago, sophisticated GIS packages could make over 1,000 different operations, and yet, not one of them related to true spatial analysis, but to "data description". This impression feels now somewhat exaggerated and dated as more and more sophisticated analytical features appear in every new version of GIS packages, but was quite appropriate at the time and underlined the problems encountered initially by users of these supposedly revolutionary new technologies. Today, the list of operations that most GIS can normally do (see Rodriguez-Bachiller and Wood, 2001, for its connection to Impact Assessment) is quite standard.

General operations

- Storage of large amounts of spatially referenced information concerning an area, in a relational database which is easy to update and use.
- Rapid and easy display of visually appealing maps of such information, be it in its original form or after applying to it database operations (queries, etc.) or map transformations.

Analysis in two dimensions

- Map "overlay", superimposing maps to produce composite maps, the most frequent use of GIS.
- "Clipping" one map with the polygons of another to include (or exclude) parts of them, for instance to identify how much of a proposed development overlaps with an environmentally sensitive area.
- Producing "partial" maps containing only those features from another map that satisfy certain criteria.
- Combining several maps (weighted differently) into more sophisticated composite maps, using so-called "map algebra", used for instance to do multi-criteria evaluation of possible locations for a particular activity.
- Calculating the size (length, area) of the individual features of a map.

- Calculating descriptive statistics for all the features of a map (frequency distributions, averages, maxima and minima, etc.).
- Doing multivariate analysis like correlation and regression of the values of different attributes in a map.
- Calculating minimum distances between features, using straight-line distances and distances along "networks".
- Using minimum distances to identify the features on one map nearest to particular features on another map.
- Using distances to construct "buffer" zones around features (typically used to "clip" other maps to include/exclude certain areas).

Analysis with a third dimension

- Interpolating unknown attribute values (a "third dimension" on a map) between the known values, using "surfaces", Digital Elevation Models (DEMs) or Triangulated Irregular Networks (TINs).
- Drawing contour lines using the interpolated values of attributes (the "third dimension").
- Calculating topographic characteristics of the 3-D terrain, like slope, "aspect", concavity and convexity.
- Calculating volumes in 3-D models (DEMs or TINs) for instance the volumes between certain altitudes (like water levels in a reservoir).
- Identifying "areas of visibility" of certain features of one map from the features of another, for instance to define the area from which the tallest building in a proposed project will be visible.
- So-called "modelling", identifying physical geographic objects from maps, like the existence of valleys, or water streams and their basins.

Many of these capabilities have been added gradually – some as "add-on" extensions, some as integral components of new versions of systems – in response to academic criticism and consumer demand. However, when these systems were being first "diffused" outside the US, to go beyond map operations to apply them to real problems tended to require considerable amount of manipulation or programming by the user, as the pioneering experience in the UK of the Regional Research Laboratories[3] suggested over ten years ago (Flowerdew, 1989; Green *et al.*, 1989; Hirschfield *et al.*, 1989; Maguire *et al.*, 1989; Openshaw *et al.*, 1989; Rhind and Shepherd, 1989; Healey *et al.*, 1990; Stringer and Bond, 1990). The bibliography of GIS applications in Rodriguez-Bachiller (1998) still showed about *half* of all GIS applications involving some degree of expert programming.

3 Funded by ESRC to set up (in the late 1980s) laboratories to research the use of geographical information, intended among other goals to help diffuse the new GIS technology.

1.5 IMPACT ASSESSMENT: RIPE FOR AUTOMATION?

Impact assessment can be said – once again – to be a US import. It has been well established in the United States since the National Environmental Policy Act (NEPA) of 1969 (Glasson *et al.*, 1999), which required studies of impact assessment to be attached to all important *government* projects. The 1970s and 1980s subsequently saw the consolidation of its institutional structure as well as its methods and procedures, and the publication of ground-breaking handbooks (e.g. Rau and Wooten, 1980) to handle the technical difficulties of this new field. Later, this nationwide approach in the US has been supplemented with additional statewide legislation ("little NEPAs") in 16 of the 52 states.

In the meantime, similar legislation, and the expertise that is needed to apply it, has been spreading around the world and has been adopted by more and more countries at a growing rate: Canada (1973), Australia (1974), Colombia (1974), France (1976), The Netherlands (1981), Japan (1984), and the European Community produced its Directive to member countries in July 1985, which has since been adopted in Belgium (1985), Portugal (1987), Spain (1988), Italy (1988), United Kingdom (1988), Denmark (1989), Ireland (1988–90), Germany (1990), Greece (1990), and Luxembourg (1990) (Wathern, 1988; Glasson *et al.*, 1999).

The European Directive 85/337 (Commission of the European Communities, 1985) structured originally the requirements for environmental impact assessment for development projects at two levels, and this approach has been maintained ever since. For certain types and sizes of project (listed in Annex I of the Directive) an "Environmental Statement" would be mandatory:

- crude oil refineries, coal/shale gasification and liquefaction;
- thermal power stations and other combustion installations;
- radioactive waste storage installations;
- cast iron and steel melting works;
- asbestos extraction, processing or transformation;
- integrated chemical installations;
- construction of motorways, express roads, railways, airports;
- trading ports and inland waterways;
- installations for incinerating, treating, or disposing of toxic and dangerous wastes.

In addition, for another range of projects (listed in Annex II of the Directive), an impact study would only be required if the impacts from the project were likely to be "significant" (the criteria for significance being again defined by the scale and characteristics of the project):

- agriculture (e.g. afforestation, poultry rearing, land reclamation);
- extractive industry;

- energy industry (e.g. storage of natural gas or fossil fuels, hydroelectric energy production);
- processing of metals;
- manufacture of glass;
- chemical industry;
- food industry;
- textile, leather, wood and paper industries;
- rubber industry;
- infrastructure projects (e.g. industrial estate developments, ski lifts, yacht marinas);
- other projects (e.g. holiday villages, wastewater treatment plants, knackers' yards);
- modification or temporary testing of Annex I projects.

In the UK, the Department of the Environment (DoE, 1988) adopted the European Directive primarily through the Town and Country Planning Regulations of 1988 ("Assessment of Environmental Effects"). These largely replicated the two-tier approach of the European Directive, classifying EIA projects into those requiring an Environmental Statement and those for which it is required only if their impacts are expected to be significant, listed in so-called Schedules 1 and 2 respectively – which broadly correspond to the Annexes I and II of the European Directive (Glasson *et al.*, 1999). In turn, the expected *significance* of the impacts was to be judged on three criteria (DoE, 1989):

1 The scale of a project making it of "more than local importance".
2 The location being "particularly sensitive" (a Nature Reserve, etc.).
3 Being likely to produce particularly "adverse or complex" effects, such as those resulting from the discharge of pollutants.

The European Directive of 1985 was updated in 1997 (Council of the European Union, 1997) with the contents of Annexes I and II being substantially extended and other changes made, including the mandatory consideration of alternatives. The new Department of Environment, Transport and the Regions (DETR) set in motion a similar process in the UK (DETR, 1997) to update not just the categories of projects to be included in Schedules 1 and 2, but also the standards of significance used, which has recently resulted in new Regulations (DETR, 1999a) with a revised set of criteria, and also in new practical guidelines in a Circular (DETR, 1999b). This represents in reality a shift to a *three-level* system, in that for Schedule 1 projects, there is a mandatory requirement for EIA, and for Schedule 2 projects there are two categories. Projects falling below specified "exclusive thresholds" do not require EIA, although there may be circumstances in which such small developments may give rise to significant environmental impacts (for example by virtue of the sensitivity of the location), and in such cases an EIA may be required. For other Schedule 2 projects, there are "indicative criteria and thresholds" which, for each category of project, indicate the characteristics which are most likely to generate significant impacts. For such projects, a case-by-case approach is normally needed, and projects will be judged on: (i) characteristics of the development (size, impact accumulation with other projects, use of natural resources, waste production, pollution, accident risks); (ii) sensitivity of the location; (iii) characteristics of the potential impacts (extent, magnitude and complexity, probability, duration and irreversibility).

1.6 THE IA PROCESS

IA can be seen as a series of processes within processes in a broader cycle that is the life of a development project. The life of a project usually involves certain typical stages:

1 decision to undertake the project and general planning of what it involves;
2 consideration of alternative designs and locations (not always);
3 conflict resolution and final decision;
4 construction;
5 operation;
6 closedown/decommissioning (not always present, some projects have theoretically an eternal life).

Within this cycle, IA is a socio-political process to add certain checks and balances to the project life, within which more technical exercises are needed to predict and assess the likely impacts of the project, sometime involving social processes of consultation and public participation. IA can be seen as a process in itself (Glasson *et al.*, 1999), with typical *stages*:

1 *Screening*: deciding if the project needs an environmental statement, using the technical criteria specified in the relevant IA legislation and guidelines, and often also involving consultation. Beyond this first stage, IA as such should be (but often is not) applied to all the main phases in the physical life of the project: *construction, operation, decommissioning*.

2 *Scoping*: determining which impacts must be studied (using checklists, matrices, networks, etc.), as well as identifying which of those are likely to be the *key* impacts, likely to be the ones that will "make or break" the chances of the project being accepted, often involving consultation with interested parties and the public. Both the "screening" and "scoping" stages require a considerable amount of work directed at the *understanding*

of the situation being considered: understanding of the project, understanding of the environment, and understanding of the alternatives involved.

3 *Impact prediction* for each of the impact areas defined previously, involving two distinct types of predictions:

3a *Baseline prediction* of the situation concerning each impact without the project.
3b *Impact prediction* as such, predicting the differences between the baseline and the project impacts using models and other expert technical means, and differentiating between:

- *direct* impacts from the project (from emissions, noise, etc.);
- *indirect* impacts derived from other impacts (like noise from traffic);
- *cumulative* impacts resulting from the project *and* other projects in the area.

4 *Assessment of significance* of the predicted impacts, by comparing them with the accepted standards, and often also including some degree of consultation.

5 *Mitigation*: definition of measures proposed to alleviate some of the adverse impacts predicted to be significant in the previous stage.

6 Assessment of the likely *residual impacts after mitigation*, and their significance.

7 After the project has been developed, *monitoring* the actual impacts from it – including monitoring the effectiveness of any mitigation measures in place – separating them from impacts from other sources impinging on the same area. Hopefully, this may lead to, and provide data for, some *auditing* of the process itself (e.g. studies of how good were the predictions).

The different stages of the IA process are "interleaved" with those of the project life and, in fact, the quality of the overall outcome often depends on how appropriately – and timely – that interleaving takes place. In general, the earlier in the design of a project the IA is undertaken, the better, because, if it throws up any significant negative impacts, it will be much easier (and cheaper) to modify the project design than applying mitigation measures afterwards. In particular, if alternative designs or locations are being considered for the project, applying IA at that stage may help identify the best options. Also, because the public should be a key actor in the whole assessment process, an earlier start will alert the public and will be more likely to incorporate their views from the beginning, thus reducing the chances of conflict later, when the repercussions of such conflicts may be far reaching and expensive for all concerned.

1.7 ENVIRONMENTAL STATEMENTS

In turn, Environmental Statements (the actual IA reports) represent a third process within IA – also interleaved with the other two – involving two main stages: (i) statement *preparation* by the proponents of the development, and (ii) statement *review* by the agency responsible. In fact, the structure of Environmental Statements should reflect all this "interleaving", which often determines also the quality of such documents. The structure and content of Environmental Statements are defined by the legislation, guidelines and "good practice" advice from the relevant agencies (Wathern, 1988; DoE, 1988, 1994 and 1995), and is usually a variation of the following list:

1 Description of the project:

- physical and operational features;
- land requirements and layout;
- project inputs;
- residues and emissions if any.

2 Alternatives considered:

- different processes or equipment;
- different layout and spatial arrangements;
- different locations for the project;
- the *do nothing* alternative (NOT developing the project).

3 Impact areas to be considered:

- socio-economic impacts;
- impacts on the cultural heritage;
- impacts on landscape;
- impacts on material assets and resources;
- land use and planning impacts;
- traffic impacts;
- noise impacts;
- air pollution impacts;
- impacts on soil and land;
- impacts on geology and hydrogeology;
- impacts on ecology (terrestrial and aquatic).

4 Impact predictions:

- baseline analysis and forecasting;
- impact prediction;
- evaluation of significance;
- mitigation measures;
- plans for monitoring.

In addition to these substantive requirements, other formal aspects can be added – in the UK for instance – including for example a non-technical summary for the layperson, a clear statement of what the objectives of the project are, the identification of any difficulties encountered when compiling the study, and others (see Chapter 11).

Early experience of Environmental Statements evidenced several problems, including the number of statements itself. All countries where IA has been introduced seem to have had a "flood" of Environmental Statements: in the US, about 1,000 statements a year were being processed during the first 10 years after NEPA, although the number of statements processed in the US dropped afterwards to about 400 each year, and this is attributed to impact assessment having become much more an integral part of the project design process and impacts being considered much earlier in the process. In France they had a similar number of about 1,000 statements per year after they started EIA in 1976, and this has subsequently risen to over 6,000. In the UK, more than 300 statements on average were processed each year between 1988 and 1998 (Glasson *et al.*, 1999; Wood and Bellanger, 1999), a much higher rate than in the US if we relate it to the population size of both countries. The number of environmental statements in the UK dropped during the 1990s to about 100–150 a year (Wood and Bellanger, 1999), probably related to a fall in economic activity, and the number of statements went back up to about 300 with the economic revival towards the end of the decade. With the implementation of the amended EU Directive in 1999, the UK figure has risen to over 600 Environmental Statements p.a., and there have been substantial increases also in other EU Member States.

The *quality* of the statements also seems to be improving after a relatively poor start: improvements were noted first from 1988/89 to 1990/91 (Lee and Colley, 1992), and also from before 1991 to after 1991 (DoE, 1996; Glasson *et al.*, 1997), even though it seems that the overall quality is still far from what would be desirable. After the teething problems in the 1980s, mostly attributed to the inexperience of all the actors involved (developers, impact assessors, local authority controllers), better impact studies seem now to be related to (i) *larger* projects of certain types; (ii) more *experienced* consultants; (iii) local authorities with customised *EIA handbooks*. Central to this improvement seems to have been (as it was in the US in the 1970s and early 1980s) the increasing dissemination of good practice and expertise – in guides by the agencies responsible (DoE, 1989 and 1995) and in technical manuals (Petts and Eduljee, 1994; Petts, 1999; Morris and Therivel, 1995 and 2001) – that show how the field can be broken down into sub-problems and the best ways of solving such sub-problems.

Glasson *et al.* (1997) believe that, had the European Community not insisted on the adoption of EIA, the then Conservative Government would not have introduced it in the UK, arguing at the time that the existing planning system was capable of dealing with the consideration of undesirable impacts from developments. This was despite their repeated attempts to streamline and in some cases dismantle the planning system as part of their general strategy of "rolling back the State" without any need for additional controls. In the UK Planning system, each development application is evaluated "on its own merits" as part of the general development control process, and the consideration of impacts could have been seen as just another set of "material considerations" to contribute to the decision, not requiring special guidelines, processes and legislation. But the European Directive *was* implemented in the UK, and there is now considerable debate in the UK about the possibility of extending the environmental impact assessment approach further, including:

- To broaden IA to include more fully under its umbrella the area of *socio-economic impact assessment*, already practised to some extent in the UK and more fully in other countries (Glasson *et al.*, 1999), but not always with the appropriate legislative recognition. Such widening of scope may lead to more integrated IA, with decisions based partly on the extent to which various biophysical and socio-economic impacts can be "traded".
- To move from a concern with projects to include "higher tiers of actions", structured into policies, plans and programmes (Wood, 1991). This has become known as *strategic environmental assessment* (SEA) with a growing literature and legislation around it. In 2001, the European Union agreed an SEA Directive for plans and programmes – although, unfortunately, not for policies (CEU, 2001). This must be implemented by Member States by 2004. Plan SEAs would define acceptable standards for an area that projects would have to adhere to, and ideally after such an assessment individual developments would not need to re-assess their impacts each time, but only to demonstrate their compliance with those standards.

1.8 INTEGRATION: THE WAY AHEAD?

The main purpose of this chapter has been to point out the potential complementarity of the three areas of IA, and the building blocks of the argument can be summarised in a final list of points:

- IA practice is growing at a fast pace, and many of the actors involved are finding it difficult to cope.
- The quality of impact assessment (although improving) is still far from satisfactory.
- One reason for the low quality of IA is still the relative scarcity of expertise.
- IA expertise is mostly legal, technical and specific, rather than "common-sensical" and diffuse.

- IA expertise and good practice exist, and are beginning to be articulated in good sources (guides, manuals, etc.) by good experts.
- The problem in many countries – including the UK – is not the existence or the quality of IA expertise and good practice, but its dissemination.
- Expert systems are particularly suited for "technology transfer" from experts to non-experts when dealing with *not-too-difficult* problems of the kind that some parts of IA pose.
- Expert systems are also increasingly becoming useful tools for interfacing in a logical and friendly way with other systems.
- Interest in the application of expert systems to IA has already started to take off as reported in the literature, and it seems timely to take a closer look at this possibility.
- Expert systems handle logical information well, but the handling of spatial information can be a problem, and GIS can help in this respect.
- GIS are efficient map-manipulation systems, and their analytical capabilities are being continually improving.
- GIS applications can require considerable programming and customising by the user.
- Expert systems technology can possibly provide the programming power and friendliness that GIS need to interface with the user, and combined together they can help IA take the leap forward that current institutional pressures are expecting of it.

These two computer technologies may be able to help with IA, each in their own different way: GIS may be able to *support* IA good practice, expert systems may be able to *spread* it. The potential for articulation of these three technologies is now clear. Chapter 2 discusses expert systems in greater detail, and the following Chapters 3–5 include a bibliographical review of IA applications of expert systems and GIS as documented in the literature.

REFERENCES

Antenucci, J.C. (1992) Product Strategies: Creative Tensions, Plenary Session March 25th, *Proceedings of the EGIS '92 Conference*, Munich (March).

Antenucci, J.C., Brown, K., Croswell, P.L.C. and Kevany, M.J. (1991) *Geographic Information Systems: A Guide to the Technology*, Van Nostrand, New York.

Campbell, H. and Masser, I. (1994) *Geographical Information Systems and Organisations*, Taylor & Francis, London.

CEU (2001) *Common Position Adopted by the Council with a View to the Adoption of a Directive of the European Parliament and of the Council on the Assessment of the Effects of Certain Plans and Programmes on the Environment*, Council of European Union, Brussels.

Chrisman, N.R. (1991) The Error Component in Spatial Data, in Maguire, D.J., Goodchild, M.F. and Rhind, D.W. (eds) *Geographical Information Systems: Principles and Applications*, Longman, London (Ch. 12).

CEC (1985) On the assessment of the effects of certain public and private projects on the environment, *Official Journal*, L175 (5 July), Council Directive 85/337/EC.

Coppock, J.T. and Rhind, D.W. (1991) The History of GIS, in Maguire, D.J., Goodchild, M.F. and Rhind, D.W. (eds) *Geographical Information Systems: Principles and Applications*, Longman, London (Ch. 2).

Council of the European Union (1997) *Council Directive 96/11/EC of 3 March 1997 amending Directive 85/337/EEC*, European Union, Brussels.

DoE (1988) *Environmental Assessment*, Department of the Environment Circular 15/88 (Welsh Office Circular 23/88), 12 July.

DoE (1989) *Environmental Assessment: A Guide to the Procedures*, HMSO, London.

DoE (1994) *Evaluation of Environmental Information for Planning Projects. A Good Practice Guide*, Report by Land Use Consultants, HMSO, London.

DoE (1995) *Preparation of Environmental Statements for Planning Projects that Require Environmental Assessment: A Good Practice Guide*, Department of the Environment, HMSO, London.

DoE (1996) *Changes in the Quality of Environmental Impact Statements for Planning Projects*, Report by the Impact Assessment Unit, School of Planning, Oxford Brookes University, HMSO, London.

DETR (1997) *Environmental Assessment (EA): Implementation of EC Directive (97/11/EC)*, Consultation Paper.

DETR (1999a) *The Town and Country Planning (Environmental Impact Assessment) (England and Wales) Regulations 1999*, DETR No. 293.

DETR (1999b) *Environmental Impact Assessment*, DETR Circular 02/99.

Edwards-Jones, G. and Gough, M. (1994) ECOZONE: A Computerised Knowledge Management System for Sensitising Planners to the Environmental Impacts of Development Projects, *Project Appraisal*, Vol. 9, No. 1 (March), pp. 37–45.

Fisher, P.F. (1991) Spatial Data Sources and Data Problems, in Maguire, D.J., Goodchild, M.F. and Rhind, D.W. (eds) *Geographical Information Systems: Principles and Applications*, Longman, London (Ch. 13).

Flowerdew, R. (1989) The North West Regional Research Laboratory, *Mapping Awareness*, Vol. 3, No. 2 (May–June), pp. 43–6.

Foresman, T.W. (1998) (ed.) *The History of Geographic Information Systems: Perspectives From the Pioneers*, Prentice Hall PTR, Upper Saddle River (New Jersey).

Geraghty, P.J. (1992) Environmental Assessment and the Application of an Expert Systems Approach, *Town Planning Review*, Vol. 63, No. 2, pp. 123–42.

Glasson, J., Therivel, R., Weston, J., Wilson, E. and Frost, R. (1997) EIA – Learning from Experience: Changes in the Quality of Environmental Impact Statements for UK Planning Projects, *Journal of Environmental Planning and Management*, Vol. 40, No. 4, pp. 451–64.

Glasson, J., Therivel, R. and Chadwick, A. (1999) *Introduction to Environmental Impact Assessment*, UCL Press, London (2nd edition, 1st edition in 1994).

Green, A., Higgs, G., Mathews, S. and Webster, C. (1989) The Wales and South West Regional Research Laboratory, *Mapping Awareness*, Vol. 3, No. 3 (July/August).

Hayes-Roth, F., Waterman, D.A. and Lenat, D.B. (1983a) An Overview of Expert Systems, in Hayes-Roth, F., Waterman, D.A. and Lenat, D.B. (eds) *Building Expert Systems*, Addison Wesley (Ch. 1).

Hayes-Roth, F., Waterman, D.A. and Lenat, D.B. (1983b) *Building Expert Systems*, Addison Wesley.

Healey, R., Burnhill, P. and Dowie, P. (1990) Regional Research Laboratory for Scotland, *Mapping Awareness*, Vol. 4, No. 2 (March).

Hirschfield, A., Barr, R., Batey, P. and Brown, P. (1989) The Urban Research and Policy Evaluation Regional Research Laboratory, *Mapping Awareness*, Vol. 3, No. 6 (December).

Jackson, P. (1990) *Introduction to Expert Systems*, Addison Wesley (2nd edition).

Jong De, W.M. (1989) *Uncertainties in Data-quality and the Use of GIS for Planning Purposes*, in "Urban Data Management Coming of Age", *Proceedings of the 13th Urban Data Management Symposium*, Lisbon (May 29–June 2), pp. 171–85.

Leary, M.E. (1989) A Spatially-based Computer System for Land Use Planning, *Ekistics*, Vol. 56, No. 338/339 (September, October, November, December), pp. 285–9.

Lee, N. and Colley, R. (1992) *Reviewing the Quality of Environmental Statements*, Occasional Paper 24, EIA Centre, Department of Planning and Landscape, University of Manchester.

Longley, P.A., Goodchild, M.F., Maguire, D.J. and Rhind, D.W. (1999) (eds) *Geographical Information Systems*, 2 Vols, John Wiley & Sons Inc. (2nd edition).

Longley, P.A., Goodchild, M.F., Maguire, D.J. and Rhind, D.W. (2001) (eds) *Geographic Information Systems and Science*, John Wiley & Sons Ltd.

Makhchouni El, M. (1987) Un Systems Graphique Intelligent D'Aide a la Conception des Plans D'Occupation des Sols, in Laurini, R. (ed.) *UDMS '87, Proceedings of the 12th Urban Data Management Symposium*, Blois (May), pp. 204–19.

Maguire, D.J., Strachan, A.J. and Unwin, D.J. (1989) The Midlands Regional Research Laboratory, *Mapping Awareness*, Vol. 3, No. 1 (March–April).

Maguire, D.J. (1991) An Overview and Definition of GIS, in Maguire, D.J., Goodchild, M.F. and Rhind, D.W. (eds) *Geographical Information Systems: Principles and Applications*, Longman, London (Ch. 1).

Maguire, D.J. and Dangermond, J. (1991) The Functionality of GIS, in Maguire, D.J., Goodchild, M.F. and Rhind, D.W. (eds) *Geographical Information Systems: Principles and Applications*, Longman, London (Ch. 21).

Maguire, D.J., Goodchild, M.F. and Rhind, D.W. (1991) (eds) *Geographical Information Systems: Principles and Applications*, Longman, London.

Masser, I. (1990a) *The Regional Research Laboratory Initiative: An Overview*, ESRC: Regional Research Laboratory Initiative, Discussion Paper No. 1.

Masser, I. (1990b) The Utilisation of Computers in Local Government in Less Developed Countries: A Case Study of Malaysia. *Proceedings of the Urban and Regional Information Systems Association (URISA) Conference*, Edmonton, Alberta, Canada (August 12–16), Vol. IV, pp. 235–45.

Morris, P. and Therivel, R. (1995) (eds) *Methods of Environmental Impact Assessment*, UCL Press, London.

Morris, P. and Therivel, R. (2001) (eds) *Methods of Environmental Impact Assessment*, Spon Press – Taylor and Francis, London (2nd edition).

Nutter, M., Charron, J. and Moisan, J.F. (1996) Geographic Information System Tool Integration for Environmental Assessment: Recent Lessons. *Improving Environmental Assessment Effectiveness: Research, Practice and Training,*

Proceedings of IAIA '96, 16th Annual Meeting (June 17–23), Centro Escolar Turistico e Hoteleiro, Estoril (Portugal), Vol. I, pp. 473–8.
Openshaw, S., Goddard, J. and Coombes, M. (1987) *Integrating Geographic Data for Policy Purposes: Some Recent UK Experience*, Research Report 87/1, North East Regional Research Laboratory, Department of Geography, University of Newcastle upon Tyne, Newcastle upon Tyne.
Openshaw, S., Gillard, A. and Charlton, M. (1989) The North East Regional Research Laboratory, *Mapping Awareness*, Vol. 3, No. 4 (September/October).
Openshaw, S. (1990) *A Spatial Analysis Research Strategy for the Regional Research Laboratory Initiative*, Regional Research Laboratory Initiative, Discussion Paper No. 3 (June).
Openshaw, S. (1991) Developing Appropriate Spatial Analysis Methods for GIS, in Maguire, D.J., Goodchild, M.F. and Rhind, D.W. (eds) *Geographical Information Systems: Principles and Applications*, Longman, London (Ch. 25).
Petts, J. (ed.) (1999) *Handbook of Environmental Impact Assessment*, Blackwell Science Ltd, Oxford (2 Vols).
Petts, J. and Eduljee, G. (1994) *Environmental Impact Assessment for Waste Treatment and Disposal Facilities*, John Wiley & Sons, Chichester .
Radwan, M.M. and Bishr, Y.A. (1994) Integrating the Object Oriented Data Modelling and Knowledge System for the selection of the Best Management Practice in Watersheds, *Proceedings of The Canadian Conference on GIS*, Ottawa (June), Vol. 1, pp. 690–9.
Rau, J.G. and Wooten, D.C. (eds) (1980) *Environmental Impact Analysis Handbook*, McGraw-Hill.
Rhind, D. and Shepherd, J. (1989) The South East Regional Research Laboratory, *Mapping Awareness*, Vol. 2, No. 6 (January–February), pp. 38–46.
Rodriguez-Bachiller, A. (1991) Expert Systems in Planning: An Overview, *Planning Practice and Research*, Vol. 6, No. 3 (Winter), pp. 20–5.
Rodriguez-Bachiller, A. and Smith, P. (1995) Diffuse Picture on Spread of Geographic Technology, *Planning* (June 14), pp. 24–5.
Rodriguez-Bachiller, A. (1998) *GIS and Decision-Support : A Bibliography*, Working Paper No. 176, School of Planning, Oxford Brookes University.
Rodriguez-Bachiller, A. (2000a) Geographical Information Systems and Expert Systems for Impact Assessment. Part I: GIS, *Journal of Environmental Assessment Policy and Management*, Vol. 2, No. 3 (September), pp. 369–414.
Rodriguez-Bachiller, A. (2000b) Geographical Information Systems and Expert Systems for Impact Assessment. Part II: Expert Systems and Decision Support Systems, *Journal of Environmental Assessment Policy and Management*, Vol. 2, No. 3 (September), pp. 415–48.
Rodriguez-Bachiller, A. and Wood, G. (2001) Geographical Information Systems (GIS) and EIA, in Morris, P. and Therivel, R. (eds) *Methods of Environmental Impact Assessment*, Spon Press – Taylor and Francis, London (2nd edition, Ch. 16).
Schibuola, S. and Byer, P. (1991) Use of Knowledge-based Systems for the Review of Environmental Impact Statements, *Environmental Impact Assessment Review*, No. 11, pp. 11–27.
Sharpe, R., Marksjo, B.S. and Thomson, J.V. (1988) Expert Systems in Building and Construction, in Newton, P.W., Taylor, M.A.P. and Sharpe, R. (eds) *Desktop Planning*, Hargreen Publications, Melbourne (Ch. 39).

Stringer, P. and Bond, D. (1990) The Northern Ireland Regional Research Laboratory, *Mapping Awareness*, Vol. 4, No. 1 (January/February).

Warner, M., Croal, P., Calal-Clayton, B. and Knight, J. (1997) Environmental Impact Assessment Software in Developing Countries: A Health Warning. *Project Appraisal*, 12(2), pp. 127–30.

Waterman, D.A. (1986) *A Guide to Expert Systems*, Addison Wesley.

Wathern, P. (1988) (ed.) *Environmental Impact Assessment: Theory and Practice*, Routledge, London.

Wood, C. (1991) EIA of Policies, Plans and Programmes, *EIA Newsletter* 5, University of Manchester, pp. 2–3.

Wood, G. and Bellanger, C. (1999) *Directory of Environmental Impact Statements: July 1988–April 1998*. Working Paper No. 179, School of Planning, Oxford Brookes University.

2 Expert systems and decision support

2.1 INTRODUCTION

This methodological – and to some extent historical – chapter focuses on the nature and potential of ES beyond the brief introduction to these systems in Chapter 1, by looking back at their early development and some of their most relevant features. It is structured into four sections: in Section 2.2, the emergence of expert systems is discussed in the context of the development of the field of Artificial Intelligence; in Section 2.3, the typical structure of expert systems is discussed; in Section 2.4 we discuss the "promise" of expert systems and the extent of its fulfillment and, in Section 2.5, we expand the discussion to cover the wider area of so-called Decision Support Systems (DSS).

2.2 EXPERT SYSTEMS AND ARTIFICIAL INTELLIGENCE

Artificial intelligence (AI) has been defined in a variety of ways, primarily by its *aims*, as reflected in a number of well-known AI manuals and textbooks:

- to *simulate* intelligent behaviour (Nilsson, 1980);
- to "study of how to make computers *do* things at which, at the moment, people are better" (Rich, 1983);
- "to *understand* the principles that make intelligence possible" (Winston, 1984);
- to *study* human intelligence by trying to simulate it with computers (Boden, 1977).

Definitions of AI such as these tend to be based on some degree of belief in the provocative statement made by Marvin Minsky (MIT) in the 1960s that "the brain happens to be a meat machine" (McCorduck, 1979) which, by implication, can be *simulated*. The main difference between these definitions is in their varying degree of optimism about the possibility of reproducing human intelligence mechanically: while the first two seem to put the emphasis on the *simulation* of intelligence (reproducing intelligent behaviour), the last two – more cautious – put the emphasis rather on *understanding* intelligence. In fact, the tension between "doing" and "knowing" has been one of the driving forces in the subsequent development of AI, and has also been one of the root causes of the birth of expert systems.

Many antecedents of AI (what can be called the "prehistory" of AI) can be found in the distant past, from the calculators of the seventeenth century to Babbage's Difference Engine and Analytical Engine of the nineteenth century, from the chess-playing machine of Torres Quevedo at the time of the First World War to the first programmable computer developed in Britain during the Second World War, together with the pioneering work of Alan Turing and his code-breaking team at Bletchley Park, part of the secret war effort only recently unveiled in its full detail and importance (Pratt, 1987) – and popularised in the recent film "Enigma". However, the consolidation of AI as a collective field of interest (and as a *label*) was very much an American affair, and AI historians identify as the turning point the conference at Dartmouth College (Hanover, New Hampshire) in the Summer of 1956, funded by the Rockefeller Foundation (McCorduck, 1979; Pratt, 1987). Jackson (1990) suggests that the history of AI after the war follows *three periods* (the classical period, the romantic period, and the modern period) each marked by different types of research interests, although most lines of research have carried on right throughout to varying degrees.

2.2.1 The classical period

This period extends from the war up to the late 1950s, concentrating on developing efficient *search* methods: finding a solution to a problem was seen as a question of searching among all possible states in each situation and identifying the best. The combinatorial of all possible states in all possible situations was conceptualised and represented as a *tree* of successive options, and search methods were devised to navigate such trees. Search methods would sometimes explore each branch in all its depth first before moving on to another branch ("depth-first" methods); some methods would explore all branches at one level of detail before moving down to another level ("breadth-first" methods). The same type of trees and their associated search methods were also used to develop game-playing methods for machines to play two-player games (like checkers or chess), where the tree of solutions includes alternatively the "moves" open to each player. The same type of tree representation of options was seen as universally applicable to both types of problems (Figure 2.1).

Efficient "tree-searching" methods can be developed independently of any particular task – hence their enormous appeal at the time as universal problem solvers – but they are very vulnerable to the danger of the so-called

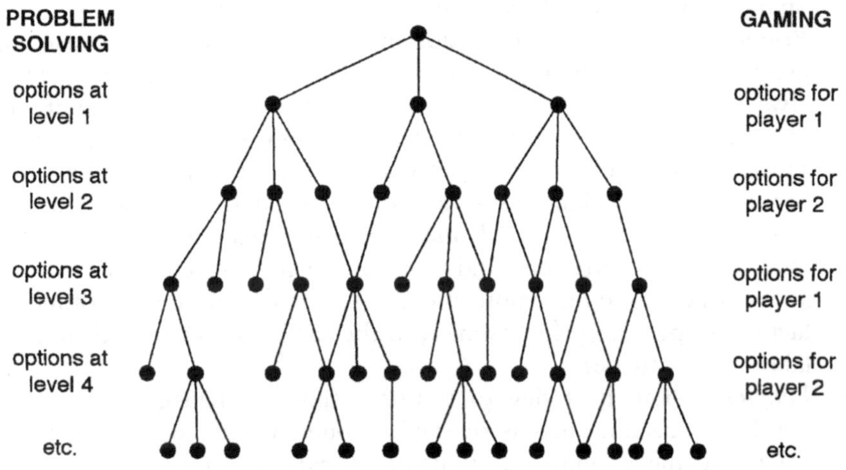

Figure 2.1 Options as trees.

"combinatorial explosion", the multiplication of possible combinations of options beyond what is feasible to search in a reasonable time. For instance, to solve a chess game completely (i.e. to calculate all 10^{120} possible sequences of moves derived from the starting position) as a *blind* tree search – without any chess-specific guiding principles – would take the most advanced computer much longer than the universe has been in existence (Winston, 1984). It is for reasons like this that these techniques, despite their aspiration to universal applicability, are often referred to as *weak* methods (Rich, 1983). On the other hand, they do provide a framework within which criteria specific to a problem can be applied. One such approach adds to the search process some form of evaluation at every step (an "evaluation function"), so that appropriate changes in the direction of search can shorten it and make it progress faster towards the best solution, following a variety of so-called "hill-climbing" methods.

2.2.2 The romantic period

This period extends from the 1960s to the mid-1970s, characterised by the interest in *understanding*, trying to simulate human behaviour in various aspects:

(a) On the one hand, trying to simulate *subconscious* human activities, things we do without thinking:

- *Vision*, usually simulated in several stages: recognising physical edges from shadows and colour differences, then reconstructing shapes (concavity and convexity) from those edges, and finally classifying the shapes identified and determining their exact position.

- *Robotics*, at first just an extension of machine tools, initially based on pre-programming the operation of machines to perform certain tasks always in the same way; but as the unreliability of this approach became apparent – robots being unable to spot small differences in the situation not anticipated when programming them – second-generation robotics started taking advantage of *feedback* from sensors (maybe cameras, benefiting from advances in vision analysis) to make small instantaneous corrections and achieve much more efficient performances, which led to the almost full automation of certain types of manufacturing operations (for instance, in the car industry) or of dangerous laboratory activities.
- *Language*, both by trying to translate spoken language into written words by spectral analysis of speech sound waves, and by trying to determine the grammatical structure ("parsing") of such strings of words leading to the understanding of the meaning of particular messages.

(b) On the other hand, much effort also went into reproducing *conscious* thinking processes, like:

- *Theorem-proving* – a loose term applied not just to mathematical theorems (although substantial research did concentrate on this particular area of development) but to general logical capabilities like expressing a problem in formal logic and being able to develop a full syllogism (i.e. to derive a conclusion from a series of premises).
- *Means-ends analysis* and planning, identifying sequences of (future) actions leading to the solution of a problem, like Newell and Simon's celebrated "General Problem Solver" (Newell and Simon, 1963).

2.2.3 The modern period

In the so-called *modern* period, from the 1970s onwards, many of the traditional strands of AI research – like robotics – carried on but, according to Jackson (1990), the main thrust of this period comes from the reaction to the problems that arose in the previous attempts to simulate brain activity and to design general problem-solving methods. The stumbling block always seemed to be the lack of criteria specific to the particular problem being addressed ("domain-specific") beyond general procedures that would apply to *any* situation ("domain-free"). When dealing with geometric wooden blocks in a "blocks world", visual analysis might have become quite efficient but, when trying to apply that efficiency to dealing with nuts and bolts in a production chain, procedures more specific to nuts and bolts seemed to be necessary. It seemed that for effective problem-solving at the level at which humans do it, more problem-specific knowledge was required than had been anticipated. Paradoxically, this need for a more domain-specific approach developed in the following years in two totally different directions.

On the one hand, the idea that it might be useful to design computer systems which did not have to be pre-programmed but which could be *trained* "from scratch" to perform specific operations led – after the initial rejection by Minsky in the late 1960s – to the development in the 1980s of *neural networks*, probably the most promising line of AI research to date. They are software mini-brains that can be trained to recognise specific patterns detected by sensors – visual, acoustic or otherwise – so that they can then be used to identify other (new) situations. Research into neural nets became a whole new field in itself after Rumelhart and McClelland (1989) – a good and concise discussion of theoretical and practical issues can be found in Dayhoff (1990) – and today it is one of the fastest growing areas of AI work, with ramifications into image processing, speech recognition, and practically all areas of cognitive simulation.

On the other hand, and more relevant to the argument here, the emphasis turned from trying to understand how the brain performed certain operations, to trying to capture and use problem-specific knowledge *as humans do it*. This emphasis on knowledge, in turn, raised the interest in methods of *knowledge representation* to encode the knowledge applicable in particular situations. Two general types of methods for knowledge representation were investigated:

(a) *Declarative* knowledge representation methods which describe a situation in its context, identifying and describing all its elements and their relationships. *Semantic networks* were at the root of this approach; they were developed initially to represent the meaning of words (Quillian, 1968), describing objects in terms of the class they belong to (which itself may be a member of another class), their elements and their characteristics, using attribute relationships like "colour" and "shape", and functional relationships like "is a", "part of" and "instance of" (Figure 2.2).

Of particular importance is the *is a* relationship which indicates class membership, used to establish relationships between families of objects and to derive from them rules of "inheritance" between them. If an object belongs to a particular class, it will inherit some of its attributes, and they do not need to be defined explicitly for that object: because a penguin is a bird, we know it must have feathers, therefore we do not need to register that attribute explicitly for penguins (or for every particular penguin), but only for the class "birds".

Other declarative methods like *conceptual dependency* were really variations of the basic ideas used in semantic networks. *Frames* were like "mini" semantic nets applied to all the objects in the environment being described, each frame having "slots" for parts, attributes, class membership, etc. even

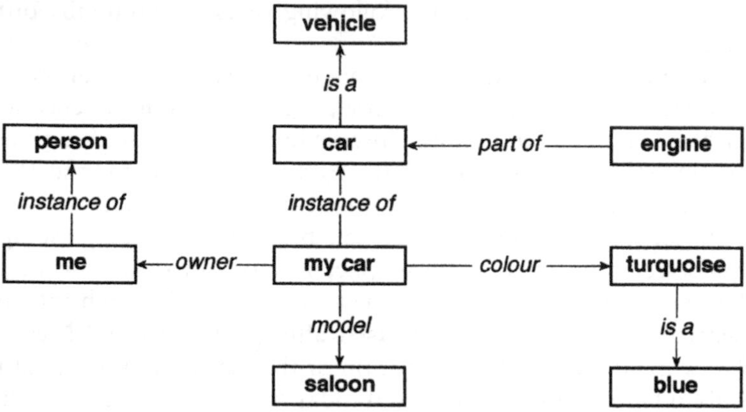

Figure 2.2 A semantic network.
Source: Modified from Rich, 1983.

for certain procedures specific to them. We can trace the current emphasis on "object-oriented" approaches to computer technology to these frames and networks of the 1970s. Also, *scripts* were proposed to represent contextual knowledge of time-related processes, standard sequences of events that common knowledge takes for granted, like the sequence that leads from entering a bar to ordering a drink and paying for it. As with the rest of these methods, the emphasis is on common-sense knowledge that we take for granted, and which acts as backcloth to any specific problem-solving situation we encounter.

(b) *Procedural* knowledge representation, on the other hand, concentrates not so much on the description of a situation surrounding a problem, but on the articulation of how to use the knowledge we have (or need to acquire) in order to solve it. The most prominent of these approaches has been the use of *production rules* to represent the logic of problem-solving, "if-then" rules which can be used to express how we can infer the values of certain variables (conclusions) from our knowledge of the values of other variables (conditions). By linking rules together graphically, we can draw chains ("trees") of conditions and conclusions leading to the answer for the question at the top. These *inference trees* do not describe the problem but simply tell us what we need to know to solve it, so that when we provide that information, the solution can be inferred automatically. For example, a rudimentary tree to work out if a project needs an impact assessment might look like Figure 2.3.

A tree like this is just a representation of a set of "if-then" rules which might be worded like this:

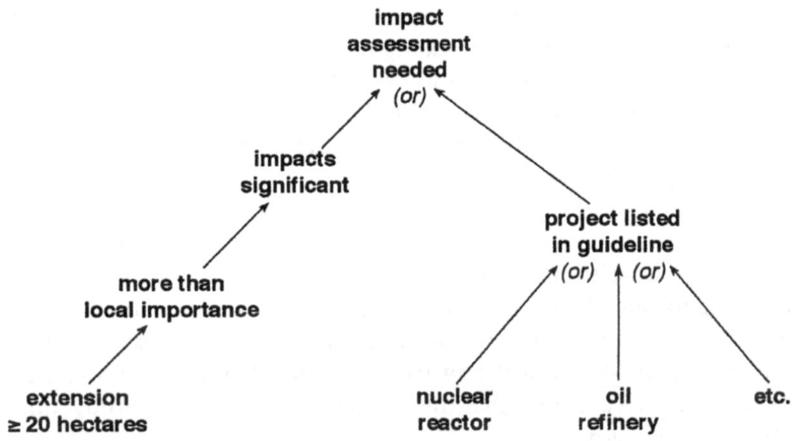

Figure 2.3 Inference tree.

Rule 1: *if the project impacts are likely to be significant*
or if the project type is included in the guidelines' list
then an impact assessment is needed

Rule 2: *if the project is a nuclear reactor*
or if the project is an oil refinery
or if the project is....
then the project type is included in the guidelines' list

Rule 3: *if the scale of the project is of more than local importance*
then the project impacts are likely to be significant

Rule 4: *if the extension of the project (in hectares) is greater than 20*
then the scale of the project is of more than local importance

As the values of the variables *at the bottom* of the tree (the "leaves") are obtained – normally by asking screen-questions about them – the appropriate production rules are "fired" sending their conclusions up the tree to activate other rules, until an answer is derived for the top question.

When queried about whether "an impact assessment is needed", the inference process will first try to find if there is any rule which has this as its *conclusion* (Rule 1 in our example), and it will try to answer it by finding if the *conditions* in that rule are true. In this case, there are two conditions (that the impacts are likely to be significant, or that the project is of a certain type) and the fact that they are linked by an "or" means that either of them will suffice. Therefore, the inference will try to evaluate each condition in turn, and stop as soon as there is enough information to determine if the rule is true.

Repeating the same logic, in order to evaluate the first condition about "the impacts being significant", the process will look for a rule that has this as its conclusion (Rule 2 in our example) and try to see if its condition(s) are true – in this case, the condition that "the scale is of more than local importance". Then, in order to conclude this, it will need to find another rule that has this as its conclusion (Rule 3 in our example) and try to evaluate its conditions, and so on.

When, at the end of this chain of conclusions and conditions, the process finds some conditions to be evaluated for which there are *no rules*, the evaluation of those conditions has to be undertaken outside the rules. The usual way will be to find the information in a database or to *ask the user*. In the latter case, the user will simply be asked to quantify *the extension of the project* (*in hectares*) and, if the answer is greater than 20, then the chain of inference will derive from it that the project needs an impact study, and this will be the conclusion.

The logic followed in this example is usually referred to as "backward-chaining" inference, which derives what questions to ask (or what conditions to check) from the conclusions being sought in the corresponding rules. Another possible approach is usually referred to as "forward-chaining" inference, by which information or answers to questions are obtained first, and from them are derived as many conclusions as possible.[4] This type of inference is also embedded in similar trees as shown above, but it can also be useful to represent it with simpler *flow diagrams* showing the succession of steps involved in the inference process. The "data-first" diagram for such approach (Figure 2.4) would look quite different from the previous tree diagram, even if both represent basically the same deductive process of deriving some conclusions from answers to certain questions, following the same logical rules.

Inference trees have the inherent appeal of having two-in-one uses: they represent the logic of analysing a problem, and at the same time they show the steps necessary to solve it. But their visual effectiveness diminishes rapidly as the complexity of the problem increases, as the number of "links" between levels increases and lines begin to cross. A clear understanding of such complex trees would require an impractical three-dimensional representation, therefore trees tend to be used only to describe relatively simple processes – or, as here, to illustrate the principle – and flow diagrams are often preferred in practical situations.

It is not by chance that the development of these methods was concurrent with the growing interest in *expert systems* in the 1970s. Semantic nets and classificatory trees were often used in the first expert systems to represent relationships between types of problems or aspects of the problem, and

4 Also, backward and forward chaining can be combined, so that, at every step of the inference, what information to get is determined by backward chaining and, once obtained, all its possible conclusions are derived from it by forward chaining.

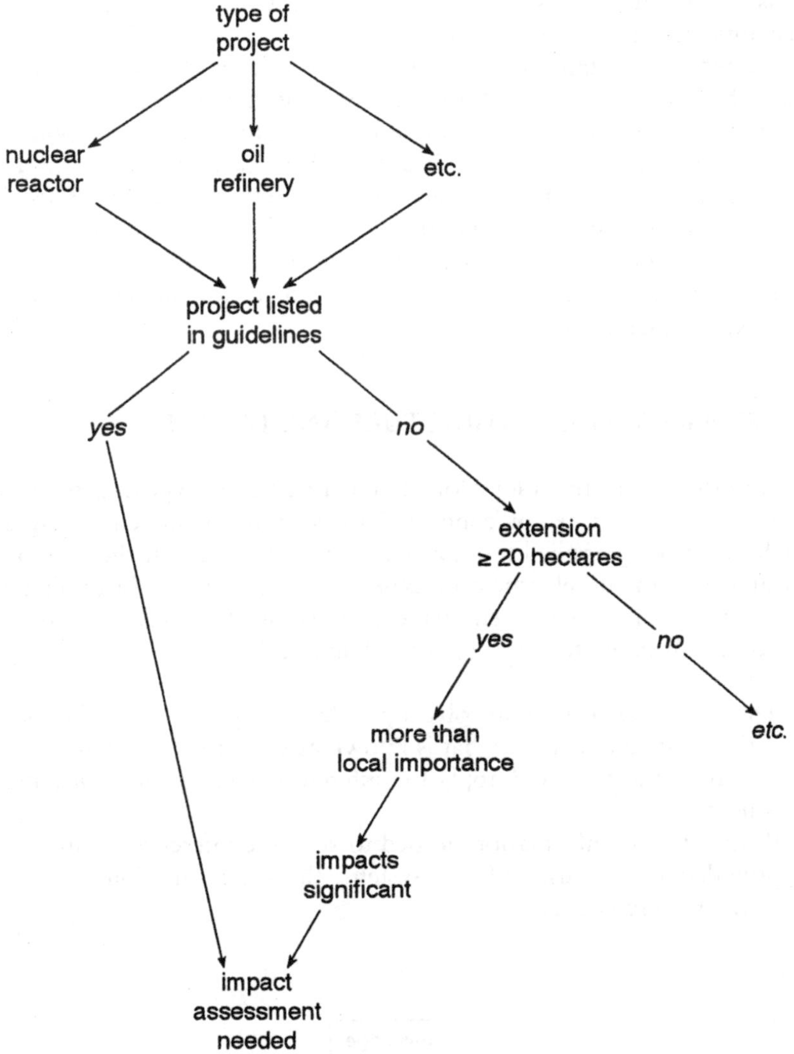

Figure 2.4 Data-first flow diagram.

production rules were used to derive conclusions to solve them. CASNET (developed in the early 1970s at Rutgers University to diagnose and treat glaucoma) used semantic nets as a basis of a model of the disease, linking observations to disease categories and these to treatment plans. INTERNIST (also known as CADUCEUS, developed at the same time at Carnegie-Mellon University in Pittsburgh for general medical diagnosis) had its central knowledge represented by a disease tree linked to sets of symptoms, to be matched to the data about the patient. PROSPECTOR (developed at Stanford University in the late 1970s to help field geologists assess geological deposits) contained a taxonomy of the geological world in a semantic net, and a series of geological "states" connected by rules. MYCIN (developed also at Stanford in the early 1970s to help doctors diagnose and treat infectious diseases) organised its substantive knowledge about types of patients, symptoms and diseases into classificatory trees, and applied the actual consultation using connected sets of rules. Although quite a few expert systems caught the attention in the 1960s and early 1970s, it is probably fair to say that PROSPECTOR and particularly MYCIN best exemplify the potential of production rules for this new approach to problem-solving and, in so doing, also provide a paradigm for the development of most expert systems today.

2.3 EXPERT SYSTEMS: STRUCTURE AND DESIGN

The idea that the methodology for solving a particular type of problem can be represented by a set of connected rules (and an inference diagram), which can then be applied to a particular case, has been at the root of the appeal and of the development of expert systems from the beginning and, to a certain extent, has given shape to what is still considered today a "standard" structure for these systems (Figure 2.5):

- The knowledge needed to solve a problem is represented in the form of if-then rules and kept in what is known as the *knowledge base*.
- To "fire" the rules and apply the inference chain, an *inference engine* is used.
- If the ultimate information needed to start the inference chain is to be provided by the user of the system, the right questions are asked through an *interface*.

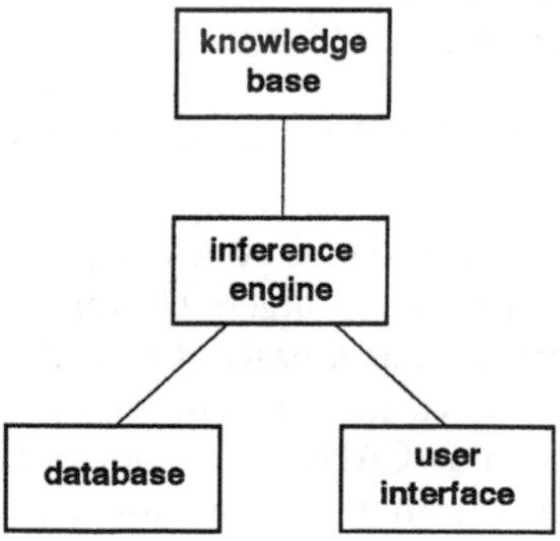

Figure 2.5 Typical structure of an expert system.

- If some of the information needed is to come from existing data instead of the user (or if the output from the system is to be stored), a *database* appropriate to the problem must be connected to the system.

MYCIN applied "backward-chaining" inference – deriving the necessary conditions from the conclusions sought and working out that way what information is needed – in what is now a well-established approach. In this context, the inference engine's role is:

- to derive what conditions need to be met for an answer to the main question to be found;
- to identify what rules may provide values for those conditions;
- to derive from those rules, in turn, what other conditions are needed to determine them;
- when no rules are found to derive information needed, to either find it in the database or ask appropriate questions of the user;
- once all the information needed has been found, to infer from it the answer to the overall question;
- finally, to advise the user about the final conclusion.

What was important and innovative at the time from the computing point of view, was that which part of the knowledge base would be used at any time, while running the system (the order of "control" evaluating the rules) was *not* pre-determined as in conventional computer programs – by writing the program as a particular sequence of commands – but would depend on how the inference was going in each case.[5] As information concerning that specific case was provided, the successive rules applicable at every stage of the inference would be "found" by the inference engine whatever their location in the knowledge base, without the need for the programmer to pre-determine that sequence and to write the rules in any particular order.

Although initially this type of inference logic was embedded in the MYCIN expert system linked to its rules about infectious diseases, it was soon realised that it could be applied to other problems as long as they could be expressed in the form of if-then rules of a similar kind. This led to the idea of *separating* the inference engine from a particular problem and giving it independence, so that it could be applied to *any* knowledge base, as long as its knowledge was expressed in the form of if-then rules. The new system developed along these lines became known as EMYCIN ("empty" MYCIN), and this idea has since been at the root of the proliferation

5 This style of program writing was taking one step further the growing preference in the computer-programming industry for so-called *structured programming*, which replaced traditional control changes using commands like "go to" by making all the parts of a computer program become integrated into one overall structure.

(commercially and for research) of a multitude of expert-system tools called "shells", empty inference engines that can be applied to any rule-based knowledge base. As these "shells" became more and more user-friendly, they contributed substantially to the diffusion of expert systems and of the idea that *anybody could build an expert system*, as long as they could express the relevant problem as a collection of linked if-then rules.

When applying an expert system to the solution of a particular problem, the inference may be quite complicated "behind the scenes" (as encapsulated in the knowledge base), but what the user sees is only a series of relatively simple questions, mostly factual. Because of this *black-box* approach, the user may be unsure about what is going on or about the appropriateness of his answers, and it is common for expert systems to include some typical additional capabilities to compensate for this:

(a) *Explanation*, the capacity of the expert system to explain its logic to the user, usually taking two forms: (i) explaining *why* a particular question is being asked, normally done by simply detailing for the user the chain of conditions and conclusions (as in the rules) that will lead from the present question to the final answer; (ii) explaining *how* the final conclusion was reached, done in a similar way, spelling out what the deductive chain was (what rules were applied) going from the original items of information to the final answer to the main question. For instance, in the example of the set of rules shown before to determine if a project needs an impact assessment, when the user is asked to quantify "the extension of the project (in hectares)" he/she could respond by asking the expert system *Why?* (why do you ask this question?) and what the system would do is to show how the answer is needed to determine a certain rule, in turn needed to evaluate another, and so on, leading to the final answer. The answer to the *Why?* question could look something like:

the *area of the project in hectares* is necessary to evaluate the rule that says that
if the extension of the project (in hectares) is greater than 20
then the scale of the project is of more than local importance

which is necessary to evaluate the rule that says that
if the scale of the project is of more than local importance
then the project impacts are likely to be significant

which is necessary to evaluate the rule that says that
if the project impacts are likely to be significant
or if the project type is included in the guidelines' list
then an impact assessment is needed

which is necessary to evaluate the final goal of whether *an impact assessment is needed.*

In a similar way, if the answer to the question was, for instance *23* (hectares), the system would conclude (and tell the user) that *an impact assessment is needed*. If the user then wanted to enquire how this conclusion was derived, he could ask *How?* and a similar chain of rules and known facts would be offered as an explanation, looking something like:

the conclusion was reached that *an impact assessment is needed* from the rule
 if the project impacts are likely to be significant
 or if the project type is included in the guidelines' list
 then an impact assessment is needed

because it was found that *the project impacts are likely to be significant* from the rule
 if the scale of the project is of more than local importance
 then the project impacts are likely to be significant

because it was found that *the scale of the project is of more than local importance* from the rule
 if the extension of the project (in hectares) is greater than 20
 then the scale of the project is of more than local importance

because it was found that *the extension of the project (in hectares) is greater than 20* from an answer to a direct question.

As we can see – and this is one of the reasons for the appeal of this approach – the rules are combined with standard phrases ("canned text" in the AI jargon) to produce text which reads almost like natural language. In the case of MYCIN, its explanation capabilities were considered so good that another system was developed from it (called GUIDON), which took advantage of these explanation facilities to be used for teaching purposes.

(b) *Uncertainty* can also be incorporated in the handling of the information: (i) there may be uncertainty associated with the user's response to a question, so he/she will need to provide a "degree of certainty" for every answer; (ii) the rules themselves may not be certain, but have a certain probability attached to their conclusion when the conditions are met, leading to the question of the *propagation* of uncertainty: if we are relatively certain of each of the conditions of a rule with varying degrees of certainty (probability), how sure can we be of its overall conclusion? MYCIN provided one of the models for many future developments in this area, by considering that, if all the conditions in a rule are necessary (they are linked by *and*) the probability of the conclusion will be the product of the probabilities of the conditions; on the other hand, if the conditions in a rule are alternative (linked by *or*), the probability of the conclusion will be equal to the probability of the *most certain* condition. PROSPECTOR used a more statistically sound approach based on Bayes' theorem, and these two ways of dealing with uncertainty have remained the most important bases for ulterior refinements (Neapolitan, 1990).

Central to expert systems is the separation between the knowledge involved in solving a problem and the knowledge involved in designing the computer software of the "inference engine". While the latter is the domain of specialised programmers, the former is the domain of *experts*, and it was essential for the development of expert systems to find ways of obtaining that knowledge from the experts. Techniques for acquiring and encoding knowledge were developed, and the field of "knowledge engineering" was born, aimed at extracting from the experts and representing the knowledge that would be at the core of these systems. Within this framework, *knowledge acquisition* became crucial to the design of expert systems (Figure 2.6).

With the popularisation and diffusion of expert systems technology in the 1980s after the first wave of pioneering projects, a variety of knowledge acquisition methods were suggested (Breuker and Wielinga, 1983; Grover, 1983; Hart, 1986; Kidd, 1987), which tend to be a combination of a few basic approaches:

- Consulting documentation like manuals, guidelines, even legislation, considered by the experts as the sources of their expertise.
- Studying past cases and the analyses experts made of them, maybe concentrating on a few key examples, or maybe looking at large numbers of them and using *automatic induction* methods to derive decision rules from their results.
- Discussing cases in person with the experts, be it current cases (although they may raise problems of confidentiality), or past cases of particular relevance, or even imaginary cases pre-prepared by the knowledge engineer.
- Watching experts apply their knowledge to current problems, maybe focusing on particular cases, maybe using comparative methods like "repertory grids".
- One variation of the last approach – a rather ingenious and probably the most productive of "case-based" approaches – is the knowledge engineer being guided verbally in the solution of a case by an expert who cannot see it (Crofts, 1988).

If more than one expert is used, the issue of consensus between experts may also need to be addressed (Trice and Davis, 1989). MYCIN also included pioneering work in knowledge acquisition, in the form of the system TEIRESIAS that was linked to it, built to allow the experts to interact

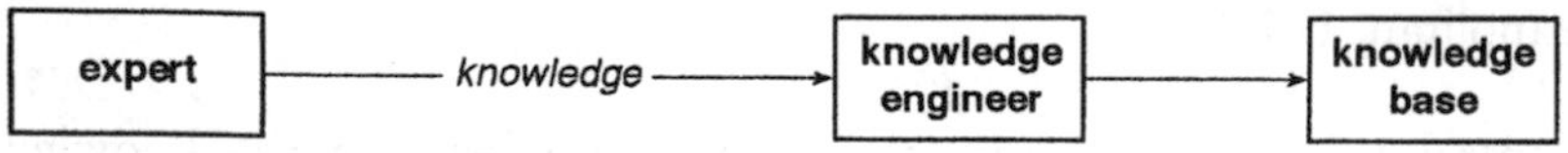

Figure 2.6 Knowledge acquisition and expert-system design.

directly with the expert system being designed and to improve its knowledge base, reducing the role of the expert system designer.

2.4 THE PROMISE OF EXPERT SYSTEMS?

One of the obvious questions to ask with respect to expert systems is about the *partial* nature of their success to date. Considering their theoretical simplicity and the universality of potential areas of application, *how is it that they are not the single most important problem-solving computer tool used in most areas of professional practice?*

In the 1980s it looked as if they were going to become the all-embracing problem-solving tools of the future, and their numbers were growing considerably, as the OVUM Reports showed (Hewett and Sasson, 1986; Hewett *et al.*, 1986). However, in the 1990s the interest seems to have faded, and expert systems are seen increasingly as no more than useful tools which can make a *partial* contribution to problem-solving, dealing with aspects that require some logical inference and some dialogue with sources of information (human or database). Also, while interest in expert systems has been apparent in traditionally technological fields, in fields more related to the social sciences – like town planning – the impact of expert systems has been minimal and research has tended to concentrate in very specific areas like building permits and development control (Rodriguez-Bachiller, 1991). This situation is not far from that identified in the US some years before (Ortolano and Perman, 1987), with city planning being among the few professions falling behind in the exploration and adoption of expert systems.

Thirty years after expert systems first came onto the scene, it is possible to look with hindsight at their emergence and growth, and identify some aspects which explain their popularity in the 1970s and 1980s but which, when put in the context of other improving computer tools and of more demanding and flexible decision-making environments, may also be at the root of their relative disappointment later on:

1 Expert systems represented at the time *a new style of interactive computing*, more personalised and friendly than the habitual "batch work" with mainframe computers. When, in the 1980s, the new trend of microcomputing (based on both PCs and Workstations) started to penetrate the market, this contributed also to this new style, reinforcing the appeal of expert systems even more. However, with these new personalised tools for communicating with computers – the screen and the keyboard, and later the "mouse" – also came a revolution in software, which started to take away the novelty that expert systems may have claimed for themselves:

- *Interactive software* started to proliferate, with menu-based interaction (we could call it "dialogue") as their backbone, much in the style in which expert systems interact with their users.
- A new generation of *interactive operating systems* – like Windows – also appeared, with "user-friendliness" as their selling pitch, based on menus of options (not unlike expert systems' questions) and with a much more "visual" approach based on icons and windows.
- *Database theory* originates from 1970, and during the 1970s and 1980s the availability of commercial database-management software became widespread, including advances such as the possibility of having programmable databases, or even so-called "intelligent" databases where the search for information can be subject to quite complicated rules, in a style again not too different from how an expert system's inference tree seeks information.

2 The *elegant logic of production rules* provided a universal framework for problem-solving so that just *one* type of structure provided for virtually all the needs of this new type of computer–user interaction. Production rules appeared as potentially universal tools capable of representing any kind of knowledge, with a logical framework providing at the same time the basis for the necessary inference to be carried out, and the basis for a sensible dialogue with the user:

- The tree of rules provided a simple mechanism for replicating human-like inference and deriving relatively complicated conclusions from answers to relatively simple questions.
- The same structure could be used to generate automatically the questions to ask the user in the dialogue (or the items of information to retrieve from databases).
- As an added bonus, the same rule structure also provided a mechanism for "why" and "how" explanation during the dialogue with the user.

Also, the easy representation of production rules in *quasi-natural language* – away from the specialised programming languages usual in computing at the time – suggested that anybody could master the use of these structures and write expert systems' knowledge bases:

- The knowledge base could be written by anybody who could articulate the expert knowledge, with no need for practically any computer expertise.
- Because the "control" of the computing process did not have to be pre-programmed explicitly, rules could be written/added into the knowledge base in any order, by anybody with sufficient knowledge of the problem but with no particular expertise in programming.
- Adding/changing knowledge in these knowledge bases would also be easy if the knowledge changed – for instance if new legislation came

about – just by adding/changing rules, by adding/changing "branches" in the inference trees.

This versatility, and the proliferation of expert system "shells" to manipulate these knowledge bases, attracted many to the idea of expert systems. It was almost too good to be true. In practice, however, all this promise proved to be more limited than at first thought when applied to larger and more complex problems in the real world.

First of all, it became increasingly clear that the process of extracting knowledge from experts was not without problems, and it has been acknowledged as a real "bottleneck" in expert system design for a long time (Davis, 1982; Buchanan *et al.*, 1983; Gaines, 1987; Cullen and Bryman, 1988; Leary and Rodriguez-Bachiller, 1988), derived from the difficulty of identifying and retrieving the expertise from experts who "don't know how much they know" and whose knowledge is often *implicit* (Berry, 1987). The expert may have forgotten the reasons why problems are solved that way, or he/she may have learned from experience without ever rationalising it properly. The difficulties for a knowledge engineer – not expert in the field in question – when interpreting the knowledge as verbalised or used by the expert (what is sometimes referred to as "shallow" knowledge) can be intractable, and suggest that the expert system designer should be at least a semi-expert in the particular domain of expertise, and not just an expert in knowledge acquisition. In the words of Gaines (1987), the solution to the knowledge acquisition bottleneck may lie, paradoxically, in "doing away with the knowledge engineer". This requirement that expert-system designers should be the experts themselves potentially solves the knowledge-acquisition problem but may create a new problem in that, given the relative scarcity of experts – this scarcity may be one of the reasons for developing expert systems in the first place – this approach may simply be replacing one bottleneck with another.

In terms of the universal applicability of production rules, Davis (1982) already pointed out how it could prove too difficult in larger expert systems to represent all the knowledge involved with just *one* type of representation like production rules. This problem could take the form of a need for some form of *control* in the middle of the inference more akin to traditional computer programming, and difficult to express in the simple syntax of if-then production rules. Sometimes it could be that complicated procedures needed to be activated when certain rules were applied, or it could be that *strategic* changes of direction were needed to change the avenue being explored when one avenue was proving fruitless, a point raised years earlier by Dreyfus (1972) against AI in general, and not just expert systems.

With respect to the explanatory capabilities of these structures (answering *why?* and *how?* questions) we have seen that what is offered as "explanation" is simply a *trace* (a "recapitulation", in the words of Davis, 1982) of the chain of inference being followed and not a proper explanation of the deeper causality involved, nor any of the many other possible elaborations which could be given (Hughes, 1987), even if this simplistic explanation seemed quite revolutionary when these systems first appeared. In terms of the user-friendliness of production rules written in quasi-natural language, it proved to be true when developing demonstration prototypes, but when building complicated knowledge bases the complexity was virtually no different from that of ordinary programming (Navinchandra, 1989). This becomes apparent very clearly, for instance, by the fact that inference "trees" become more and more difficult to draw on a piece of paper as the problem becomes more complex, as multiple connections become the norm rather than the exception, and trees become "lattices". One of the implications of this was that, against what was anticipated, adding to or modifying an existing knowledge base proved to be as difficult as in traditional programming – where it is often impossible to change a program by anyone other than the person who wrote it originally – and the idea of incremental modifications to the knowledge base as the knowledge evolved, started to appear much less practical than at first thought. And, once the user-friendliness of expert-system design disappears, these systems become similar to other computer tools, with their specific programming language requiring considerable expertise for their design and maintenance.

From this discussion, some of the possible reasons for the relative loss of appeal that expert systems have suffered in the last ten years become apparent. First, their innovative interactive approach to computing is not the novelty it once was. Second, the user-friendliness of expert-system design is questionable, as these systems can be almost as difficult to design and modify as other computer tools, except in the simplest cases. Third, the universal applicability and the durability of expert systems is also put into question, and these systems are at their best when applied to relatively small problems whose solution methods are well established and are unlikely to change. It is for these reasons that expert systems, which started offering great promise as universal problem-solving tools for non-experts, have been gradually reduced either to research prototypes in academic departments or to the role of tools – albeit quite elegant and effective – to solve relatively small and specific problems within wider problem-solving frameworks, which are dealt with by other means and with different computer tools.

2.5 FROM EXPERT SYSTEMS TO DECISION SUPPORT SYSTEMS

As problems become bigger and more complex, the simple rule-based logic of ES begins to prove inadequate, and needs a framework within which to perform its problem-solving. Also, as problems become more complex, their aims and solution approaches often become more tentative and open-ended, and for such exploratory problem-solving ES are less suitable. For ES to be applicable to a problem, the solution procedures for that problem must be "mapped out" in all their possibilities, so that the system can guide the user when confronted with any combination of circumstances. The problem-solving process is embedded in the expert system, and the user is "led" by the system which, in this respect, is not very different from traditional models or algorithms.

A Decision Support System (DSS), on the other hand, is designed to support problem-solving in less well-defined situations, when the decision-maker has to find his/her way around the problem by performing some form of interactive evaluation of possibilities. DSS are "interactive computer-based systems which help decision-makers utilise data and models to solve unstructured problems" (Sprague, 1980). The one-way evaluation implicit in traditional models and, to a certain extent, in expert systems, changes into an open-ended interactive evaluation (Janssen, 1990) where *the user* guides the system (instead of being led by it) through a process which is at the same time a problem-solving process and a *learning* process. There is a link between how a problem is defined and how its evaluation is performed: if a problem is completely defined – and there is consensus on its solution method – then a one-way evaluation approach using a model or an ES is appropriate. DSS are useful when the definition of a problem is open-ended, and therefore the evaluation required to solve it is also incompletely defined. Such "ill-defined" problems are characterised by (Klein and Methlie, 1990):

- the search for a solution involving a mixture of methods;
- the sequence of their use cannot be known in advance, as in ES;
- decision criteria which are numerous and largely dependent on the perspective of the user;
- the need for support not at predetermined points in a decision process, but on an *ad hoc* basis.

The fact that the phrase "decision support system" is quite meaningful and self-explanatory has contributed to its excessive use, with a tendency to apply it to any system used to support decision-making – which could potentially be applied to virtually all computer applications – but DSS developed historically as a quite specific and new approach to computer-aided decision-making. DSS research started in the late 1960s (in the 1970s they were called "Management Decision Systems") at several business schools: the Sloane School of Management at MIT, the Harvard Business School, the Business School HEC in France, and the Tuck School of Business Administration at Dartmouth College (Klein and Methlie, 1990). They came from the academic tradition of management science, and were seen as the culmination of an evolutionary process followed by successive generations of increasingly sophisticated computerised information systems for management (Sprague, 1980; Thierauf, 1982; Bonczek *et al.*, 1982; Ghiaseddin, 1987):

1 At the lowest level of sophistication, *non-real-time* Information Systems (IS) were based largely on "electronic data processing" (EDP) routines, and were oriented mostly towards "reporting the past".
2 Next, *real-time* Management Information Systems (MIS) were geared to "reporting the present", so that data were put into the system as soon as available, and summary reports were generated regularly to help decision-making.
3 Decision Support Systems (DSS) were designed to "explore the future" using interactive computing facilities to help the decision-maker, but not taking the decision for one.

Although there is no theory for DSS (Sprague and Watson, 1986), a conceptual framework evolved out of the IBM Research Laboratories in San Jose (California) in the late 1970s. DSS typically consist of (Bonczek *et al.*, 1982; Sprague and Watson, 1986):

- a set of *data* sources;
- a set of models and *procedures* (ES can be part of these);
- a set of display and report *formats*;
- a set of *control* mechanisms to "navigate" between the other three, which is the most important element, since in these systems it is the user who steers the system instead of being led by it.

If spatial information and/or spatial analysis are included, another set of spatial procedures and data may have to be added to the list above, and we are talking about a so-called Spatial Decision Support System (SDSS) (Densham, 1991), where GIS can – and often does – play an important role, as we shall see. What is also crucial as a complement to the navigation possibilities in DSS is that at the core of these systems there are:

- some kind of evaluation function to assess the quality of the options being considered and some criteria for "satisfying" (not necessarily optimising);
- "what-if" capabilities to test alternative combinations of procedures and data;
- some learning capability, so that when certain combinations or "routes" are proven particularly successful, the system can "remember" them for next time.

Within such systems, ES can play an important role (like GIS, models or other procedures) being called by the user to apply their problem-solving capabilities to particular aspects of the (large) problem (Figure 2.7).

In a way, DSS can be seen as complementary to ES, also helping with decision-making but in a very different way (Turban and Watkins, 1986):

	ES	*DSS*
Objectives	to replicate humans	to assist humans
Who decides	the system	the user
Orientation	expertise transfer	decision-making
Query	machine queries human	human queries machine
Client	individual user	possible group-user
Problem area	narrow	complex, wide

The most important feature of DSS is their flexibility, the user's control of the process, and the most important part of the DSS structure is the "navigator", which embodies that flexibility in the form of a range of choices of data, procedures, displays, etc. available to the user. Because of this inherent flexibility and open-endedness, the emphasis when discussing DSS structure has shifted towards discussing "DSS generators" (Sprague, 1980) rather than DSS themselves:

- The DSS is seen as a collection of problem-solving tools (models, data, etc.).
- The *DSS generator* is seen as a flexible framework for the user to construct the DSS over time; these "generators" can be seen as "empty" DSS, as DSS "shells", not too different from the expert systems shells we have already mentioned.

Because of the open-ended nature of the problems these systems are applied to, a standard linear design approach (analysis, design, implementation) cannot be used, but instead an *iterative* design cycle is used. The idea is

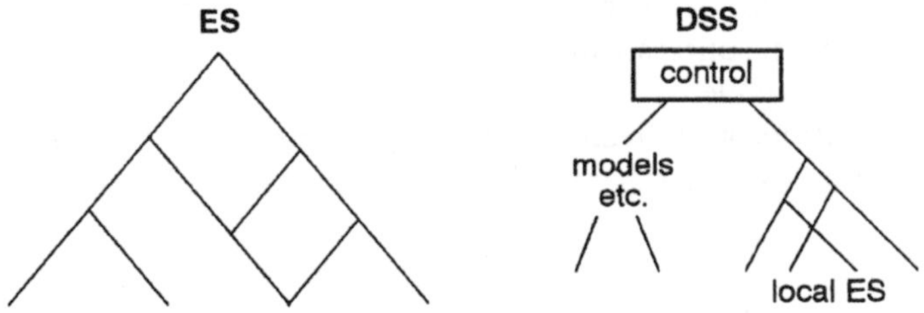

Figure 2.7 ES and DSS.

that the system will become modified (will "learn") with use; learning is integrated in the design process. In traditional linear design, looking back is seen as a failure, in DSS design it is seen as essential.

A survey by Hogue and Watson (1986) found that DSS design took less time when DSS generators were used, and also when the designers were people already working in the domain area of the problem, which finds parallels with the field of ES. Also, in comparison with ES, where one typical problem in expert system design is how to determine when a system is finished, in the case of DSS this does not present theoretical or practical problems. The aims of DSS are themselves open-ended, and the objective is not to develop a "finished" DSS, but to help decision-making in a cumulative learning process of successive changes and improvements.

2.6 CONCLUSION: EXPERT SYSTEMS ARE DEAD, LONG LIVE EXPERT SYSTEMS!

The conclusions from the discussion in this chapter are "mixed": on the one hand, the technical potential of expert systems to be good vehicles for the dissemination of good practice is clear. They represent precisely what is needed, the extraction of the expertise from those who know and making that knowledge available to those who don't know, with very positive additional connotations of top-down technology transfer within organisations. Expert systems represent a very powerful *enabling* technology. On the other hand, their association with specific forms of representation of the knowledge – like if-then rules and their associated inference trees – or with specific technologies – like the universal expert systems "shells" – can be limiting beyond the simplest demonstration prototypes. Such negative aspects suggest that the greatest contribution of "pure" expert systems is likely to be in relation to specific tasks within the overall problem-solving framework, rather than as "master-controllers" of the whole process. On the other hand, at a more general level, the basic principles of what we could call the expert systems "approach" are perfectly appropriate to what is needed:

- The whole approach is based on the know-how extracted from experts and accepted sources, and only to the extent that these exist, the approach is viable.
- The operation of the technology is highly interactive and user-friendly for the non-expert, relying all the time on natural language and feedback.

The paradox is that these traits – initially pioneered by expert systems – increasingly characterise most computer applications and, in this sense, while pure expert systems become relegated to being just another specialist computer technique, the expert systems "approach" has become mainstream and pervades practically all modern computer applications.

REFERENCES

Berry, D.C. (1987) The Problem of Implicit Knowledge, *Expert Systems*, Vol. 4, No. 3 (August), pp. 144–50.

Boden, M.A. (1977) *Artificial Intelligence and Natural Man*, The MIT Press.

Bonczek, R.H., Holsapple, C.W. and Winston, A.B. (1982) The Evolution from MIS to DSS: Extension of Data Management to Model Management, in Ginzberg, M.J., Reitman, W. and Stohr, E.A. (eds) *Decision Support Systems*, North Holland.

Breuker, J.A. and Wielinga, B.J. (1983) *Analysis Techniques for Knowledge Based Systems. Part 2: Methods for Knowledge Acquisition*, Report 1.2, ESPRIT Project 12, University of Amsterdam.

Buchanan, B.G., Barstow, D., Bechtal, R., Bennett, J., Clancey, W., Kulikowski, C., Mitchell, T. and Waterman, D.A. (1983) Constructing an Expert System, in Hayes-Roth, F., Waterman, D.A. and Lenat, D.B. (eds) *op. cit.* (Ch. 5).

Crofts, M. (1988) *Expert Systems in Estate Management*, Working Paper, Surrey County Council.

Cullen, J. and Bryman, A. (1988) The Knowledge Acquisition Bottleneck: Time for Reassessment, *Expert Systems*, Vol. 5, No. 3 (August), pp. 216–25.

Davis, R. (1982) Expert Systems: Where Are We? And Where Do We Go From Here?, *The AI Magazine* (Spring), pp. 3–22.

Dayhoff, J. (1990) *Neural Network Architectures*, Van Nostrand Reinhold, New York.

Densham, P.J. (1991) Spatial Decision Support Systems, in Maguire *et al.* (eds) *op. cit.* (Ch. 26).

Dreyfus, H.L. (1972) *What Computers Can't Do: The Limits of Artificial Intelligence*, Harper & Row, New York.

Gaines, B.R. (1987) Foundations of Knowledge Engineering, in Bramer, M.A. (ed.) *Research and Development in Expert Systems III*, Proceedings of "Expert Systems '86" (Brighton, 15–18 December 1986), Cambridge University Press, pp. 13–24.

Ghiaseddin, N. (1987) Characteristics of a Successful Decision Support System: User's Needs versus Builder's Needs, in Holsapple, C.W. and Whinston, A.B. (eds) *Decision Support Systems: Theory and Applications*, Springer Verlag.

Grover, M.D. (1983) A Pragmatic Knowledge Acquisition Methodology, *International Journal on Computing and Artificial Intelligence*, Vol. 1, pp. 436–8.

Hart, A. (1986) *Knowledge Acquisition for Expert Systems*, Kogan Page, London.

Hayes-Roth, F., Waterman, D.A. and Lenat, D.B. (1983a) An Overview of Expert Systems, in Hayes-Roth, F., Waterman, D.A. and Lenat, D.B. (eds) *op. cit.* (Ch. 1).

Hayes-Roth, F., Waterman, D.A. and Lenat, D.B. (1983b) (eds) *Building Expert Systems*, Addison Wesley.

Hewett, J. and Sasson, R. (1986) *Expert Systems 1986*, Vol. 1, USA and Canada, Ovum Ltd.

Hewett, J., Timms, S. and D'Aumale, G. (1986) *Commercial Expert Systems in Europe*, Ovum Ltd.

Hogue, J.T. and Watson, R.H. (1986) Current Practices in the Development of Decision Support Systems, in Sprague Jr., R.H. and Watson, R.H. (eds) *op. cit.*

Hughes, S. (1987) Question Classification in Rule-Based Systems, in Bramer, M.A. (ed.) *Research and Development in Expert Systems III*, Proceedings of "Expert Systems '86" (Brighton, 15–18 December 1986), Cambridge University Press, pp. 123–31.

Jackson, P. (1990) *Introduction to Expert Systems*, Addison Wesley (2nd edition).

Janssen, R. (1990) Support System for Environmental Decisions, in Shafer, D. and Voogd, H. (eds) *Evaluation Methods for Urban and Regional Planning*, Pion.

Kidd, A.L. (1987) (ed.) *Knowledge Acquisition for Expert Systems. A Practical Handbook*, Plenum Press, New York, London.

Klein, M. and Methlie, L.B. (1990) *Expert Systems: A Decision Support Approach*, Addison-Wesley Publishing Co.

Leary, M. and Rodriguez-Bachiller, A. (1988) The Potential of Expert Systems for Development Control in British Town Planning, in Moralee, D.S. (ed.) *Research and Development in Expert Systems IV*, Proceedings of "Expert Systems '87" (Brighton, 14–17 December 1987), Cambridge University Press.

McCorduck, P. (1979) *Machines Who Think*, W.H. Freeman and Co.

Navinchandra, D. (1989) *Observations on the Role of A.I. Techniques in Geographical Information Processing*, paper given at the First International Conference on Expert Systems in Environmental Planning and Engineering, Lincoln Institute, Massachusetts Institute of Technology, Boston (September).

Neapolitan, R.E. (1990) *Probabilistic Reasoning in Expert Systems. Theory and Algorithms*, John Wiley & Sons Inc., New York.

Newell, A. and Simon, H.A. (1963) GPS, A Program That Simulates Human Thought, in Feigenbaum, E.A. and Feldman, J. (eds) *Computers and Thought*, McGraw-Hill, New York.

Nilsson, N. (1980) *Principles of Artificial Intelligence*, Springer Verlag.

Ortolano, L. and Perman, C.D. (1987) Expert Systems Applications to Urban Planning: An Overview, *Journal of the American Planners Association*, No. 1, pp. 98–103; also in Kim, T.J., Wiggins, L.L. and Wright, J.R. (1990) (eds) *Expert Systems: Applications to Urban Planning*, Springer Verlag.

Pratt, V. (1987) *Thinking Machines*, Basil Blackwell.

Quillian (1968) Semantic Memory, in Minsky, M. (ed.) *Semantic Information Processing*, The MIT Press.

Rich, E. (1983) *Artificial Intelligence*, McGraw-Hill Inc.

Rodriguez-Bachiller, A. (1991) Expert Systems in Planning: An Overview, *Planning Practice and Research*, Vol. 6, Issue 3, pp. 20–5.

Rumelhart, D.E., McClelland, J.L. and the PDP Research Group (1989) *Parallel Distributed Processing*, The MIT Press, Cambridge (Massachusetts), 2 Vols.

Sprague Jr., R.H. (1980) A Framework for the Development of Decision Support Systems, in *MIS Quarterly*, Vol. 4, No. 4 (June).

Sprague Jr., R.H. and Watson, R.H. (1986) (eds) *Decision Support Systems: Putting Theory into Practice* (Introduction, by the editors), Prentice-Hall.

Thierauf, R.J. (1982) *Decision Support Systems for Effective Planning and Control*, Prentice-Hall.

Trice, A. and Davis, R. (1989) *Consensus Knowledge Acquisition*, AI Memo No. 1183, Massachusetts Institute of Technology Artificial Intelligence Laboratory.

Turban, E. and Watkins, P.R. (1986) Integrating Expert Systems and Decision Support Systems, in Sprague Jr., R.H. and Watson, R.H. (eds) *op. cit.*

Winston, P.H. (1984) *Artificial Intelligence*, Addison Wesley.

3 GIS and impact assessment

3.1 INTRODUCTION

This chapter reviews GIS applications concerning only the "natural" environment and Impact Assessment in particular, as they have been reported in the published literature.[6] One of the striking features of the literature is the relatively small proportion of accounts of GIS use that reaches the public domain in books or research journals, with the vast majority appearing as papers given at conferences – often sponsored at least partially by GIS vendors – with no follow-up publications afterwards, or as short articles in magazines heavily dependent on GIS advertising (*GisWorld, GeoWorld, GisEurope, Mapping Awareness, GeoEurope* are typical examples). In such accounts, often the interest does not lie in theoretical or technical issues raised by the particular application, but in the very fact that it happened, in the fact that GIS technology was used. This is typical of the current stage in GIS development, where much of the interest is in the *diffusion* of this technology – who is adopting it and how fast – just as with other technologies before. The proliferation of such outlets for the monitoring of GIS diffusion also provides very useful market research for the industry itself.

The chapter starts by putting Impact Assessment (IA) in the wider context of impact management – to be discussed in Chapter 4 – and the use of GIS for IA is discussed in its different levels of complexity: GIS just for mapping, GIS linked to external models, GIS using its own functionality, and combinations of the three.

3.2 IMPACT ASSESSMENT AND ENVIRONMENTAL MANAGEMENT

The introduction to GIS in Chapter 1 indicated how much of the functionality of these systems is more directed to the solution of cartographic problems

6 Rodriguez-Bachiller (2000) includes an earlier version of this bibliographical review.

than to solving substantive analytical problems, even if the situation is changing as this technology evolves. It is not surprising therefore that the relatively complex technical operations involved in the core of Impact Assessment have made in the past only limited use of GIS. In the UK, GIS has been absent from virtually all Environmental Statements up until the end of the 1990s[7] and, even afterwards, GIS use has been limited to displaying a few maps without any analytical manipulation of them. In terms of published references worldwide, Joao (1998) already pointed out in her brief review the paradox that, while environmental applications of GIS are very numerous, IA applications of this technology represent only a fraction, quoting as an indication the fact that in the Database GEOBASE (covering usage between 1990 and 1996) she found only 1.2 per cent of all GIS-related references being concerned with IA, and only about 6 per cent of the references related to IA involving GIS. The bibliography in Rodriguez-Bachiller (1998) also showed this apparent contradiction: more than half (53 per cent) of all GIS applications recorded were concerned with the environment, but only 8.4 per cent were concerned with IA as normally defined.

Over time, the relative importance of different areas of GIS application has changed considerably. Updating the information in Rodriguez-Bachiller (2000),[8] Figure 3.1 shows the relative "share" of various areas of GIS application, not in absolute numbers of publications – this would only be accurate if the bibliographical reviews had covered the same or equivalent sources every year, which they do not – but in percentages of all the publications recorded each year. We can see that the share of environmental applications – the sum of "rural", "environmental" and EIA – seems to be declining over time, as GIS use in transport and various services (public, private, "utilities") increases, although this is probably not an indication of a decline in environmental GIS use, but a reflection of a fast increase in the diffusion of GIS in these other growing sectors. The low share of IA applications does not seem to vary much over time.

Undoubtedly, this apparent anomaly is partially due to the mentioned mismatch between the relatively simple analytical functionality of GIS and the technical complexity of impact prediction and assessment. However, it is suggested here that it is also due to the relatively narrow definition of IA that is normally used, which tends to include only the technical core of IA consisting of impact scoping, prediction and mitigation. On the other hand,

7 Judging from the collection held at the Impacts Assessment Unit at Oxford Brookes University – a sample of about 25 per cent of all EIS produced in the UK covering the complete period since EIA was formally introduced – only since 1998 have some statements contained GIS (Arc View) maps (Wood, 1999c, personal communication).

8 That publication updates an earlier bibliography in Rodriguez-Bachiller (1998), which looked at GIS magazines (of the type already mentioned), books, articles and conference proceedings from the late 1980s.

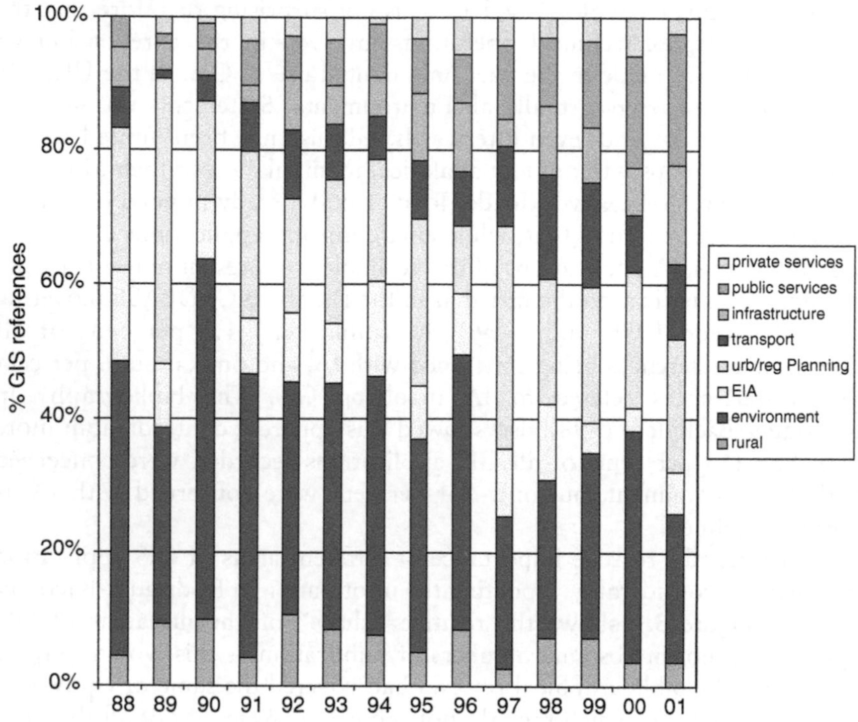

Figure 3.1 Areas of GIS applications during the 1990s.

if we broaden our view, even from the abbreviated description of IA in Section 1.6 we can appreciate the wide range of environment-related operations that *really* constitute IA:

1 Appraising the environment and assessing its quality and sensitivity, needed for the determination of the key impacts (scoping) which need investigation.
2 Identification of all potential impacts from a project to determine if it requires an impact study (screening) and which future impacts ought to be studied (scoping).
3 Consulting the public and specific interest groups about the significance of impacts, about alternative locations for the project and about possible mitigation measures.
4 Modelling and forecasting the evolution of the environment without the project, to establish the various baselines for comparison with the impact predictions.

5 Forecasting the impacts on that environment of the particular project, the impact prediction as such which is included in all IA reports.
6 Forecasting impacts from other projects likely to add their influence to that of the project, to determine possible cumulative impacts.
7 Assessing the significance of the likely impacts on the environment by comparison with the relevant standards.
8 Establishing possible mitigation measures to counteract any significant effects on the environment identified in the previous stages.
9 Monitoring the actual impacts once the project is under way for correction and mitigation or for reassessment.

What is normally considered IA constitutes the central part of this list, but the wider definition of IA also must include other tasks (in particular 1, 4 and 9) which serve the purpose of general environmental management but are also essential to good IA. One of the reasons why the relatively narrow definition of IA is normally used as opposed to the wider definition is probably that the two involve not only different sets of operations, but they are usually performed by *different actors*:

1 Identifying, forecasting and assessing project impacts with varying degrees of public consultation (tasks 2, 3, 5, 7 and 8, sometimes also 6) – what we can call IA as such – are project-specific and usually the responsibility, in the US and Europe, of those agencies or actors behind the project being assessed, the "developers".
2 On the other hand, monitoring, assessing and auditing the environment (tasks 1, 4 and 9 above) – what we can call *environmental management* – are also essential to IA but are not necessarily associated with any project in particular, and are usually carried out by large organisations (sometimes in the public sector) or environmental agencies.

In relation to this distinction between Impact Assessment and environmental management, one particular environmental management task, *environmental modelling* and *forecasting* (unrelated to any particular future project) is crucial to the baseline part of IA, but tends to "fall between two stools" and not be systematically performed by anyone. Developers do not have the data and resources to undertake it for an area where they are involved in just one project, and larger organisations and environmental agencies very rarely consider it part of their terms of reference to keep the kind of *ongoing simulation* of the environment in all areas of the country that this would entail. It is therefore not surprising that this part of IA is very rarely done, or done well, and baseline studies usually confine themselves to the presentation of the environmental situation at the time of the study, but with little or no forecasting.

If one considers IA as the project-based process mainly carried out by developers, it is not surprising to find that GIS is scarcely used, given the considerable costs not only of the expertise and the hardware/software (important bottlenecks years ago but gradually becoming less of an obstacle) but of the *data*, as Joao and Fonseca (1996) found in their small survey of environmental consultants. Even if that survey had a low number of respondents, it is interesting that the time and cost of setting up a GIS database to be used only for *one* project was quoted as the most important drawback of GIS, while the more traditional problem of start-up costs of hardware and software was the second most important, followed by lack of digital data and training requirements for the staff – not all IA consultants can afford to have up-to-date GIS experts. Nutter *et al.* (1996) also pointed out the difficulties in IA with GIS data managers, as well as the conflicts between the rapidly changing GIS technology and the staff involved. Although average training and hardware/software costs diminish within an organisation as GIS is applied to more projects, data problems are usually specific to only one project, unless an organisation specialises in IA in the same geographical area – of which there is no evidence, at least in the UK – and it is these very high one-off costs which are likely to be the strongest deterrent against GIS. In less developed countries, resource-related problems are likely to be even greater (Masser, 1990), and Warner *et al.* (1997) repeat the "health warning" about GIS data accuracy in developing countries, where data are collected only sporadically (and often from remote sensing without "ground-truthing"), not reflecting fast-changing seasonal situations which can make all the difference for IA.

This chapter concentrates on reviewing GIS applications which are related more to those tasks listed above linked to the technical core of Impact Assessment as such. Those concerned with environmental management will be reviewed in the next chapter.

3.3 THE ROLE OF GIS

Whether GIS is used for environmental management or IA, an aspect which is crucial to our understanding of the contribution of GIS is the *role* that these systems play and the sophistication of their contribution. We can express this by the degree to which GIS is used just as "provider" of information (maps or data for a technical task), or as a true analytical instrument:

1 At the lowest level of sophistication, GIS may be used *just for mapping*, for the production of maps of the environment, of the project, or of particular impacts from it, to provide visual aids to researchers or managers who will use this information in a non-technical way and externally to the system.
2 At the next level, GIS can itself be involved in *technical analytical tasks*, which can be internalised to different degrees into the GIS:

(i) The GIS can provide data (more or less prepared or "pre-processed") to an *external model*, programmed outside the GIS and "coupled" in some way to it. In a similar way, GIS can be used to display output (more or less manipulated or "post-processed") from such models.
(ii) The *internal functionality* of GIS – buffering, overlay, map algebra, visibility analysis, etc. – can be used for the task in question. In such cases, it is also useful to distinguish whether the GIS is set up to be operated *hands-on* by a relatively expert user, or has been *pre-programmed* so that a non-expert user can apply it.

3 Finally, the pre-programmed approach just mentioned can reach the sophistication of the GIS being integrated with an interactive system (an *Expert System* for example), so that the operation of the GIS and its links with other tools – if any – are guided by the user's choices in "dialogue" with the system, used as a *decision-support* tool. The focus in this chapter and the next is on GIS applications *not* involving decision-support tools. Expert Systems and other decision-support tools will be discussed in Chapter 5.

Updating the information in Rodriguez-Bachiller (2000) – that classifies GIS references using similar categories – we still see a fairly balanced distribution between these different levels of complexity in GIS use over the whole period 1988–2001: GIS for mapping is – somewhat unexpectedly – quite frequent, amounting to 27 per cent of all cases; GIS linked to external models accounts for 18 per cent; and more than half of the cases involve some degree of expert pre-programming, be it to handle GIS' internal functionality or to link these systems to external models or to wider systems. Over time (Figure 3.2) the share of the most sophisticated approaches ("decision support") seems to be declining, as the relatively simpler level of use ("mapping") seems to be on the increase. This seems to contradict a natural expectation of increased sophistication with time, although, as pointed out earlier when discussing Figure 3.1, it is probably *not* due to a decline in more sophisticated GIS use, but to a fast increase in its use at the lower end of the scale, as this technology is diffused to more and more countries and to more and more new areas of application.

3.4 GIS FOR IMPACT ASSESSMENT

As with general environmental modelling, reviewed in the next chapter, a series of conferences mark the evolution of the interest in the use of GIS for Impact Assessment. The difference with modelling, however, is that with the passage of time and the increase in skills and knowledge, the interest in the use of GIS does not seem to have increased – taking research into deeper and deeper layers as we would expect – but rather the opposite.

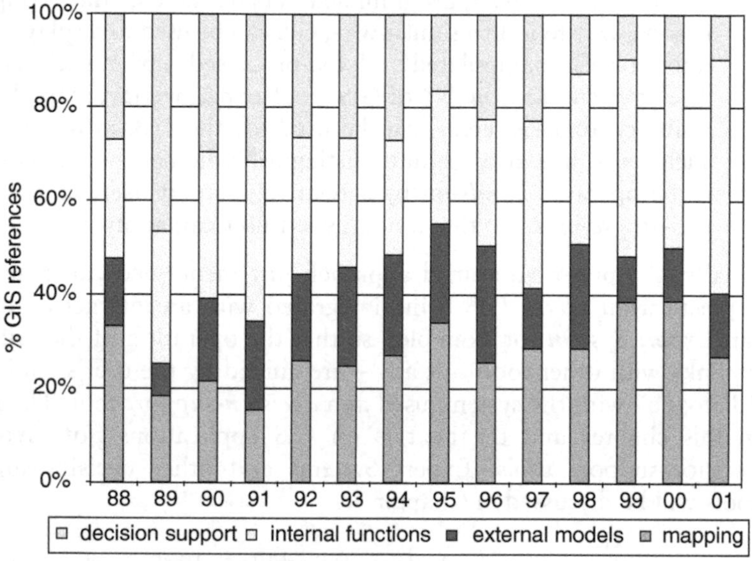

Figure 3.2 Complexity in GIS use from the late 1980s.

As one of the first steps on that road, Guariso and Page (1994) report on a conference in 1993 on Information Technology (not only GIS) for IA, where GIS features prominently and arguments about its potential abound. Around that time, Eedy (1995) lists the potential of GIS for various aspects of Environmental Assessment, based on their capacity for storing information in "real time", providing data for models, and performing map overlay, buffering, viewshed analysis, etc. The World Bank (1995) provides a similar argument, pointing out the different needs of project-based GIS and institution-based GIS at national/regional level. The conferences of the International Association for IA (IAIA) – meeting annually since 1980 – also take notice, with a "peak" of interest in GIS in 1996 when, at the conference held in Portugal, a whole section was devoted to "GIS for EIA", with seven papers in it and many more on the same subject in other sections. Then, GIS in later IAIA conferences gradually fades away: in 1997 (in New Orleans) there are only five papers mentioning rather unsophisticated uses of GIS, in 1998 (in New Zealand) there are six papers mentioning the potential of new technologies like GIS, and in 1999 (in Glasgow) there are just a couple of papers underlining the use of GIS, even if other papers mention its use. It seems as if the novelty of GIS as the object of research has been exhausted rather rapidly – for IA experts at least – and it only remains as a tool to be used. On the other hand, when we looked at the relative importance of different areas of GIS application over the years (Figure 3.1), we saw that the share of EIA remained fairly constant. This contradiction seems to suggest an *asymmetrical* relationship between GIS and IA: while IA maintains – and even increases – its appeal in the field of GIS application over the years, the latter has kept its use in IA mainly as a practical instrument, in ways which we shall now go on to review.

3.4.1 GIS mapping for impact assessment

When GIS are first adopted by organisations (and by professional and political cultures), the simple production of maps is often their most frequent use – as in British local authority Planning departments (Rodriguez-Bachiller and Smith, 1995) – and it makes sense to expect that only as confidence and experience grows, GIS functionality is increasingly used in its more technical aspects. For this reason, it is often over-simplistic to put a GIS application in the "just mapping" category, as it is likely to evolve over time into more sophisticated uses. The extent to which this applies to some of the applications classified here in the mapping category is difficult to tell from the published references, and could only be determined with "longitudinal" studies following the development of these applications over time, a task well beyond this review, or the project it developed from. On the other hand, the large number of applications which seem to be aimed mainly or exclusively at the production of maps[9] – at least at the time of publication – makes it impossible to ignore this level of rather superficial GIS use, even at the risk of under-rating the real depth and complexity of some of these applications.

Accordingly, we start by looking at GIS uses at this level, when these systems seem to be applied (maybe only temporarily) just to the production of maps for external visual analysis. Mapping impact-related information can play an important role in IA, simply by displaying the information and letting the viewer make the connections. Collins *et al.* (1986) give an early example of producing maps to assess habitat risk from a proposed new town using satellite data, Henzel *et al.* (1990) relate in the same way ground water pollution to different farming practices, and Dodge (1996) maps fire incidents in South Wales *animating* them over time to see if there are any patterns the eye can detect. This approach has been effectively used for anticipating and assessing visually the probable impacts from the *siting of new facilities*:

- Siderelis and Tribble (1988) used GIS maps to support a bid for the location of a particle accelerator in North Carolina, and Oliver (1988)

9 In Rodriguez-Bachiller (1998), 25 per cent of all references, by far the most numerous, are in this category.

did the same for Illinois (at the time, 25 States were submitting bids to the US Department of Energy for the location of such facilities).

- Pereira and Mourab (1999) use GIS maps to illustrate visually the potential impact of different locations for the planned new bridge over the river Tagus in Lisbon.
- Roper (1996) reports how the duration of a road project in Florida was cut down from the expected eight years to four by using GIS to carry out a study of probable impacts and producing maps of the population in the areas likely to be affected.

Ex-post impact *monitoring* is a typical area where simple mapping can provide very useful insights: Friel *et al.* (1993) use satellite data to monitor an oil spill in Tampa Bay, and Allen (1995) does the same to monitor the Komi oil spill. Wagner (1994) maps the impacts of a new car-manufacturing plant, Corbley (1995) uses GIS maps to monitor the aftermath of a hurricane in Florida, Meldrum (1996) monitors in this way abnormal levels of radiation (after Chernobyl) in the UK, and Evers and Most (1996) localise and map emissions from landfills in the Netherlands for their Emission Inventory System of that country, and Longhorn and Moreira Madueno (1998) monitor toxicity from an open-cast mine in the Coto de Dona Ana (Spain). Also, Brown (1994) uses mapping to help with impact *mitigation* in South Carolina, to identify and assess wetland areas and the opportunities they offer, helping to detail mitigation categories.

It is often the input–output links between the GIS and the outside that attract attention: satellite imagery is mentioned frequently as an ideal source of information, and Rodbell (1993) discusses the potential of using GPS for accurate mapping of impacts. In particular, *multimedia* GIS for impact recording and display is quite prominent since the early 1990s, when video information (Shiffer, 1991, 1993) and noise simulation (Dubbink, 1991) were used to improve interaction and collective participation in environmental decision-making. And then *hypertext* (or "hypermedia", a hierarchical way of storing information that allows "nested" zooming in and out of different items) came into the scene to link it all up: Fonseca *et al.* (1994, 1996) discuss a system integrating GIS maps, photographs, videos, 3D graphics, text and sound – all in hypertext mode – for the Expo '98 in Lisbon, to help with impact scoping, prediction and visualisation (including "walk through" effects) for public participation. In fact, *participatory* IA is one of the themes at the forefront of Impact Assessment: Patindol (1996) discusses a system for participatory IA using a GIS database of environmental risk information and also economic evaluation of enhancement measures for a hydroelectric power project, and Richardson (1999) indicates the potential of GIS to help participation in areawide Strategic Environmental Assessment, another emergent area attracting increasing interest in IA.

3.4.2 GIS linked to external models for IA

As expected, this type of GIS use is central to IA, with GIS providing data for models and then being used to display the results from those models. The linkage of a GIS to a model can be related to any of the typical stages in a modelling exercise, which in turn will define the type of use made of the GIS:

- The model can be at the *design* stage, when its form and the intervening variables are being defined, and the GIS can provide the base data to be analysed and modelled.
- The model may have been designed already and it only requires *estimation* – calculating its parameters and their statistical significance – for the particular area or case study, as is often the case with environmental models; the GIS provides the data to the external model, which is estimated by statistical means.
- Finally, a model may have been already designed and estimated for a particular situation and it only requires *application*, using it for the purpose it was designed for, be it predicting environmental events, predicting impacts, or any other simulation; the GIS provides the data and registers the results, maybe also in map form.

Sometimes the distinction between these stages is blurred, and several stages are involved. For example, with some types of models (like regression models, widely used in environmental modelling) design and estimation are combined, as the estimation of the significance of parameters is used at the same time to include or exclude intervening variables in the model (design).

Linking these two technologies raised, from the early days, a number of methodological issues (Nyerges, 1993). In particular, the question of *how* models are connected to GIS is not trivial, and is reflected in the number of references on the subject that appeared in the first half of the 1990s. Mandl (1992) identifies three ways in which the connection between GIS and external models can be organised: (i) so-called "loose coupling" of GIS and models, where the two exchange data and results through files; (ii) "tight coupling" where not only data but other information is shared between the two "tool boxes" of the GIS and the model; and (iii) full integration of all the modelling and spatial operations into one software product, which is very rare. Fedra (1993, 1996) identifies the broad alternative approaches to integration very much along the same lines: (i) as two separate systems exchanging files, or (ii) through deeper integration sharing memory space with transparent transfer between the two, either by using a higher-level language built into the GIS or by using a tool kit that talks to the GIS functionality and to the models. Raper and Livingstone (1996) argue that integration should take place at the highest level, using "object orientation" as the integrating approach. This issue is still today one of the stumbling blocks of off-the-shelf GIS – which are not particularly easy to link in either of these ways – which has been periodically revisited in the research literature (Jankowski, 1995; Dragosits *et al.*, 1996), be it dealing with the simulation of impacts or with wider environmental aspects.

The simulation of impacts can be applied to help at the *planning* stage of projects. For example, Ladha and Robertson (1988) describe an early GIS system (conceived as early as 1972) for route planning of power lines based on impact prediction and its perception by the population; Schaller (1995) describes how landscape analysis models and water flow models were used to predict ecological transition and danger to species, for different alternative ways of building the Rhine-Main-Danube connection in Southern Bavaria. Guimaraes Pereira and Antunes (1996) use genetic algorithms to search for alternative sites for facilities using map algebra with IDRISI; Wu (1998) uses GIS and a cellular automata model to simulate urban encroachment on rural land, and Jones *et al.* (2000) combine Map-Info with an environmental prediction model to help "planning for a sustainable city". However, the most common use of impact simulation is in a later stage in the development of projects, and examples incorporating GIS are numerous, covering a wide range of impact types.

Water pollution in different forms is one of the areas of impact that has attracted more interest than most in simulation modelling linked to GIS:

- Craig and Burnette (1996) linked a GIS to a water quality simulation model; Kuhlman *et al.* (1994) modelled non-point pollution, and Bennett and Vitale (2001) use GIS in a similar way linked to a model.
- Simpson (1990) and Leipnik (1993) applied this approach to *groundwater contamination*, and Gauthier *et al.* (1992) simulated the contamination of water resulting from the use of pesticides in agriculture; Jankowski and Haddock (1996) discuss a seamless integration of Arc-Info and an agricultural pollution model; Bhaduri *et al.* (2000) use a model with GIS to assess the hydrological impacts of land-use changes, and Harman *et al.* (2001) use GIS and a model to assess potential contamination sources.
- Harris *et al.* (1991) describe research at the university of Madison (Wisconsin) into a system to simulate non-point pollution of *run-off* water in urban areas after rain, looking at all pollution sources in urban environments (including the roofs of buildings, and the like); water run-off impacts linked to urbanisation are predicted by Mattikalli and Richards (1996) using a run-off model and Spot satellite data to simulate surface water quality changes resulting from land-use changes in the river Glen watershed (South Lincolnshire), and Brun and Brand (2000) use a simulation model for the Gwynns Falls watershed in Baltimore; in the related area of *soil pollution*, Schou *et al.* (2000) use modelling and

GIS to estimate the economic effects of applying different pollution-tax policies in the Vejle Fjord in Denmark.

- An area of water pollution that has attracted particular attention is related to the occurrence of *oil spills* (Roth, 1991; Green, 1996). Belore *et al.* (1990) discuss an early *interactive* system (using SPANS) where the co-ordinates of an area are input and the pollution situation around a spill area is simulated, and at-risk populations of existing species are given; French and Reed (1996) use a similar model for oil and chemical spills, and Li *et al.* (2000) discuss issues of GIS data quality related to the simulation of coastal oil spills.

On a different aspect of water modelling, Wu and Xia (1991) use a flood-simulation model linked to a GIS for Bangladesh, and Rodda (2001) discusses a similar approach for Europe. From another angle – this time related to *snow* – GeoWorld's News Link reports attempts to predict avalanches using GIS technology (Geo World, 2001). *Earth movements* are modelled by Boggs *et al.* (2000) combined with a GIS, to estimate geological impacts derived from mining in the Northern Territory of Australia.

On *air pollution*, Anderson and Taylor (1988) provide an early PC-based mapping system (not a fully developed GIS) which maps the output of a model simulating air pollution from traffic. Osborne and Stoogenke (1989) link GIS and air-dispersion models for EIA, Moore (1991) uses a model to simulate point pollution and construct health-risk maps for California, and Rinaldi *et al.* (1993) add up air pollution simulations from all sources in an area to produce overall maps of cumulative impacts. Noe (1993) uses a plume model linked to a GIS to simulate air pollution for urban populations, Fouda *et al.* (1993) applies a similar approach in the Sixth of October city in Egypt, and Kim *et al.* (1996) use Arc-Info to see the population impact of air pollution odour.

On a different note, Goncalves Henriques *et al.* (1992) model forest fire impacts. Related to other types of pollution, Krasovskaia and Tikunov (1991) report on a system which calculates pollution potential from records of existing concentrations of pollutants in the Kola peninsula (Siberia), and Van der Perk *et al.* (2001) discuss a system to estimate the transfer of radio-active agents through the food chain after the Chernobyl accident.

On *noise*, Schaller (1992) reports on a system to predict the effects on the environment of increased traffic in Munich, and the system is also applied to the new international airport in that city. Bilanzone *et al.* (1993) simulate noise levels for urban areas in Ancona (Switzerland) to be compared with tolerance to different land uses in order to predict social conflicts and suggest mitigations, and Lam *et al.* (1999) assess road traffic noise impacts. *Aircraft noise* in particular has been the subject of considerable interest: Reddingius and Finegold (1990) use a model (with the raster-GIS GRASS) to predict noise effects of aircraft at the planning stage of an Air Force base; Zhuang and Burn (1993) suggest using a similar system to map areas of various noise levels and the populations in them (including the property owners) to help in the management of an airport.

The simulation of *visibility* impacts can be done relatively easily and efficiently using GIS' internal functionality – as we shall see in the next section – and, when this type of analysis is done with GIS linked to *external software*, it is usually linked to CAD tools used as visualisation aids. Most applications of this kind try to simulate the visual impact of urban development but, occasionally, extensions to the natural environment are also considered: Mayall *et al.* (1994) discuss the use of such links for landscape visualisation in an urban context (showing urban developments in perspective) discussing how it could be extended to the natural environment, and Shang (1995) describes a similar combination of computer tools to assess the visual impact of different silvicultural systems.

Just as in environmental management, there is a type of modelling activity for IA which is often borderline with impact prediction, and that is *model building*. Mallants and Badji (1991) use satellite data to construct a model of the rainfall–runoff relationships for EIA, and Bernardo *et al.* (1994) discuss a system for Setubal (Portugal) where local perceptions of flood hazards were modelled for different social groups and a hydrological model was integrated with a GIS so that impacts could be assessed using population's perceptions.

3.4.3 Using GIS' own functionality for IA

Already in the late 1980s, Kramer (1989) sketched out the methodology for impact assessment with the standard functions available in GIS at the time, and applied it to the case of large scale commercial developments. Similarly, Gray (1993) developed a system for environmental assessment of industrial land and defined some of the requirements which he believed "sophisticated" GIS (purpose built for environmental assessment) should have – including collections of ready-made tables with the most common features and attributes used in environmental evaluation – as opposed to "simple" GIS where the user must define the data models to be used: the entities and their attributes, the links between them, etc.

Urban development can be used in IA-related GIS both as the recipient of impacts or as their cause: for instance, Katz (1993) uses GIS to simulate the impacts *caused by* different development scenarios on the natural landscape in Ottawa (Canada). On the other hand, Van Slagmaat and Van der Veen (1990) describe a research prototype (the REMIS project) for Dutch municipalities to map potentially hazardous buildings in a built-up area from the point of view of noise, and buffer zones around them, sensitive areas, etc. Dale *et al.* (1998) use GIS map algebra to assess the environmental impact of land uses in Tennessee, Le Lay *et al.* (2001) produce a risk map for wildlife species in urban areas using GIS map algebra, and Young and Jarvis (2001) use a similar approach to predict risks for urban habitats.

Andrews (2001) discusses the application of the "industrial ecology" approach to urban development, with an example application using GIS for Trenton (New Jersey).

This type of general *risk assessment* – overall vulnerability to a whole range of impacts derived from the area's location and characteristics – can also be applied to non-urban areas: for example, Kooistra *et al.* (2001) use GIS to study spatial variability in ecological risk in Dutch river floodplains. Also related to population settlements as recipients of impacts, the field of *disaster planning* has shown obvious potential for the application of GIS technology: Smith and Greenway (1988) discuss an early computer system (not a fully fledged GIS) for the assessment of flood damages to property, also used for the evaluation of possible mitigation measures; Watson (1992a,b) applies GIS to the evaluation of the potential damage from hurricanes in South Carolina, and Hill *et al.* (2001) discuss a similar approach to tornadoes and floods.

The impact of *transport facilities* (traffic impacts as well as direct impingement of the facilities on the environment) is a well-established area of GIS application: already Moreno and Siegel (1988) lay down the methodology for corridor siting based on analysis of their potential impacts, Paoli *et al.* (1992) discuss a system to decide transportation routes on the basis of their impacts on natural and man-made environments, and Appelman and Zeeman (1992) review IA for highways in the Netherlands. Bina *et al.* (1995) illustrate how buffering around projected transport links could be used at a European scale, using the Oresund link (mentioned in the next section) and the Via Egnatia motorway in Greece as examples of Strategic EIA; Lee and Tomlin (1997) use a more sophisticated system based on map algebra for automatic siting of transport corridors, and Klungboonkrong and Taylor (1998) apply a multi-criteria approach to simulate the environmental impacts of road networks.

As might be expected, *pollution* impacts constitute a major area of application of GIS' functionality:

- Related to *water*, Albertson *et al.* (1992) discuss a system which simulates groundwater contamination from different land uses, Merchant (1994) reviews the DRASTIC model to calculate water pollution potential given the characteristics of the water system and of the soil, and Secunda *et al.* (1998) use a similar approach to develop a groundwater vulnerability index in Israel's Sharon region. Mertz (1993) reports on a system to simulate contaminating sediments in waterways, and Giupponi *et al.* (1999) use a GIS map algebra approach to map the risks of agricultural pollution of the water in the Venice Lagoon; Spence *et al.* (1995) describe the SLURP model to calculate (with the raster-GIS SPANS) run-off water volumes depending on soil and land cover types, and Weng (2001) combines GIS and remote sensing to model the run-off effects of urban growth.

- For *soil* pollution-risk analysis, Pires and Santos (1996) use IDRISI with satellite and other data to construct a risk model for São Paulo (Brasil) using the GIS' internal map algebra facilities; Turner *et al.* (1997) use GIS to compare indicators of pollution-hazard risk, and Bennett (2000) uses GIS to assess the risks of land contamination in Huntingdonshire District in the UK; Trevisan *et al.* (2000) use GIS map algebra to assess the risk of water contamination from agriculture in the province of Cremona (Italy).
- For *air* pollution, Bocco and Sanchez (1997) measure the potential impact of lead contamination using GIS, and Briggs *et al.* (1997) use a GIS' internal statistical capabilities, applying regression analysis (in Arc-Info's GRID) to model and simulate NO_2 concentrations as a function of traffic and land use characteristics. Shivarama *et al.* (1998) discuss the integration of air pollution models with GIS to help with land-use planning in Bombay (India).
- In the area of *radiation*, Van der Heiligenberger (1994) describes a system for monitoring and mapping emission sources and radiation effects to produce risk maps.
- On impacts from *mining*, Asabere (1992) uses GIS to simulate and map such impacts, and Suri and Venkatachalam (1994) relate copper mining to air quality, damage to vegetation and cumulative impacts on human health in Bihar State (India).

The potential impacts of *hazardous waste* have been studied as a source of pollution – related to the last set of applications discussed – and the siting of waste facilities is a typical area of application (Siderelis, 1991); also dangerous waste has been studied as a dangerous product to transport, needing careful route planning – and this relates to the next set of applications below – as in the study by Brainard *et al.* (1996) using GIS to assess hazardous waste transport risks for Southeast England in order to select the best routes. Fatta *et al.* (1998) use GIS map algebra to identify the best locations for industrial waste facilities in Cyprus, and Basnet *et al.* (2001) use a similar approach to find suitable locations for animal-waste facilities.

Visibility analysis is probably the most popular impact area using GIS, simply because most GIS incorporate these days a "viewshed analysis" function using 3D terrain-modelling capabilities to define the areas from where certain features (like the structures in a project) will be visible, a quite impressive facility that can also incorporate the effects of barriers. Kluijtmans and Collin (1991) incorporate the "cartooning" of viewshed analysis views (from a Digital Terrain Model) to give the impression of walking through; Fels (1992) describes an interactive system (for an Apple Macintosh computer) to define the type of visibility analysis wanted, and Davidson *et al.* (1992) review the usefulness of GIS to assess visual and environmental impacts of four projects for rural planning in

Scotland. Howes and Gatrell (1993) try to *quantify degrees of visibility* as applied to wind farms; Boursier *et al.* (1994) use the same type of analysis to decide a location in the Languedoc-Roussillon, and Hebert and Argence (1996) use Digital Elevation Modelling to assess the visibility impact of electricity pylons for the French national power company. On a slightly different note, Gracia and Hecht (1993) describe a visualisation GIS system applied to the evaluation of restoration projects for military areas. Wood (2000) approaches the issue of GIS and visibility impacts from a different angle – much needed in all areas of IA: he undertakes an *audit* of visibility impacts as predicted by GIS, providing an interesting evaluation of the technology. Wood (1999a) applies a similar auditing approach to simple noise-prediction models; he also discusses the issues raised by impact auditing and applies the approach to air pollution (Wood, 1999b).

An area of IA (and of GIS use) which is attracting increasing attention is *cumulative* impact assessment, which can have two meanings: (i) it can refer to the *prediction* of all the impacts likely to affect an area and how a particular project can add to them, as explored by Parker and Coclin (1993) with examples in New Zealand; and (ii) it can also refer to the assessment of all the impacts already affecting an area, and in that sense it becomes synonymous with environmental *monitoring*. Johnston *et al.* (1988) argue the potential of GIS for this latter purpose using satellite imagery to classify wetlands, showing the effects of cumulative pollution of the water, and Li (1998) integrates GIS and remote sensing to monitor the loss of agricultural land in the Pearl River Delta; Roose (1994) uses GIS to model cumulative impacts of industrial pollution to derive pollution surfaces.

Sometimes the GIS' own functionality is sophisticated enough to be used for *model development* internal to the system (as opposed to using external models, already discussed), for example, when this functionality includes statistical capabilities. We have already mentioned Briggs *et al.* (1997) using regression capabilities internal to Arc-Info; Johannesen *et al.* (1997) use IDRISI in a similar way to make the statistical analysis necessary to build a model of marine transport of radioactive contamination, applied to the northern seas of Kava and Barents.

As can be seen from our discussion, only a few types of GIS functions are used most frequently for IA-related purposes:

- *map overlay*, to detect/measure direct impingement between projects and sensitive areas;
- *buffering*, to detect impingement "at a distance" by radiation, emissions, etc.;
- *map algebra*, when it is a combination of factors that needs to be calculated;
- *3D modelling* to simulate terrains, visibility, etc.;

- sometimes, if available, *statistical analysis* (like regression) for model-building purposes.

Beyond such functions, innovation in the use of GIS functionality tends to be associated with *input* and *output* devices more than with the GIS functionality itself. For example, on the input side the potential of *satellite imagery* was identified from the 1980s; followed by the *Global Positioning System* (GPS) for accurate location of point events like fires, etc.; and the growing availability of *Internet* access to data and tools that can be used with GIS for IA. On the output side, *multimedia* interfaces are at the forefront of innovation, usually linked to an increase of the level of interactivity in these systems.

The discussion of IA applications of GIS' functionality, concludes with a look at some applications where GIS functionality is in some way *pre-programmed*, making it possible for non-GIS experts to use them. Sometimes they are planned this way from the start, sometimes they start as "hands-on" applications and, as staff confidence and experience increase, they start adding some pre-programming, in a way similar to mapping applications evolving into more technical ones, as mentioned earlier. The areas of interest and the approaches used are virtually the same as for the hands-on versions just discussed, the only difference being that the sequences of operations have been automated by encapsulating them into a programme which decision-makers and managers can activate themselves. Moreno (1990) describes a quite sophisticated pre-programmed system in Nevada which is an example of an early hands-on system (Moreno and Siegel, 1988) that evolved, to undertake route selection for power lines and then estimate the impacts (ground impacts, accessibility impacts and visibility impacts) of a specific route. Gardels *et al.* (1990) use pre-programmed GIS functions (buffering and overlay) for modelling impacts of land uses on water quality in the San Francisco Bay estuary, and Cova and Church (1997) use AML (Arc-Info's Macro language) to define emergency planning zones around possible catastrophe points for the preparation of catastrophe-evacuation plans. When it comes to pre-programming, the most popular approach is to use the GIS' own internal *macro language* (like Arc-Info's AML) if it has one, probably reflecting the considerable difficulties of accessing GIS from external programmes.

3.4.4 Multi-purpose GIS systems

As already mentioned, applications are sometimes difficult to classify in the groupings used above because they develop over time, but in some cases the difficulty is that they fit into *all* the groups, usually because they are set up for multi-purpose management and require the complete range of technical capability, from simple operations like mapping to links with models or map manipulation using GIS functions. Such systems are more akin to so-called Decision Support Systems (DSS) to be discussed in Chapter 5, but it is worth mentioning here some relatively simple examples that do not describe themselves as DSS. Grieco (1992) describes a system that integrates all stages of impact evaluation and clean-up of contaminated land using a whole range of approaches; Antunes *et al.* (1996) discuss a system (with IDRISI) used to integrate impact predictions from models and evaluate their significance to produce indices maps, applied to some case studies (a highway and a tourist development) in Portugal. Boulmakoul *et al.* (1999) discuss a project for the general management of the transportation of hazardous materials in the city of Mohammedia (Morocco) combining GIS and GPS. Andersen (1996) and Baumann (1998) describe different stages in the development of a system for *instant* monitoring and mitigation of impacts "as they happen" on the Oresund link between Denmark and Sweden involving rail tunnels and bridges and an artificial island; it is the EAGLE information support system – with simulation models about movement of the sediment, etc. – to evaluate various construction scenarios using information about the state of the ecosystem. It monitors closely eel grass and sediments and currents to control the maximum load of sediment spills that dredging contractors can reach in each area, reporting immediately on any "incident" of excess spills, its impact being simulated with hydrologic models. The system also allows on-line monitoring by contractors and other interested parties.

3.5 CONCLUSIONS

The spread of GIS in environmental work in the last 15 years has been phenomenal, as shown by the vast number of accounts of environmental usage of GIS. Many such reports research the use of this new technology in research articles and books, but many – in conference papers and magazines – are just accounts of GIS being used, simply monitoring its diffusion. After the initial enthusiasm of the 1990s, it seems that the research appeal of GIS for IA has "levelled off", even if the diffusion of GIS as such continues at a fast pace, more as practical tools than as an innovation requiring theoretical discussion. This is probably the result – at least partially – of the progressive realisation of the relative unsophistication of GIS functionality, illustrated in our review by the relatively narrow range of operations that they are called to perform:

- map display;
- map overlay and intersection;
- buffering around given features;
- multi-factor map algebra;
- visibility analysis derived from terrain modelling.

In addition to the technical power of GIS as databases, the purely "visual" appeal of their outputs (maps) has been and still is a major contributor to their success, as reflected in the large proportion of GIS applications (27 per cent overall) whose main aim seems to be map production, a proportion which seems to be on the increase. There are some applications that are becoming more sophisticated over time, but they represent a decreasing proportion compared to the growing number of new and simpler applications aimed at map output. In fact, interest and innovation in GIS environmental applications seem to be focussing more and more on the *external links* of these systems:[10] on the input side, links with the Internet as a source of environmental data, satellite imagery, GPS for accurate location; on the output side, multimedia (and hypermedia) interfaces are at the forefront of innovation, usually linked to an increase in the interactivity of these systems associated with a growing interest in *public participation*. Internet connections are also seen as a step towards more participatory decision support.

GIS continue to "diffuse" to more and more agencies in more and more countries, but the sophistication of their use seems to have reached a plateau, and further improvements seem to come from the way these systems are linked to the outside more than from developments in their own functionality. It is probably right to say that "partial" technologies like GIS (or modelling) maximise their usefulness when operating within the framework of other decision-support tools (like expert systems) that structure and focus their performance, and this will be explored further in Chapter 5. But first, the role of GIS in the broader area of environmental management is now discussed in Chapter 4.

REFERENCES

Albertson, P.E., Bourne, S.G. and Hennington, G.W. (1992) An Integrated Groundwater and Land-use GIS for Impact Assessment, *Photogrametric Engineering and Remote Sensing*, Vol. 58 (August), pp. 1203–7.

Allen, P. (1995) Monitoring, the Environment From Space: Observations of the 1994 Komi oil spill, *GIS Europe* (September), pp. 42–3.

Andersen, R. (1996) EAGLE – A GIS for Environmental IA, in Rumor, M. *et al.* (eds) *op. cit.*, Vol. 1, pp. 311–17.

Anderson, M. and Taylor, M.A.P. (1988) Estimating the Environmental Impacts of Road Traffic at the Local Level: A PC-Based Modelling System, in Newton, P.W., Taylor, M.A.P. and Sharpe, R. (eds) *Desktop Planning*, Hargreen Publishing Company, Melbourne, Australia, Ch. 27.

10 We have already seen how important external *functional* links are for GIS: external models were present in 18 per cent of applications, and decision-support tools were present in 24 per cent.

Andrews, C.J. (2001) Overcoming the Open System Problem in Local Industrial Ecological Analysis, *Journal of Environmental Planning and Management*, Vol. 44, No. 4 (July), pp. 491–508.

Antunes, P., Santos, R., Jordao, L., Goncalves, P. and Videira, N. (1996) A GIS Based Decision Support System for Environmental IA, in Guariso, G. and Page, B. (eds) *op. cit*., pp. 451–6.

Appelman, K. and Zeeman, M. (1992) Landscape Ecology and GIS in Environmental IA for Highways, *Proceedings of the EGIS '92 Conference*, Munich (March 23–6), Vol. 2, pp. 398–407.

Asabere, R.K. (1992) Environmental Assessment in the Mining Industry using Geographical Information System, *Mapping Awareness*, Vol. 6, No. 10 (December), pp. 41–5.

Basnet, B.B., Apan, A.A. and Raine, S.R. (2001) Selecting Suitable Sites for Animal Waste Application Using a Raster GIS, *Environmental Management*, Vol. 28, No. 4 (October), pp. 519–31.

Baumann, J. (1998) The Oresund Goes Against the Flow, *GIS Europe* (March), pp. 32–4.

Belore, R.C., Trudel, B.K., Jessiman, B.J. and Ross, S.L. (1990) An Automated Oil Spill IA System Using a Microcomputer Based GIS, *Proceedings of the "GIS for the 1990s" National Canadian Conference*, Ottawa, Canada (March 5–8), pp. 87–102.

Bennett, P. (2000) Contaminated Land: Assessing the Risk, *GeoEurope*, Issue 12 (December), pp. 33–4.

Bennett, D.A. and Vitale, A.J. (2001) Evaluating Nonpoint Pollution Policy Using a Tightly Coupled Spatial Decision Support System, *Environmental Management*, Vol. 27, No. 6 (June), pp. 825–36.

Bernardo, F., Almeida, R., Ramos, I. and Marques, Z. (1994) GIS in Flood Risk Management, *Proceedings of the EGIS/MARI '94 Conference*, Paris (March 29–April 1), Vol. 1, pp. 508–13.

Bhaduri, B., Harbor, J., Engel, B. and Grove, M. (2000) Assessing Watershed-Scale, Long-Term Hydrologic Impacts of Land-Use Change Using a GIS-NPS Model, *Environmental Management*, Vol. 26, No. 6 (December), pp. 643–58.

Bilanzone, G., Chini, P. and Salis, A. (1993) A Case Study for a Noise Reduction Planning System for the City of Ancona, *Proceedings of the EGIS '93 Conference*, Genoa, Italy (March 29–April 1), Vol. 1, pp. 96–105.

Bina, O., Briggs, B. and Bunting, G. (1995) *The Impact of Trans-European Networks on Nature Conservation: A Pilot Project*, The Royal Society for the Protection of Birds, Bedfordshire.

Bocco, G. and Sanchez, R. (1997) Identifying Potential Impact of Lead Contamination Using a Geographic Information System, *Environmental Management*, Vol. 21, No. 1 (January/February), pp. 133–8.

Boggs, G.S., Evans, K.G., Devonport, C.C., Moliere, D.R. and Saynor, M.J. (2000) Assessing Catchment-wide Mining-related Impacts on Sediment Movement in the Swift Creek Catchment, Northwestern Territory, Australia, Using GIS and Landform-evolution Modelling, *Journal of Environmental Management*, Vol. 59, No. 4 (August – Special issue), pp. 321–34.

Boulmakoul, A., Laurini, R., Servigne, S. and Idrissi, M.A.J. (1999) First Specifications of a Telegeomonitoring System for the Transportation of Hazardous Materials. *Computers, Environment and Urban Systems*, Vol. 23, No. 4, pp. 259–70.

Boursier, P., Dbouk, M., Germa, J.-M., Dugue, C., Geffray, F., Souris, M. and Mullon, C. (1994) The Use of Geographic Information and Digital Terrain Models for Evaluating Wind Energy, *Proceedings of the EGIS/MARI '94 Conference*, Paris (March 29–April 1), Vol. 2, pp. 1464–70.

Brainard, J., Lovett, A. and Parfitt, J. (1996) Assessing Hazardous Waste Transport Risks Using a GIS, *International Journal of Geographical Information Systems*, Vol. 10, No. 7 (October–November), pp. 830–49.

Briggs, D.J., Collins, S., Elliott, P., Fischer, P., Kingham, S., Lebret, E., Pryl, K., Van Reeuwijk, H., Smallbone, K. and Van der Veen, A. (1997) Mapping Urban Air Pollution Using GIS: A Regression-based Approach, *International Journal of Geographical Information Science*, Vol. 11, No. 7 (October–November), pp. 699–718.

Brown, C.R. (1994) Toward No Net Loss: A Methodology for Identifying Potential Wetland Mitigation Sites Using a GIS, *Proceedings of the Urban and Regional Information Systems Association (URISA) Conference*, Milwaukee, Wisconsin (August 7–11), Vol. I, pp. 594–607.

Brun, S.E. and Brand, L.E. (2000) Simulating runoff behavior in an Urbanizing Watershed. *Computers, Environment and Urban Systems*, Vol. 24, No. 1, pp. 5–22.

Collins, W.G., Kelsey, K.G. and Benson, G. (1986) A Remote Sensing Evaluation of Habitat Resources in a New Town Site, *The Planner*, Vol. 72, No. 5, pp. 20–5.

Corbley, K.P. (1995) South Florida Fine-Tunes GIS in Hurricane's Aftermath, *GIS World* (September), pp. 40–3.

Cova, T.J. and Church, R.L. (1997) Modelling Community Evacuation Vulnerability Using GIS, *International Journal of Geographical Information Science*, Vol. 11, No. 8 (December), pp. 763–844.

Craig, P.M. and Burnette, G.A. (1996) Basinwide Water-Quality Planning Using the QUAL2E Model in a GIS Environment, in Goodchild, M.F., Steyaert, L.T., Parks, B.O., Johnston, C., Maidment, D., Crane, M. and Glendinning, S. (eds) *op. cit.*, Ch. 53, pp. 283–6.

Dale, V.H., King, A.W., Mann, L.K., Washington-Allen, R.A. and McCord, R.A. (1998) Assessing Land-Use Impacts on natural Resources, *Environmental Management*, Vol. 22, No. 2 (March/April), pp. 203–11.

Davidson, D.A., Selman, P.H. and Watson, A.I. (1992) The Evaluation of a GIS for Rural Environmental Planning, *Proceedings of the EGIS '92 Conference*, Munich (March 23–6), Vol. 1, pp. 135–44.

Dodge, M. (1996) The Visualisation and Analysis of Fire Incidents Using GIS, in Rumor, M. *et al.* (eds) *op. cit.*, Vol. 1, pp. 592–601.

Dragosits, U., Place, C.J. and Smith, R.I. (1996) Potential of GIS and Coupled/Conventional Systems to Model Acid Deposition of Sulphur Dioxide, in Goodchild, M.F. *et al.* (1996b) *op. cit.*

Dubbink, D. (1991) Noise Simulations as a Tool for Planning Decisions, in Klosterman, R.E. (ed.) *op. cit.*, Vol. 1, pp. 23–7.

Eedy, W. (1995) The Use of GIS in Environmental Assessment, *Impact Assessment*, Vol. 13 (Summer), pp. 199–206.

Evers, C.W.A. and Most, P.F.J. (1996) Calculation and Localization of Emmissions from Landfills, in Rumor, M. *et al.* (eds) *op. cit.*, Vol. 1, pp. 343–5.

Fatta, D., Saravanos, P. and Loizidou, M. (1998) Industrial Waste Facility Site Selection using Geographical Information System Techniques, *International*

Journal of Environmental Studies, Vol. 56, No. 1 (Section A: Environmental Studies), pp. 1–14.

Fels, J.E. (1992) Viewshed Simulation and Analysis: An Interactive Approach, *Proceedings of the Urban and Regional Information Systems Association (URISA) Conference*, Washington DC (July 12–6), Vol. I, pp. 267–76.

Fedra, K. (1993) Clean air, *Mapping Awareness*, Vol. 7, No. 6 (July/August), pp. 24–7.

Fedra, K. (1996) Distributed Models and Embedded GIS: Integration Strategies and Case Studies, in Goodchild, M.F., Steyaert, L.T., Parks, B.O., Johnston, C., Maidment, D., Crane, M. and Glendinning, S. (eds) *op. cit.*, Ch. 74, pp. 413–17.

Fonseca, A., Gouveia, C., Camara, A. and Ferreira, F. (1994) Environmental IA Using Multimedia GIS, *Proceedings of the EGIS/MARI '94 Conference*, Paris (March 29–April 1), Vol. 1, pp. 416–25.

Fonseca, A., Fernandes, J.P., Gouveia, C., Silva, J.P., Pinheiro, A., Sousa, I., Aragao, D. and Concalves, C. (1996) Environmental Multimedia Exploratory Systems, in Rumor, M. *et al.* (eds) *op. cit.*, Vol. 1, pp. 147–66.

Fouda, Y.E., El-Kadi, H. and El-Raey, M. (1993) A Geographical Information System for Environmental Auditing of the Sixth of October City, Egypt, in Klosterman, R.E. and French, P.S. (eds) *op. cit.*, Vol. 2, pp. 121–43.

French, D.P. and Reed, M. (1996) Integrated Environmental Model and GIS for Oil and Chemical Spills, in Goodchild, M.F., Steyaert, L.T., Parks, B.O., Johnston, C., Maidment, D., Crane, M. and Glendinning, S. (eds) *op. cit.*, Ch. 35, pp. 197–8.

Friel, C., Leary, T., Norris, H., Warford, R. and Sargent, B. (1993) GIS Tackles Oil Spill in Tampa Bay, *GIS World* (November), pp. 30–3.

Gardels, K.D., McCreary, S.T. and Huse, S. (1990) Land Use Policy Analysis With GIS: San Francisco Bay Estuary, *Proceedings of the GIS/LIS '90 Conference*, Anaheim, California (November 7–10), Vol. 1, pp. 191–9.

Gauthier, E., Dupuis, P., Banton, O. and Villeneuve, J.P. (1992) L'Utilisation de la Modelisation Mathematique et des Systemes d'Information Geographique Dans la Gestion du Risque de Contamination des Eaux Souterraines par les Pesticides, *Proceedings of the Canadian Conference on GIS '92*, Ottawa, Canada (March 24–6), pp. 519–35.

GeoWorld (2001) News Link: Researchers Use Geospatial Technology to Predict Avalanches, *GeoWorld*, No. 1 (January), pp. 12.

Giupponi, C., Fiselt, B. and Ghetti, P.F. (1999) A Multicriteria Approach for Mapping Risks of Agricultural Pollution for Water Resources: The Venice Lagoon Watershed Case Study, *Journal of Environmental Management*, Vol. 56, No. 4 (August), pp. 259–69.

Goncalves Henriques, R., Pereira Reis, R., Joao, E., Fonseca, A. and Gouvela, C. (1992) From Forest Fire Simulation to the Assessment of Environmental Impacts, *Mapping Awareness*, Vol. 6, No. 5 (June), pp. 13–17.

Goodchild, M.F., Parks, B.O. and Steyaert, L.T. (1993) (eds) *Environmental Modelling with GIS*, Oxford University Press, Oxford.

Goodchild, M.F., Steyaert, L.T., Parks, B.O., Johnston, C., Maidment, D., Crane, M. and Glendinning, S. (1996a) (eds) *GIS and Environmental Modeling: Progress and Research Issues*, GIS World Books.

Goodchild, M.F., Steyaert, L.T., Parks, B.O., Johnston, C., Maidment, D., Crane, M. and Glendinning, S. (1996b) *Proceedings of the Third International*

Conference/Workshop on Integrating GIS and Environmental Modelling, National Centre for Geographic Information and Analysis, Santa Fe (New Mexico) (in CD format).

Gracia, J.M. and Hecht Jr., L.G. (1993) GIS Improves Visualization, Evaluation Capabilities in Superfund Cleanup, *GIS World* (February), pp. 36–41.

Gray, C. (1993) The Application of a Simple and Sophisticated GIS in the Environmental Assessment of Industrial Land, *Proceedings of the EGIS '93 Conference*, Genoa, Italy (March 29–April 1), Vol. 2, pp. 871–5.

Green, D.R. (1996) Between the Desktop and the Deep Blue Sea, *Mapping Awareness*, Vol. 9, No. 6 (July), pp. 19–22.

Grieco, D.K. (1992) Integrated Solutions for the Evaluation and Clean-Up of Contaminated Land, *Proceedings of the EGIS '92 Conference*, Munich (March 23–6), Vol. 2, pp. 1059–65.

Guariso, G. and Page, B. (eds) (1994) *Computer Support for Environmental IA*, The IFTP TC5/WG5.11 Working Conference on Computer Support for Environmental IA – CSEIA 93, Como, Italy (6–8 October 1993), IFIP Transactions B: Applications in Technology (B-16), North holland: Elsevier Science BV.

Guimaraes Pereira, A. and Antunes, M.P. (1996) Extending the EIA Process: Generation ad Evaluation of Alternative Sites for Facilities Within a GIS: A Multi-Criteria Genetic Algorithm Based Approach, in *Improving Environmental Assessment Effectiveness: Research, Practice and Training*, Proceedings of IAIA '96, 16th Annual Meeting (June 17–23), Centro Escolar Turistico e Hoteleiro, Estoril (Portugal), Vol. I, pp. 479–84.

Harman, W.A., Allan, C.J. and Forsythe, R.D. (2001) Assessment of Potential Groundwater Contamination Sources in a Wellhead Protection Area, *Journal of Environmental Management*, Vol. 62, No. 3 (July), pp. 271–82.

Harris, P.M., Kim, K.H., Ventura, S.J., Thum, P.G. and Prey, J. (1991) Linking a GIS with an Urban Nonpoint Source Pollution Model, *Proceedings of the GIS/LIS '91 Conference*, Atlanta, Georgia (October 28–November 1), Vol. 2, pp. 606–16.

Hebert, M.-P. and Argence, J. (1996) Virtual Pylons into Geographic Reality, *GIS Europe* (August), pp. 28–30.

Henzel, W.M., Thum, P.G., Ventura, S.J., Niemann Jr., B.J., Nowak, P.J., McCallister, R.B. and Shelley, K. (1990) Understanding Relationship Between Farm Management Practices and Groundwater Contamination Using a GIS, *Proceedings of the GIS/LIS '90 Conference*, Anaheim, California (November 7–10), Vol. 2, pp. 919–27.

Hill, J.M., Cotter, D.M., Dodson, R. and Graham, L.A. (2001) Winds of Change, *GeoWorld*, Vol. 14, No. 11 (November), pp. 38–41.

Howes, D. and Gatrell, T. (1993) Visibility Analysis in GIS: Issues in the Environmental IA of Windfarm Developments, *Proceedings of the EGIS '93 Conference*, Genoa, Italy (March 29–April 1), Vol. 2, pp. 861–70.

Jankowski, P. (1995) Integrating Geographical Information Systems and Multiple Criteria Decision-Making Methods, *International Journal of Geographical Information Systems*, Vol. 9, No. 3, pp. 251–73.

Jankowski, P. and Haddock, G. (1996) Integrated Nonpoint Source Pollution Modeling System, in Goodchild, M.F., Steyaert, L.T., Parks, B.O., Johnston, C., Maidment, D., Crane, M. and Glendinning, S. (eds) *op. cit.*, Ch. 39, pp. 209–11.

Joao, E. (1998) Use of Geographic Information Systems in IA, in Porter, A. and Fittipaldi, J. (eds) *Environmental Methods Review: Retooling IA for the New Century*, The Army Environmental Policy Institute (Atlanta), pp. 154–63.

Joao, E. and Fonseca, A. (1996) *Current Use of Geographical Information Systems for Environmental Assessment: A Discussion Document*, Research Papers in Environmental and Spatial Analysis No. 36, Department of Geography, London School of Economics.

Johannessen, O.M., Pettersson, L.H., Bobylev, L.P., Rastoskuev, V.V., Shalina, E.V. and Volkov, V.A. (1997) The Slumbering Bear of the Kara Sea, *GIS Europe* (May), pp. 20–2.

Johnston, C.A., Detembeck, N.E., Bonde, J.P. and Niemi, G.J. (1988) Geographic Information Systems for Cummulative IA, *Photogrametric Engineering and Remote Sensing*, Vol. LVI, No. 11 (November), pp. 1609–15.

Jones, P., Williams, J. and Lannon, S. (2000) Planning for a Substainable City: An Energy and Environmental Prediction Model, *Journal of Environmental Planning and Management*, Vol. 43, No. 6 (November), pp. 855–72.

Katz, G.E. (1993) Cartographic Modelling of Development Impacts in Natural Areas in Ottawa, Canada, *Proceedings of the Canadian Conference on GIS '93*, Ottawa, Canada (March 23–5), pp. 670–81.

Kim, M.-J., Han, E.-J. and Kang, I.-G. (1996) Integration of Geographic Information System and Air Dispersion Model, in *Improving Environmental Assessment Effectiveness: Research, Practice and Training*, Proceedings of IAIA '96, 16th Annual Meeting (June 17–23), Centro Escolar Turistico e Hoteleiro, Estoril (Portugal), Vol. I, pp. 467–72.

Klosterman, R.E. (1991) (ed.) *Proceedings of the 2nd. International Conference on Computers in Urban Planning and Urban Management*, Oxford Polytechnic (July 6–8).

Klosterman, R.E. and French, S.P. (1993) (eds) *Proceedings of the 3rd. International Conference on Computers in Urban Planning and Urban Management*, Atlanta (Georgia).

Kluijtmans, P. and Collin, C. (1991) 3D Computer Technology for Environmental Impact Studies, *Proceedings of the EGIS '91 Conference*, Brussels (April 2–5), Vol. I, pp. 547–50.

Klungboonkrong, P. and Taylor, M.A.P. (1998) A Microcomputer-based System for Multicriteria Environmental Impacts Evaluation of Urban Road Networks, *Computers, Environment and Urban Systems*, Vol. 22, No. 5 (September), pp. 425–46.

Kooistra, L., Leuven, R.S.E.W., Nienhuis, P.H., Wehrens, R. and Buydeus, L.M.C. (2001) A Procedure for Incorporating Spatial Variability in Ecological Risk Assessment of Dutch River Flood Plains, *Environmental Management*, Vol. 28, No. 3 (September), pp. 359–73.

Kramer, M.R. (1989) A GIS Application Design Methodology to Determine the Impact of Large-Scale Commercial Development, *Proceedings of the Urban and Regional Information Systems Association (URISA) Conference*, Boston (August 6–10), Vol. II, pp. 210–6.

Krasovskaia, T.V. and Tikunov, V.S. (1991) Geographical Information Systems for Environmental Impact Assessment, *Proceedings of the EGIS '91 Conference*, Brussels (April 2–5), Vol. I, pp. 574–80.

Kuhlman, K., Hart, D., Ventura, S. and Prey, J. (1994) GIS and Nonpoint Pollution Modeling: Lessons Learned from Three Projects, *Journal of the Urban and Regional Information Systems Association*, Vol. 6, No. 2 (Fall), pp. 69–72.

Ladha, N. and Robertson, A.G. (1988) Technology and the Power Line Planner: The Impact of the Computer on Environmental IAs for New Transmission Facilities, *Proceedings of the GIS/LIS '88 Conference*, San Antonio, Texas (November 30–December 2), Vol. 2, pp. 498–506.

Lam, K.C., Li, B., Ma, W. and Yu, S. (1999) GIS-Based Road Traffic Noise Assessment System, paper presented at the conference *Forecasting the Future: IA for a New Century*, IAIA '99, 19th Annual Meeting (June 15–19), The University of Strathclyde (John Anderson Campus), Glasgow.

Lee, B.D. and Tomlin, C.D. (1997) Automate Transportation Corridor Allocation, *GIS World* (January), pp. 56–60.

Leipnik, M.R. (1993) GIS and 3-D Modeling Fight Subsurface Contamination at Federal Site, *GIS World* (May), pp. 38–41.

Le Lay, G., Clergeau, P. and Hubert-Moy, L. (2001) Computerized Map of Risk to Manage Wildlife Species in Urban Areas, *Environmental Management*, Vol. 27, No. 3 (March), pp. 451–61.

Li, X. (1998) Measurement of Rapid Agricultural Land Loss in the Pearl River Delta with the Integration of Remote Sensing and GIS, *Environment and Planning B: Planning and Design*, Vol. 25, No. 3 (May), pp. 447–61.

Li, Y., Brimicombe, A.J. and Ralphs, M.P. (2000) Spatial Data Quality and Sensitivity Analysis in GIS and Environmental Modelling: The Case of Coastal Oil Spills, *Computers, Environment and Urban Systems*, Vol. 24, No. 2 (March), pp. 95–108.

Longhorn, R. and Moreira Madueno, J.M. (1998) Toxic shock. *GIS Europe* (November), pp. 24–7.

Mallants, D. and Badji, M. (1991) Integration of GIS and Deterministic Hydrologic Models – A Powerful Tool for Environmental IA, *Proceedings of the EGIS '91 Conference*, Brussels (April 2–5), Vol. I, pp. 671–80.

Mandl, P. (1992) A Conceptual Framework for Coupling GIS and Computer-Models, *Proceedings of the EGIS '92 Conference*, Munich (March 23–6), Vol. 1, pp. 431–7.

Masser, I. (1990) The Utilisation of Computers in Local Government in Less Developed Countries: A Case Study of Malaysia. *Proceedings of the Urban and Regional Information Systems Association (URISA) Conference*, Edmonton, Alberta, Canada (August 12–16), Vol. IV, pp. 235–45.

Mattikalli, N.M. and Richards, K.S. (1996) Estimation of Surface Water Quality Changes in Response to Land Use Change: Application of The Export Coefficient Model Using Remote Sensing and Geographical Information Systems, *Journal of Environmental Management*, Vol. 47, No. 3 (November), pp. 263–82.

Mayall, K., Hall, G.B. and Seebohm, T. (1994) Integrate GIS and CAD to Visualize Landscape Change, *GIS World* (September), pp. 46–9.

Meldrum, A. (1996) Coping with another Chernobyl, *GIS Europe* (June), pp. 50–1.

Merchant, J.W. (1994) GIS-Based Groundwater Pollution Hazard Assessment: A Critical Review of the DRASTIC Model, *Photogrametric Engineering and Remote Sensing* (September), pp. 1117–27.

Mertz, T. (1993) GIS Targets Agricultural Nonpoint Pollution, *GIS World* (April), pp. 41–3.

Moore, T.J. (1991) Application of GIS Technology to Air Toxics Risk Assessment: Meeting the Demands of the California Air Toxics "Hot Spots" Act of 1987, in Klosterman, R.E. and French, P.S. (eds) *op. cit.*, Vol. 2, pp. 694–714.

Moreno, D. and Siegel, M. (1988) A GIS Approach for Corridor Siting and Environmental Impact Analysis, *Proceedings of the GIS/LIS '88 Conference*, San Antonio, Texas (November 30–December 2), Vol. 2, pp. 507–14.

Moreno, D.D. (1990) Advanced GIS Modelling Techniques in Environmental IA, *Proceedings of the GIS/LIS '90 Conference*, Anaheim, California (November 7–10), Vol. 1, pp. 345–56.

Noe, S.V. (1993) Using GIS to Predict Urban Growth Patterns and Risk form Accidental Release of Industrial Toxins, Woodfin, T.M. and Fendley, M.S. (1993) Supporting Decisions in the International Zone: Transboundary Planning in the Texas/Mexico Borderlands, in Klosterman, R.E. and French, P.S. (eds) *op. cit.*, Vol. 2, pp. 197–226.

Nyerges, T.L. (1993) Understanding the Scope of GIS: Its Relationship to Environmental Modeling, in Goodchild, M.F., Parks, B.O. and Steyaert, L.T. (eds) *op. cit.*, Ch. 8, pp. 75–93.

Nutter, M., Charron, J. and Moisan, J.F. (1996) Geographic Information System Tool Integration for Environmental Assessment: Recent Lessons, in *Improving Environmental Assessment Effectiveness: Research, Practice and Training*, Proceedings of IAIA '96, 16th. Annual Meeting (June 17–23), Centro Escolar Turistico e Hoteleiro, Estoril (Portugal), Vol. I, pp. 473–8.

Oliver, S.G. (1988) A GIS Automation of County Tax Assessor Maps and their Application for Siting the Superconducting Super Collider in Illinois, *Proceedings of the GIS/LIS '88 Conference*, San Antonio, Texas (November 30–December 2), Vol. 1, pp. 448–58.

Osborne, S. and Stoogenke, M. (1989) Integration of a Temporal Element Into a Natural Resource Decision Support System, *Proceedings of the GIS/LIS '93 Conference*, Minneapolis, Minnesota (November 2–4), Vol. 1, pp. 10–18.

Paoli, M., Giles, W. and Hagarty, J. (1992) GIS in Environmental Assessment: A Linear Utility Case Study, *Proceedings of the Canadian Conference on GIS '92*, Ottawa, Canada (March 24–6), pp. 180–90.

Parker, S. and Coclin, C. (1993) The Use of Geographical Information Systems for Cummulative IA, *Computers, Environment and Urban Systems*, Vol. 17, No. 5 (September–October), pp. 393–407.

Patindol, S.L. (1996) Environmental IA for Kanan B1 Hydroelectric Power Project, in *Forecasting the Future: IA for the New Century*, in *Improving Environmental Assessment Effectiveness: Research, Practice and Training*, Proceedings of IAIA '96, 16th Annual Meeting (June 17–23), Centro Escolar Turistico e Hoteleiro, Estoril (Portugal), Vol. II, pp. 665–70.

Pereira, M. and Mourab, D. (1999) Environmental Mapping in the Tagus Estuary, *GeoEurope* (October), pp. 36–8.

Pires, J.S.R. and Santos, J.E. (1996) Preliminary Analysis of Environmental Impacts Applied to a Rural Area of the State of São Paulo (Luiz Antonio, SP-Brasil), in *Improving Environmental Assessment Effectiveness: Research, Practice and Training*, Proceedings of IAIA '96, 16th Annual Meeting (June 17–23), Centro Escolar Turistico e Hoteleiro, Estoril (Portugal), Vol. II, pp. 969–74.

Raper, J. and Livingstone, D. (1996) High-Level Coupling of GIS and Environmental Process Modeling, in Goodchild, M.F., Steyaert, L.T., Parks, B.O.,

Johnston, C., Maidment, D., Crane, M. and Glendinning, S. (eds) *op. cit.*, Ch. 70, pp. 387–90.

Reddingius, N.H. and Finegold, L.S. (1990) Integrating GIS With Predictive Models, *Proceedings of the GIS/LIS '90 Conference*, Anaheim, California (November 7–10), Vol. 1, pp. 289–98.

Richardson, J. (1999) Can GIS Help Facilitate Public Participation Within SEA?, Paper Presented at the Conference *Forecasting the Future: IA for a New Century*, IAIA '99, 19th Annual Meeting (June 15–19), The University of Strathclyde (John Anderson Campus), Glasgow.

Rinaldi, G., Cavallone, G., Stanghellini, S. and Vestrucci, P. (1993) Air Quality Assessment Using Geographical Information System, *Proceedings of the EGIS '93 Conference*, Genoa, Italy (March 29–April 1), Vol. 1, pp. 284–93.

Rodbell, S. (1993) GPS/GIS Mapping Accurately Assesses Environmental Impact, *GIS World* (December), pp. 54–6.

Rodda, H. (2001) Insuring against disaster, *GeoEurope*, Issue 1 (January), pp. 48–9.

Rodriguez-Bachiller, A. and Smith, P. (1995) Diffuse Picture on Spread of Geographic Technology, *Planning* (June 14), pp. 24–5.

Rodriguez-Bachiller, A. (1998) *GIS and Decision-Support : A Bibliography*, Working Paper No. 176, School of Planning, Oxford Brookes University.

Rodriguez-Bachiller, A. (2000) Geographical Information Systems and Expert Systems for Impact Assessment. Part I: GIS, *Journal of Environmental Assessment Policy and Management*, Vol. 2, No. 3 (September), pp. 369–414.

Roose, A. (1994) Compounding Effects of Industrial Change: A GIS Application, *Proceedings of the EGIS/MARI '94 Conference*, Paris (March 29–April 1), Vol. 1, pp. 988–96.

Roper, J.J. (1996) Digital Imagery Promotes Road Project Acceptance, *GIS World* (October), pp. 58–60.

Roth, A. (1991) Oil spill cleanup and GIS, *Mapping Awareness*, Vol. 5, No. 4 (May), pp. 19–23.

Schaller, J. (1992) GIS Helps Measure Impact of New Munich II Airport, *GIS Europe* (June), pp. 20–1.

Schaller, J. (1995) Landscape Analysis Modelling for an Environmental Impact Study of the Danube River Construction, *Proceedings of the Joint European Conference and Exhibition on Geographic Information JEC-GI '95*, The Hague (March 26–31), Vol. 1, pp. 279–84.

Schou, J.S., Skop, E. and Jensen, J.D. (2000) Integrated Agri-environmental Modelling: A Cost-effectiveness Analysis of Two Nitrogen Tax Instruments in the Vejle Fjord Watershed, Denmark, *Journal of Environmental Management*, Vol. 58, No. 3 (March), pp. 199–212.

Secunda, S., Collin, M.L. and Melloul, A.J. (1998) Groundwater Vulnerability Assessment Using a Composite Model Combining DRASTIC with Extensive Agricultural Land Use in Israel's Sharon Region, *Journal of Environmental Management*, Vol. 54, No. 1 (September), pp. 39–57.

Shang, H. (1995) Visualisation in Urban and Environmental Projects Using Computerised Image Setting, *Proceedings of the 4th International Conference on Computers in Urban Planning and Urban Management*, Melbourne, Australia (July 11–14), Vol. 2, pp. 623–31.

Shiffer, M.J. (1991) Towards a Collaborative Planning System, in Klosterman, R.E. (ed.) *op. cit.*, Vol. 2, pp. 465–82.

Shiffer, M.J. (1993) Environmental Review with Hypermedia Systems, in Klosterman, R.E. and French, P.S. (eds) *op. cit.*, Vol. 1, pp. 587–606.

Shivarama, M.S., Patil, R.S. and Venkatachalam, P. (1998) Integrating Air Quality Models With Geographical Information System for Regional Environmental Planning, *International Journal of Environmental Studies*, Vol. 54, No. 3–4 (Section A: Environmental Studies), pp. 195–204.

Siderelis, K.C. and Tribble, T.N. (1988) Using a Geographical Information System to Prepare a Site Proposal for the Superconducting Super Collider: A Case Study in North Carolina, *Proceedings of the GIS/LIS '88 Conference*, San Antonio, Texas (November 30–December 2), Vol. 1, pp. 459–68.

Siderelis, K.C. (1991) Land resource information systems, in Maguire, D.J., Goodchild, M.F. and Rhind, D.W. (eds) *Geographical Information Systems: Principles and Applications*, Longman, London, Vol. 2, pp. 261–73.

Simpson, H. (1990) Simulating Groundwater and Contaminant Flow Using Geographical Information Systems, *Proceedings of the EGIS '90 Conference*, Amsterdam (April 10–13), Vol. 1, pp. 581–5.

Smith, D.I. and Greenway, M.A. (1988) The Computer Assessment of Urban Flood Damages: ANUFLOOD, in Newton, P.W., Taylor, M.A.P. and Sharpe, R. (eds) *Desktop Planning*, Hargreen Publishing Company, Melbourne (Australia), Ch. 26.

Spence, C., Dalton, A. and Kite, G. (1995) GIS Supports Hydrological Modeling, *GIS World* (January), pp. 62–5.

Suri, J.K. and Venkatachalam (1994) Environmental Impact Analysis Using GIS – A Case Study, *Proceedings of the Canadian Conference on GIS '94*, Ottawa, Canada (June 6–10), Vol. 2, pp. 944–53.

Trevisan, M., Padovani, L. and Capri, E. (2000) Nonpoint-Source Agricultural Hazard Index: A Case Study of the Province of Cremona, Italy, *Environmental Management*, Vol. 26, No. 5 (November), pp. 577–84.

Turner, G.W., Ruffio, R.M.C. and Robert, M.W. (1997) Comparing Environmental Conditions Using Indicators of Pollution Hazard, *Environmental Management*, Vol. 21, No. 4 (July/August), pp. 623–34.

Van der Heiligenberg, H.A.R.M. (1994) GIS in Radiation Research: Environmental Modelling and Risk Assessment, *Proceedings of the EGIS/MARI '94 Conference*, Paris (March 29–April 1), Vol. 1, pp. 217–22.

Van der Perk, M., Burema, J.R., Burrough, P.A., Gillett, A.G. and Van der Meer, M.B. (2001) A GIS-based Environmental Decision Support System to Assess the Transfer of Long-livedradiocaesium Through Food Chains in Areas Contaminated by the Chernobyl accident, *International Journal of Geographical Information Science*, Vol. 15, No. 1 (January–February), pp. 43–64.

Van Slagmaat, M.J.M. and Van der Veen, A.A. (1990) Building an Environmental Information System for Local Authorities. The REMIS – Project, *Proceedings of the EGIS '90 Conference*, Amsterdam (April 10–13), Vol. 2, pp. 1044–53.

Wagner, M.J. (1994) Controlling Prosperity: Monitoring the Environmental Impact of a New Auto Plant, *GIS Europe* (June), pp. 24–7.

Warner, M., Croal, P., Calal-Clayton, B. and Knight, J. (1997) Environmental Impact Assessment Software in Developing Countries: A Health Warning, *Project Appraisal*, Vol. 12, No. 2, pp. 127–30.

Watson Jr., C.C. (1992a) GIS and Hurricane Planning at the Town of Hilton Head island, SC, *Proceedings of the Urban and Regional Information Systems Association (URISA) Conference*, Washington DC (July 12–16), Vol. I, pp. 180–7.

Watson, C.C. (1992b) GIS Aids Hurricane Planning, *GIS World* (July), pp. 46–52.

Weng, Q. (2001) Modeling Urban Growth Effects on Surface Runoff with the Integration of Remote Sensing and GIS, *Environmental Management*, Vol. 28, No. 6 (December), pp. 737–48.

World Bank (1995) *Implementing Geographic Information Systems in Environmental Assessment*, Environmental Assessment Sourcebook Update, No. 9 (January), Environment Department, The World Bank.

Wood, G. (1999a) Assessing Techniques of Assessment: Post-development Auditing of Noise Predictive Schemas in Environmental Impact Assessment, *Impact Assessment and Project Appraisal*, Vol. 17, No. 3 (September), pp. 217–26.

Wood, G. (1999b) Post-development Auditing of EIA Predictive Techniques: A Spatial Analysis Approach, *Journal of Environmental Planning and Management*, Vol. 42, No. 5 (September), pp. 671–89.

Wood, G. (1999c) *Personal Communication with the author.*

Wood, G. (2000) Is What You See What You Get? Post-development Auditing of Methods Used For Predicting the Zone of Visual Influence in EIA, *Environmental Impact Assessment Review*, Vol. 20, No. 5 (October), pp. 537–56.

Wu, B. and Xia, F. (1991) Flood Damage Evaluation System Design for Pilot Area on Bangladesh Floodplain Using Remote Sensing & GIS, *Proceedings of the EGIS '91 Conference*, Brussels (April 2–5), Vol. I, pp. 115–20.

Wu, F. (1998) Simulating Urban Encroachment on Rural Land with Fuzzy-logic-controlled Cellular Automata in a Geographical Information System, *Journal of Environmental Management*, Vol. 53, No. 4 (August), pp. 293–308.

Young, C.H. and Jarvis, P.J. (2001) A Simple Method for Predicting the Consequences of Land Management in Urban Habitats, *Environmental Management*, Vol. 28, No. 3 (September), pp. 375–87.

Zhuang, X. and Burn, M. (1993) Integrating GIS for Management of Sound Insulation Programs, *Proceedings of the GIS/LIS '93 Conference*, Minneapolis, Minnesota (November 2–4), Vol. 2, pp. 820–8.

4 GIS and environmental management

4.1 INTRODUCTION

Chapter 4 provides a structured discussion of the application of GIS in the wider area of environmental management,[11] with the dual role of being a bibliographical review and a "taxonomy" of different types and areas of GIS application. It uses a similar general framework to Chapter 3, grouping GIS applications into four types of approach corresponding to different levels of sophistication:

- GIS just for mapping;
- GIS linked to external models;
- using GIS' own functionality;
- multi-purpose GIS systems.

As with IA, the literature on GIS applications to environmental management is characterised by the high proportion of cases reported in conferences and magazines, as opposed to research journals or books. This chapter draws particularly on the latter type of publication,[12] and conference papers and magazine articles are only referred to when they provide particularly interesting cases.

4.2 GIS FOR ENVIRONMENTAL MAPPING AND MANAGEMENT

The framework starts at the lowest level of sophistication in GIS use within environmental management, looking at GIS applications where these systems seem to be used just for the *production of maps* for visual use by

11 Rodriguez-Bachiller (2000) includes an earlier version of this review.

12 A full review of conference papers and magazine articles would require too much space and, also, it can be said that there is a "natural selection" with the best of those items going further and getting converted into research articles.

decision-makers or researchers. Sometimes these systems may evolve into all-purpose management systems using GIS in more sophisticated ways, as was the case, for example with the fully integrated information system for New South Wales developed at the CSIRO research institute in Australia (Walker and Young, 1997). Taking this as a valid – albeit temporary – category, one of the typical uses of such mapping systems is to provide *areawide information systems*, to service a varied range of needs in a particular area:

1 Prominent in this class is what we can call general *environmental inventories* used for monitoring the environment, like the early Massachusetts environmental database (Taupier and Terner, 1991), or similar systems for North Estonia (Meiner *et al.*, 1990), for Hungary (Scharek *et al.*, 1995), for the ecological regions of the Netherlands (Klijn *et al.*, 1995), for the Rif mountains in Morocco (Moore *et al.*, 1998), for the National Wilderness Preservation System in the US (Lomis and Echohawk, 1999), for the Antarctic Treaty area (Cordonnery, 1999), or for the Papua New Guinea Resource Information System (Montagu, 2000).
2 Also typical is the monitoring of *land cover* in an area, often using satellite data, which can range from covering a whole country, like the Land Cover Map of Great Britain (Fuller and Groom, 1993a,b), or even a continent – like the CORINE Land Cover project for Europe (GIS Europe, 1992) – to a specific region, maybe to identify land use changes (Adeniyi *et al.*, 1992, for North Western Nigeria; Ringrose *et al.*, 1996, for North Central Botswana; Baldina *et al.*, 1999, for the Lower Volga Delta in Russia). Haack (1996) combines GIS and satellite data for monitoring wetland changes in East Africa. Priya and Shibasaki (1997) use Landsat data simply to classify land uses in a region in India, Haak and Bechdol (1999) use radar satellites for the same purpose, Scott and Udouj (1999) use the GRASS GIS for spatial and temporal characterisation of land uses in a watershed in Arkansas, and Brown and Shrestha (2000) use GIS mapping to study market-driven land-use changes in the mountains of Nepal.
3 Some mapping systems can be integrated with general *regional planning* to provide environmental information to be combined with other information, as in Botswana (Nkambwe, 1991), or in the Mediterranean area (Giavelli and Rossi, 1999) for the promotion of sustainable tourism.
4 Sometimes, just the production of certain maps is worth reporting, as in the project to map the whole world in *3D* using new satellite technology (Chien, 2000); Thomas *et al.* (2000) discuss different mapping systems for Ghana and, on a different note, Rhind (2000) discusses the *problems* involved in global mapping.

Considering more specific uses of GIS mapping for *environmental management* as such, the range of environmental aspects addressed is quite varied:

- *Ecology* is typical, in that interest in GIS mapping arose in the 1980s and early 1990s linked to the perceived potential of using the Landsat satellite technology combined with GIS, and the issues raised by this new combination (Davis *et al.*, 1991; Tappan *et al.*, 1991), although a few years later the "novelty shock" appears to be wearing off, and articles of this type become less frequent in research publications. This is partly linked to the development of newer technologies like the Global Positioning System (GPS) (Havens *et al.*, 1997; McWilliam, 1999), and the application of satellite data becomes almost routine, as for example Phinn *et al.* (1996), who used this type of data to map the biomass distribution in Southern New Mexico; Lammert and Allan (1999) use GIS to relate land-cover and habitat structure to the ecology of fresh water, Geist and Dauble (1998) study in a similar way salmon habitats in large rivers, McMahon and Harned (1998) study the Albemarle-Pamlico drainage basin in North Carolina and Virginia (USA), and Sarch and Birkett (2000) apply it to detecting lake-level fluctuations to manage fishing and farming practices in Lake Chad. Cruickshank *et al.* (2000) use the CORINE database to estimate the carbon content of vegetation in Ireland, and Akcakaya (2000) integrates fieldwork and GIS to the management of multiple species and, on a different note, Bowker (2000) discusses the *problems* involved in using GIS to map ecological diversity.

- *Landscape* mapping and monitoring is also typical: Higgs *et al.* (1994) develop a "demonstrator" system of common lands in England and Wales, Isachenko and Reznikov (1994) map the landscapes of the Ladoga region in Russia, and Taylor (1994) does it for the Niagara region in the US; Clayson (1996) monitors landscape change in the Lake District (UK) using remote sensing, Kirkman (1996) also combines GIS and remote sensing to monitor seagrass meadows, and Macfarlane (1998) applies a "landscape-ecology" perspective to the Lake District in the UK.

- Environmental planning of *heritage* sites is reported by Wagner (1995) using GIS for a case study in Cambodia.

- The monitoring and management of *forestry* – a particularly important component of the landscape – also shows a number of applications: Tortosa and Beach (1993) use "desk-top" portable GIS with GPS to map forest fire hot-spots and lightning strikes on the ground; Dusart *et al.* (1994) combine GIS with remote sensing in a river valley in Senegal, Thuresson *et al.* (1996) use GIS to visualise landscape changes in the Gulkal forest (Sweden), Jang *et al.* (1996) use a similar approach to assess global forest changes over time, and Johnson *et al.* (1999) use the same approach for mapping freshwater wetlands and forests in Australia; Bateman and Lovett (2000) use GIS to estimate the carbon content of forests in Wales.

- *Soil/agriculture* management: Price (1993) reports on a project to help customers of the Department of Agriculture in the US, Girard *et al.* (1994)

use remote sensing to map fallow land, and Allanson and Moxey (1996) map agricultural land-use changes in England and Wales; Pratt *et al.* (1997) discuss the use of GIS to estimate the extension of areas under irrigation in North East Nigeria, where soil is at a premium – as it is in Japan (Kato, 1987) – or also for soil-protection organisations as in Baden-Wurthemberg, where Wolf (1996) reports on a project mapping hazardous sites. On a related note, Ackroyd (2000) reports on "precision farming" as a growing area of GIS use, and Knox *et al.* (2000) use GIS to map the financial benefits of sprinkler irrigation in the Anglian Region in the UK.

- Related to *geology*, Knight *et al.* (1999) use GIS to map the sand and gravel resources in Northern Ireland.
- *Water* quality monitoring: Beaulac *et al.* (1994) report on a project for the State of Michigan, Ford and Lahage (1996) report on Massachusetts, Cambruzzi *et al.* (1999) propose a system for the Venetian coastal ecosystem using GPS on boats; on other related aspects, Belknap and Naiman (1998) use GIS to map groundwater streams in Western Washington State, and Shivlani and Suman (2000) use GIS to study the distribution of diving operations in the Florida Keys.
- *Air*, as inventories of air pollution (Trozzi and Vaccaro, 1993; Sifakis *et al.*, 1999).

4.3 GIS LINKED TO EXTERNAL MODELS FOR ENVIRONMENTAL MANAGEMENT

The next level of sophistication in GIS application, where these systems are linked with the use (or development) of analytical/simulation models, is one of the most popular uses of GIS. Its development was marked in the 1990s by a succession of conferences on the subject, starting with the IBM-sponsored meeting on computer-assisted environmental modelling in the summer of 1990 (Melli and Zanetti, 1992), followed by a series of conferences – every two years approximately – specifically on GIS and environmental modelling (Goodchild *et al.*, 1993, 1996a,b).

4.3.1 Water modelling

Fedra (1993) reviews a set of systems dealing with a wide range of environmental issues like Impact Assessment or site suitability, but the most popular area where GIS and simulation models are linked is probably that of water-related modelling: Van der Heijde (1992) provided an early "eye-opener" article about the potential of new computer technologies like GIS to help water modelling, Maidment (1993) and Moore *et al.* (1993) review comprehensively the linking of hydrologic models and GIS. Both Maidment

(1996a,b) and Moore (1996) provide a second review of GIS and hydrologic modelling three years later, and Sui and Maggio (1999) provide another comprehensive review three years on. At a less ambitious level, Srinivasan *et al.* (1996) give a specific example of GIS and modelling in the Texas Gulf Basin, while Harris *et al.* (1993), D'Agnese *et al.* (1996) and Vieux *et al.* (1998) show the application to three-dimensional *groundwater* modelling. Freeman and Fox (1995) use IDRISI with models of watershed analysis for Hawaii, and DePinto *et al.* (1996) use a similar approach, showing a characteristic example of GIS in its typical *dual role* with respect to models: GIS is used first for pre-processing data to be fed into the models, and then for post-processing and displaying the results from the models. Murray and Rogers (1999) simulate groundwater vulnerability to "brownfield" development in the Rouge river watershed, and Aspinall and Pearson (2000) integrate landscape ecology, hydrologic modelling and GIS to assess conditions in water catchment areas.

Water modelling is present also in various other areas of GIS use. For example, *flood risk* modelling has attracted considerable attention, for obvious practical reasons, from the early real-time flood warning system of Johnson *et al.* (1990), to Lanza and Conti (1994) forecasting flood hazards using remote sensing data. Burlando *et al.* (1994) illustrate the use of a GIS Digital Elevation Model (DEM) with a flood-risk model, using climatic, soil and land-use data for the Sausobbia river basin in Liguria (Italy), Brimicombe and Bartlett (1996) use a simulation model to assess flood risk in Hong Kong, and Thumerer *et al.* (2000) discuss a similar system related to climate change for the east coast of England. Related to this – insofar as flood risks are mainly associated to rainfall – is the major water-related theme of *rainfall* in its various aspects:

- Hay *et al.* (1993, 1996) and Lakhatakia *et al.* (1996) integrate GIS with water and *climate change* models.
- Gao *et al.* (1993) use a DEM with a "raster" GIS (GRASS) for Arizona to simulate *runoff* water, and Battaglin *et al.* (1996) use a precipitation-runoff model for a river in Colorado.
- As another effect of rainfall, the simulation of *soil erosion* also attracts considerable attention, for instance, De Roo *et al.* (1994) link GIS to a simulation model to predict runoff soil erosion in the Limburg province of the Netherlands. These areas of water simulation are all related, and Wilson (1996) reviews critically the performance of six models covering the whole range of runoff, soil erosion and subsurface pollution.

Finally, for *water pollution*: Rogowski (1996) and Cronshey *et al.* (1996) report on the use of water pollution models with GIS, Sham *et al.* (1995, 1996) concentrate on modelling septic nitrogen levels in particular, and Xiang (1993) combines GIS with models to define potential impact-mitigation measures, testing the width of vegetal buffer zones needed to protect against water pollution in the Mountain Island Lake Basin (North Carolina). Garnier *et al.* (1998) combine GIS and the GLEAMS model to simulate groundwater pollution resulting from agricultural disposal of animal waste.

4.3.2 GIS and other environmental modelling

Modelling *air* – be it air pollution or atmospheric conditions – has also been combined with GIS: Lee *et al.* (1993) use satellite maps and atmospheric models to show how different landscapes influence the atmosphere in the US, and Novak and Dennis (1993) combine a range of air pollution simulation models and use GIS to show their cumulative results. Fedra (1999) reviews a range of systems combining GIS and simulation models for environmental monitoring (mostly of air quality) in various countries of Europe. On a different note, Chang and Wei (1999) combine GIS with a multi-objective programming model to plan the location of *recycling* stations in Taiwan.

Modelling in *terrestrial ecology* is more rare due to the intrinsic difficulties of such models – which are still more the subject of research and development than application – but the discussion of such models linked to GIS is also developing: Lyon and Adkins (1995) link a raster-GIS (ERDAS) to a model for the identification of wetlands, and Mackey (1996) reviews the issues raised by habitat modelling with GIS. Church *et al.* (1996) discuss an ecological optimisation model for California, Van Horssen (1996) uses regression analysis with GIS for landscape ecological modelling in the Netherlands, Akcakaya (1996) links GIS with models of ecological risk for endangered species, and Kittel *et al.* (1996) assess terrestrial ecological vulnerability to climate change. Bian (2000) combines GIS and component modelling to represent wildlife movements. In the related area of *water ecology*, Pierce *et al.* (2001) combine modelling and GIS and apply the approach to fisheries in the North-East Atlantic.

Various aspects of *forestry* have also attracted interest: Malanson *et al.* (1996) try to anticipate forest response to climate change, Acevedo *et al.* (1996) simulate forest dynamics, Mladenoff *et al.* (1996) extend the simulation into forest management, and Mayaux *et al.* (1998) combine GIS and modelling techniques to measure the extension of tropical forests. Almeida (1994) uses a model to classify *fire risk* areas in Portugal and their ecological relevance, also an area of obvious practical importance. In the related area of *agriculture*, Liao and Tim (1994a) link a GIS (Arc-Info) to external modules to predict soil loss, sediment yield and phosphorus loading, Collins *et al.* (1998) link GIS to the simulation of nitrogen leaching from agriculture, and Quiel (1995) uses satellite data to assess (and model) local conditions and water needs for different soils.

4.3.3 GIS for model design and development

The last example mentioned in the previous section goes beyond applications using existing models, into the equally important area of *model development*. GIS data can be used to help construct models – sometimes at the design stage, sometimes at the estimation stage – of different aspects of the environment, including:

- *Ecology*: Lowell (1991) uses a discriminant analysis to model ecological succession between species, Johnston *et al.* (1996) use GIS to model ecological processes, Ortega-Huerta and Medley (1999) use GIS to construct a map algebra model of the jaguar habitats in Mexico, and Khaemba and Stein (2000) combine GIS with Principal Component and Regression analyses to do spatial and temporal analysis of wildlife in Kenya.
- *Forestry*: Arsenau and Lowell (1992) build a monitoring model for forests, Mackey *et al.* (1996) model boreal forest ecosystems in the Rinker Lake.
- *Landscape* dynamics (Krummel *et al.*, 1996) in the Cadiz township in Wisconsin.
- *Soil* classification from a Spot satellite image of the Misiones province in Argentina (Lardon *et al.*, 1994).
- *Rainfall*: Ardiles-Lopez *et al.* (1996) estimate a rainfall-runoff model, and Jaagus (1996) uses the IDRISI GIS to estimate the impact of climate change on snow cover and river runoff in Estonia.
- *Solar radiation*: on a related aspect, McKenney *et al.* (1999) calibrate a model of solar radiation using data from DEMs, to be used in Canadian forests.
- *Hazard risks* modelling: in geology, Hao and Chugh (1993) model mine-subsidence risks using contour maps; in soils, Jones *et al.* (1994) use a raster-based GIS to evaluate and model soil risks for the National Soil Inventory in the UK, and Johnston and Sales (1994) construct a model to predict erosion in Lake Superior.

4.3.4 GIS and other modelling approaches

All the models mentioned so far are analytical or statistical but, to finish this discussion, mention must also be made of occasional links of GIS to very different computer tools that do not fit precisely into this category, to help with environmental management. Two types of models in particular are becoming increasingly popular:

1 *Process-simulation* models which, instead of using formulae to predict a situation, seek to replicate the process that leads to the prediction.

For example, Bergamasco *et al.* (1996) use a "cellular automata" model to simulate the dispersion of particles in water.

2 *Computer Aided Design* (CAD) packages applied to the natural environment. GIS-CAD combinations are used most commonly to visualise *urban* applications, but they can also be used to visualise the natural environment, as Nelson (1995) does for Alaska.

3 *Virtual Reality* packages combined with GIS, as in the example that Bishop and Gimblett (2000) apply to the management of recreational areas.

4.4 USING GIS' OWN FUNCTIONALITY FOR ENVIRONMENTAL MANAGEMENT

Kinsley (1995) lists the possible contributions of the functionality of GIS to natural resource planning and management in the areas of "communication", "inventory" or "monitoring", while he also identifies "analysis" and "synthesis" as areas where he thought these systems were weaker, as Anselin and Getis (1993) had also identified earlier. Even with the limited analytical capabilities that GIS have, their standard functions can be used to good effect to perform some environmental management tasks, as reviewed by Albrecht (1996) and Maidment (1996a), who examined the requirements of environmental modelling in comparison to GIS functionality. The focus here is not the more basic information-handling functions that GIS can perform (see the list in Chapter 1), but analytical functions – albeit simple ones – to help with decision-making, such as:

- superimposing maps (map "overlay") to identify and measure overlaps;
- combining several maps into composite maps ("map algebra");
- using distances to construct "buffer" zones around certain features;
- drawing contour maps from the point values of variables;
- building a Digital Elevation Model of a terrain;
- identifying "areas of visibility" of certain features on one map.

As in the case of systems used just for mapping (see Section 4.2), it has been common from the early days to develop systems using more complex GIS functionality whose purpose is not necessarily to perform a specific technical function but to coordinate and apply information on an *area-wide* basis. Dippon *et al.* (1989) describe the project to build the Western Oregon Database for forest management, Weber (1990) discusses a GIS for municipal environmental management in Virginia, and Ahearn and Osleeb (1993) want to demonstrate to the Department of Environmental Protection of New York – using as an example an area of Brooklyn – the advantages of GIS to integrate all information to manage sensitive areas. Campbell and Hastie (1993) describe a system to manage the 2300 Indian Reserves in

Canada to resolve conflicts of land uses and interests, and Hutchinson (1993) proposes a continentwide DEM for climate analysis in Canada. Rybaczuk (2001) proposes a similar areawide system to help the management of the Negril Watershed (Jamaica) and to encourage *public participation*, another growth area in GIS applications: Goncalves Henriques (2000) report on a nationwide information system for Portugal, and Ahlenius and Langaas (2000) discuss a GIS-based *interactive* information system for the Baltic region. Jankowski and Nyerges (2001) discuss "Public Participation GIS" in depth, Craig *et al.* (2002) bring the discussion up to date in a variety of areas of application, and Harrison and Hacklay (2002) discuss its potential related to environmental matters in an urban setting, based on an experiment in the London borough of Wandsworth.

The majority of applications of GIS' own functionality do not mention explicitly whether these functions are to be operated step-by-step by the user or whether they are pre-programmed, and it can only be assumed that a *hands-on* approach is expected, except in those cases (less numerous) where pre-programming is explicitly mentioned, which will be reviewed later in this section.

Johnston (1993) reviews methods of *ecological* modelling, arguing that GIS functionality can answer questions about "where", while remote sensing answers questions of "how much". Lajeunesse *et al.* (1995) apply map algebra to the management of a regional park in Montreal, Chang *et al.* (1995) use GIS for habitat analysis in Alaska, and Duguay and Walker (1996) use GIS to monitor an ecological research site. Chou and Soret (1966) study bird distributions in Navarre (Spain), Skidmore *et al.* (1996) use GIS to classify kangaroo habitats in Australia, Healey *et al.* (1996) use satellite data for locust forecasting and monitoring, and Kernohan *et al.* (1998) apply kernel analysis in a GIS to calculate habitat use. Bernert *et al.* (1997) use GIS map algebra to help define "eco-regions" in the Western Corn Belt plains of the USA, and Harding and Winterbourn (1997) use a similar approach in the South Island (New Zealand). Smallwood *et al.* (1998) use map algebra to assess habitat quality for a conservation plan for Yolo County (California), Clarke *et al.* (1999) model re-vegetation strategies for Western Australia, and Carriquiry *et al.* (1998) use GIS to devise sampling schemes for environmental policy analysis. From a different angle, Carver *et al.* (1995) evaluated the usefulness of *portable* field-based GIS for environmental characterisation.

In *forestry*, Davidson (1991) reviews the various methods and GIS technologies available, and Chou (1992) develops an index for fire rotation in the San Bernardino National Forest (California). Hussin *et al.* (1994) use remote sensing for land cover change detection, and Taylor *et al.* (1966) apply GIS to test the health of a eucalyptus forest in New South Wales (Australia). Hunter *et al.* (1999) assess the prospects of riparian forests in Sacramento (California), Bojorquez-Tapia *et al.* (1999) use the map algebra facility in GRASS to define suitability maps for different types of forest land uses in Mexico, Mertens *et al.* (2001) predict the impact of logging on forests in Cameroon, Gustafson *et al.* (2001) assess the impact on terrestrial salamanders of different forest-management approaches, and Velazquez *et al.* (2001) study forest quality in an indigenous community in Mexico. Hogsett *et al.* (1997) assess ozone risks in forests, Kovacs *et al.* (2001) combine GIS and Landsat data to study forest disturbances, Cassel-Gintz and Petschel-Held (2000) assess the threat to world forests from non-sustainable developments, and Ochoa-Gaona (2001) uses GIS to study forest fragmentation in Chiapas (Mexico). On a different note, Wing and Johnson (2001) use GIS to quantify forest visibility in McDonald Forest (Oregon).

In the more general area of *landscape* and land cover, Cihlar *et al.* (1989) combined satellite pictures with other maps and variables to analyse their correspondence in the growth season (by overlay, using Arc Info), Amissah-Arthur *et al.* (2000) use a similar approach to assess land degradation and farmland dynamics in Nigeria, and Petit and Lambin (2001) combine GIS and multi-source remote sensing information to detect land-cover changes in Zambia. Peccol *et al.* (1996) use GIS to assess the influence of planning policies on landscape change, and Namken and Stuth (1997a,b) analyse and model (using map algebra) the effects on landscape of grazing pressures on land. Mendonca-Santos and Claramunt (2001) use a similar map algebra approach to integrate landscape and local analysis of land-cover changes. Gustafson and Crow (1996) use ERDAS to simulate the effects of different landscape-management strategies in Hoosier National Forest (Indiana), and Baskent and Yolasigmaz (1999) review the literature concerning *forest landscape* management.

Applying GIS technology to *farming* is also an area of growing interest (Berry, 1998; Charvat, 2001), and Brown *et al.* (2000) combine GIS and remote sensing to model the relationships between land-use and land-cover in the Upper Midwest of the USA. Also, Smith *et al.* (2000) use the ArcView GIS to assess the sustainability of agriculture.

General *environmental evaluation* has been approached using GIS in New Zealand (Watkins *et al.*, 1997) and Brainard *et al.* (1999) suggest an interesting variation, using GIS and visitor information to assess the "worth" of environmental features by travel-cost analysis. Kliskey (1998) and Kliskey *et al.* (1994) apply buffering to analyse "wilderness perception" in North-West Nelson (New Zealand), and Merrill *et al.* (1995) evaluate "wilderness planning" options in Idaho (US). Swetnam *et al.* (1998) do a risk assessment of the relationship between hydrology and grassland in Somerset, Zalidis and Gerakis (1999) use map algebra to evaluate the sustainability of watershed resources in Karla (Greece), and Hawks *et al.* (2000) apply GIS to fisheries management in the Meramec river basin (Missouri). Scott and Sullivan (2000) use GIS to help select and design habitat preserves, Iverson *et al.* (2001) apply a similar approach to evaluate riparian habitats, and Eade and Moran (1996) use GIS to estimate the environmental economic benefits in a conservation area in Belize.

Burley and Brown (1995) apply GIS Principal Component Analysis to construct more "understandable" models of the environment and, on a slightly different note, Gumbricht (1996) uses GIS for *training* environmental managers.

The analysis of *visibility* areas (one of the most sophisticated GIS functions) has also been put to good use, usually for landscape assessment (not linked to IA): Uchida *et al.* (1997) analyse the visual potential of woodlands as seen from the city of Yamada (Japan), Sato *et al.* (1995) use this type of analysis to characterise the landscape views into the natural environment from 76 City Halls in Japan. On a related note, O'Sullivan and Turner (2001) develop a methodology to combine "visibility graphs" with GIS for landscape-visibility analysis.

Various aspects of *water* are also studied using GIS functionality, often using map algebra to apply multivariate models developed previously by other means. For *surface* water, Webber *et al.* (1996) study the role of wetlands in reducing water pollution in the Lake Champlain basin (Canada), Mitasova *et al.* (1996) study erosion potential in Illinois using GRASS, and Vieux *et al.* (1996) also use GRASS for storm runoff modelling. Thapa and Weber (1995) use map algebra to model the vulnerability of *watersheds* in Upper Pokhara Valley (Nepal), Wickman *et al.* (1998) use GIS cluster-analysis to identify watersheds in the US Mid-Atlantic region, and Liang and Mackay (2000) use GIS terrain-modelling capabilities to identify and define local watersheds. On a variation of the theme, Etzelmuller and Bjornsson (2000) apply GIS techniques to *glaciological* analysis and glacier flow in Iceland, and Chang and Li (2000) use GIS to model (by multiple regression) snow accumulation. Knox and Weatherfield (1999) discuss the application of GIS to the management of irrigation water in England and Wales. For *groundwater*, Canter *et al.* (1994) discuss GIS as a management tool, and McKinney and Tsai (1996) use raster GIS with multi-criteria map algebra. For water *pollution*, GIS is used from the Boston Harbour (Ardalan, 1988) to Lake Balaton in Hungary (Cserny *et al.*, 1997), and Osborn and Cook (1997) use GIS to discuss groundwater protection policies for England and Wales. Wang (2001) relates water quality management and land-use planning in watersheds.

Air quality is also monitored using GIS as described by Dev *et al.* (1993), who construct contour maps of air-quality indices by interpolation (with "Kriging", a technique which takes into account the spatial autocorrelation of data) for environmental monitoring in India. Modelling atmospheric data has also been undertaken using Digital Elevation Models (Lee and Pielke, 1996).

The area of *geology* has been particularly attractive in aspects with potential for immediate financial returns: for example, Memmi (1995) discusses an application of GIS to diamond exploration, and Fry (1995) reports on the search for gold. Related to more traditional aspects of geology, Hart and Zilkoski (1994) study subsidence in the New Orleans region, and Giles (1995) explores geological layers in the London Basin and their suitability for tunnelling for the Underground. In the area of hydrogeology, Fritch *et al.* (2000) use GIS map algebra to assess aquifer vulnerability in Texas.

The assessment of *hazard risks* has always been – for obvious practical reasons – a major area of study and GIS application, focusing on a wide range of hazards:

- *Pollution* risks are assessed in an interactive system for the Netherlands in Stein *et al.* (1995), and Heywood *et al.* (1989) provide an early use of GIS for radiation analysis and modelling in Cumbria.
- *Floods*: Emani *et al.* (1993) produce maps of vulnerability indices in Massachusetts, and Hickey *et al.* (1997) use a similar approach to assess coastal risk in the Gulf of Mexico.
- *Landslides*: after the early work of Bender and Bello (1990) on the potential of GIS for landslide assessment and monitoring through landslides inventories (they argued the case for Latin America), Wang and Unwin (1992) use a similar approach to develop a landslide potential model for central China, and Guzetti *et al.* (2000) use GIS to compare landslide maps in the Tiber basin (Italy); on a different note, Tang and Montgomery (1995) apply GIS buffering around rivers to define potentially unstable ground.
- *Avalanches*: Martin *et al.* (1999) use map algebra with terrain features like slopes, etc. calculated from a DEM.
- *Forest fire* risks: Chou (1992) uses his fire-rotation index (already mentioned) for the development of a fire-probability map for the San Bernardino National Forest in Southern California, and Chuvieco and Salas (1996) use GIS to assess fire risks for the Sierra de Gredos near Madrid (Spain).

The evaluation of rural and ecological *land suitability* (a similar application to urban land-use planning is also quite common) makes typical use of GIS functions like overlay and map algebra: Pereira and Duckstein (1993) use multi-criteria evaluation to measure land suitability for the red squirrel in Arizona, Bertozzi *et al.* (1994) produce soil vulnerability indices in the Padamo plain (Italy), Davidson *et al.* (1994) apply the approach to land evaluation in Greece, and Schmidt *et al.* (1995) evaluate forest soil fertility in Nepal. Also, the approach can be extended to land-use planning: Hallett *et al.* (1996) use GIS to plan "sustainable" land uses, Xia (1997) combines GIS and remote sensing to allocate land uses in Dongguan (China), and Ramirez-Sanz *et al.* (2000) suggest a methodology for environmental planning based on GIS map algebra. This same approach can be extended and applied to the *location* of certain activities or facilities, for example the location of sewage sludge in areas where its nutrients can be recycled (Francek *et al.*, 1999, 2001), or the location possibilities around a "dammed" river in Arizona (Graf, 2000), or the location of evaporation basins for saline irrigation schemes (Jolly *et al.*, 2001).

As can be seen, by far the most used GIS functions are *map overlay* and *map algebra*, usually in the context of some form of *multi-criteria evaluation* (see Malczewski, 1999 for a good discussion of the issues involved in this methodology), be it for land-suitability analysis or to map model results, like hazard risks. While overlay is predominantly a function used in vector-based GIS, map algebra is mostly used in raster-based GIS (or in raster-transformations of vector maps), with obvious potential for data sources like satellite imagery, already working in raster format. Beyond relatively simple functions like these, innovation in the use of GIS for environmental management tends to be associated with *input* and *output* devices more than with GIS functionality: the potential of *satellite imagery* for environmental description and monitoring has been identified since the 1980s; the potential of *Global Positioning System* (GPS) for quick and accurate location of point events (fires, etc.) and, linked to GPS, the potential of *portable* GIS for field work have also been identified. On the output side, *multimedia* interfaces are at the forefront of innovation, usually linked to an increase in the level of interactivity in these systems and, finally, it is worth mentioning that the last of the conferences on GIS and environmental modelling quoted above (Goodchild *et al.*, 1996b) contained a whole section and several other isolated papers devoted to the obvious growth area of the *Internet*, as a possible depository of environmental data, as a vehicle for the diffusion of software, and as an aid and encouragement to public participation in local environmental decision-making (Kingston *et al.*, 2000).

4.4.1 Pre-programmed GIS applications

As in IA, some applications of GIS for environmental management are *pre-programmed*, sometimes because they were planned that way from the start, sometimes because they have matured that way. The areas of interest and the approaches used (often map algebra) are virtually the same as for the hands-on versions just discussed, the only difference being that the sequences of operations have been automatised by encapsulating them into a program which decision-makers and managers can activate themselves.

For *ecology*, Lankhorst (1992) uses a pre-programmed map-algebra model to assess suitability and accessibility indices for habitats, Power and Barnes (1993) use algorithms (in the PC-based GIS SPANS) to transform forest-inventory data into habitat suitability indices for different species in New Jersey, and Parrish *et al.* (1993) evaluate an ecological risk index in Region 6 of the US. Yarie (1996) uses Arc-Info's Macro language AML to program a model of forest ecosystems, Woodhouse *et al.* (2000) use AML routines to model species-richness and select priority areas for conservation, and Cedfeldt *et al.* (2000) use a similar approach to identify wetlands in North-eastern USA.

A very common modality of "land suitability" studies is *site-selection* for a private or public facility, and such studies can be automatised for non-expert users. For example, Carver (1991) adds external Fortran routines to a GIS to combine multi-criteria evaluation with map overlay for waste site selection in the UK, and Carver (1999) extends the argument to the integration of GIS and the Internet to help with more participatory decision support. Gupta and Sahai (1993) report on a menu-driven system programmed internally to the Arc-Info GIS to evaluate the suitability of land for the location of aquaculture facilities in West Bengal (India). An extension of site-selection – by generalising its methodology to a whole range of uses – is *land use planning* for agricultural and rural management, and GIS has been suggested for this purpose from quite early on (Riezebos *et al.*, 1990).

In rural *land* management, Ventura (1988) provides an early system combining land records and environmental information with AML (the macro-language of Arc-Info) for land management in Wisconsin, Johnson *et al.* (1991) report on a system programmed to classify habitats for land management by the US Forest Service, and Eaton (1995) discusses a project developing models for the US Forest Service to predict vegetation type, so that when the models are ready they will be incorporated as "macros" (using AML) into the GIS. As an important aspect of land management, modelling forest *fire* risk, is reported in Thivierge (1994), using AML to get the data and produce indicators for various forest management and planning agencies in British Columbia, and Condes *et al.* (1996) describe the CARDIN forest-fire propagation model programmed also in AML.

Concerning *water*, Wang *et al.* (2000) integrate the ROUT water quality model with the ArcView GIS (a "friendly" relative of Arc-Info) using its internal macro-language "Avenue", for purposes of river-watershed planning. Programming GIS functionality to help *develop* water models has also been attempted successfully in a variety of aspects:

- *Tide* and *wave propagation* is modelled in Liebig (1996) using an external model, but the links between the GIS and the model are pre-programmed in Arc-Info's AML.
- For *ground water*, Saghafian (1996) uses a program for a hydrologic model written *inside* GRASS (this GIS is written in "C" which makes programming the model inside it much easier).
- For surface *runoff* water, Samulski (1991) shows an early discussion of the potential of a program (using AML) to simulate storm water flows so that drainage needs (and sewers) can be calculated later, Lehman (1994) describes a storm water flow simulation program (also linked to CAD software) for the Los Angeles Public Works Department, and Liao

and Tim (1994b) describe an interactive model to simulate soil erosion in Lake Icaria (Iowa).
- On the borderline between water and geology modelling, *landslide risk* assessment has also been programmed into a GIS, as reported by Noguchi *et al.* (1991) on a project for the Japanese railways.

As can be seen, the most common GIS function being pre-programmed is *map algebra*. With respect to the tools used, by far the most popular approach to GIS programming is – as in IA – to use the GIS' own *macro language*, AML in the case of Arc-Info. The exception to this rule is the rare case where the GIS itself is written in a language that lends itself to external connections, like "C". The problem is that not many GIS have a macro language incorporated, or are written in such accessible languages.

4.5 GENERAL-PURPOSE ENVIRONMENTAL MANAGEMENT SYSTEMS

As already mentioned, applications are sometimes difficult to classify in the groupings used above because they are not reported in sufficient detail or because they develop over time, but in some cases the difficulty is that they fit into *all* the groups, usually because they are set up for multi-purpose management and require the complete range of technical capability, from simple operations like mapping to linking with models (and other sophisticated tools) or map manipulation using GIS functions. In the field of industrial environmental management, Douglas (1995) explores the whole range of GIS environmental applications from a practical point of view (it is almost a "cook book" of how to incorporate GIS into this area). Examples of such environmental systems can be found in Strobol (1992) for managing forest resources, Moreira *et al.* (1994) describe the environmental information system for Andalucia (Spain), Ljesevic and Filipovic (1995) describe a similar system for environmental protection in Serbia, Ernst *et al.* (1995) discuss a system to help the American Environmental Monitoring and Assessment Program with wetland management, Leggett and Jones (1996) discuss a flood-defence system in the Anglian coast from the Thames to the Humber, and Bettinetti *et al.* (1996) discuss an integrated system for the restoration of the Venice lagoon. Wickham *et al.* (1999) use GIS for the management of salmon fisheries in Scotland, in a *pre-programmed* system using AML.

4.6 CONCLUSIONS

Even with the limited analytical capabilities that GIS have, their standard functions can be used to good effect in environmental management. As in

IA, one of the most popular uses of GIS is based on linking them with simulation models for the environment, be it for simulation or for model design and/or estimation, for which GIS can provide the data. When it is the GIS' own functionality that is used, the most common GIS functions are map overlay, buffering and map algebra, often in the context of some form of multi-criteria evaluation. As in IA, a lot of interest is generated by the potential of new input and output devices linked to GIS: the Internet, satellite imagery, GPS (also linked to the idea of portable GIS for field work), multimedia and hypermedia interfaces, virtually overtaking the interest in GIS functionality itself. As noted when reviewing GIS applications to IA in the previous chapter, it seems as if GIS maximise their potential when operating within a wider framework of other decision-support tools (like expert systems) that structure and focus their performance, and it is this area of GIS application that is covered in Chapter 5.

REFERENCES

Acevedo, M.F., Urban, D.L. and Ablan, M. (1996) Landscape Scale Forest Dynamics: GIS, Gap, and Transition Models, in Goodchild, M.F., Steyaert, L.T., Parks, B.O., Johnston, C., Maidment, D., Crane, M. and Glendinning, S. (eds) *op. cit.*, Ch. 33, pp. 181–5.

Ackroyd, N. (2000) GPS: Getting a Return on the Investment, *GeoEurope*, Issue 8 (August), p. 35.

Adeniyi, P.O., Omojola, A. and Soneye, A.S.O. (1992) Application of Remote Sensing and GIS in the Mapping, Evaluation and Monitoring of Agricultural Resources in Northwestern Nigeria, *Proceedings of the Canadian Conference on GIS '92*, Ottawa, Canada (March 24–6), pp. 803–20.

Ahearn, S. and Osleeb, J.P. (1993) Greenpoint/Williamsburg Environmental Benefits Program: Development of a Pilot Geographic Information System, *Proceedings of the GIS/LIS '93 Conference*, Minneapolis, Minnesota (November 2–4), Vol. 1, pp. 10–18.

Ahlenius, H. and Langaas, S. (2000) Baltic Interactive: Crating a Spatial Resource for the Region, *GeoEurope*, Issue 12 (December), pp. 38–9.

Akcakaya, H.R. (1996) Linking GIS with Models of Ecological Risk Assessment for Endangered Species, in Goodchild, M.F. *et al.* (1996b) *op. cit.*

Akcakaya, H.R. (2000) Conservation and Management for Multiple Species: Integrating Field Research and Modeling into Management Decisions, *Environmental Management*, Vol. 26 (Supplement), pp. S75–S83.

Albrecht, J.H. (1996) Universal GIS Operations for Environmental Modeling, in Goodchild, M.F. *et al.* (1996b) *op. cit.*

Allanson, P. and Moxey, A. (1996) Agricultural Land Use Change in England and Wales 1892–1992, *Journal of Environmental Planning and Management*, Vol. 39, No. 2 (June), pp. 243–54.

Almeida, R. (1994) Forest Fire Risk Areas and Definition of the Prevention Priority Planning Actions Using GIS, *Proceedings of the EGIS/MARI '94 Conference*, Paris (March 29–April 1), Vol. 2, pp. 1700–6.

Amissah-Arthur, A., Mougenot, B. and Loireau, M. (2000) Assessing Farmland Dynamics and Land Degradation on Sahelian Landscapes Using Remotely Sensed and Socioeconomic Data, *International Journal of Geographical Information Science*, Vol. 14, No. 6 (September), pp. 583–99.

Anselin, L. and Getis, A. (1993) Spatial Statistical Analysis and GIS, in Fischer, M.M. and Nijkamp, P. (eds) *GIS, Spatial Modelling and Policy Evaluation*, Springer-Verlag, Ch. 3, pp. 36–49.

Ardalan, N. (1988) A Dynamic Archival System (DAS) for the Clean-Up of Boston Harbour, *Proceedings of the Urban and Regional Information Systems Association (URISA) Conference*, Los Angeles (August 7–11), Vol. II, pp. 97–103.

Ardiles-Lopez, L., Ferre-Julia, M. and Rodriguez-Chaparro, J. (1996) The Use of GIS to Estimate Hydrological Parameters in a Rainfall-Runoff Model, in Rumor, M. *et al.* (eds) *op. cit.*, Vol. 1, pp. 408–17.

Arsenau, G. and Lowell, K. (1992) Elements de Gestion Pour les Suivis de la Planification Forestiere au Quebec a l'Aide de GIS, *Proceedings of the Canadian Conference on GIS '92*, Ottawa, Canada (March 24–6), pp. 303–16.

Aspinall, R. and Pearson, D. (2000) Integrated Geographical Assessment of Environmental Condition in Water Catchments: Linking Landscape Ecology, Environmental Modeling and GIS, *Journal of Environmental Management*, Vol. 59, No. 4 (August – Special issue), pp. 299–319.

Baldina, E.A., De Leeuw, J., Gorbunov, A.K., Labutina, I.A., Zhivogliad, A.F. and Kooistra, J.F. (1999) Vegetation Change in the Astrakhanski Biosphere Reserve (Lower Volta Delta, Russia) in Relation to Caspian Sea Level Fluctuations, *Environmental Conservation*, Vol. 26, No. 3 (September), pp. 169–78.

Baskent, E.Z. and Yolasigmaz, H.A. (1999) Forest Landscape Management Revisited, *Environmental Management*, Vol. 24, No. 4 (November), pp. 437–48.

Bateman, I.J. and Lovett, A.A. (2000) Estimating and Valuing the Carbon Sequestered in Softwood and Hardwood Trees, Timber Products and Forest Soils in Wales, *Journal of Environmental Management*, Vol. 60, No. 4 (December), pp. 301–23.

Battaglin, W.A., Kuhn, G. and Parker, R. (1996) Using GIS to Link Digital Spatial Data and the Precipitation Runoff Modeling System: Gunnison River Basin, Colorado, in Goodchild, M.F., Steyaert, L.T., Parks, B.O., Johnston, C., Maidment, D., Crane, M. and Glendinning, S. (eds) *op. cit.*, Ch. 29, pp. 159–63.

Beaulac, M.N., Businski, S. and Forstat, D. (1994) The Evolution of Michigan's Geospatial Data Infrastructure, *Journal of the Urban and Regional Information Systems Association*, Vol. 6, No. 1 (Spring), pp. 63–8.

Belknap, W. and Naiman, R.J. (1998) A GIS and TIR Procedure to Detect and Map Wall-base Channels in Western Washington, *Journal of Environmental Management*, Vol. 52, No. 2 (February), pp. 147–60.

Bender, S.O. and Bello, E.E. (1990) GIS Applications for Natural Hazards Management in Latin America and the Caribbean, *Proceedings of the Urban and Regional Information Systems Association (URISA) Conference*, Edmonton, Alberta, Canada (August 12–16), Vol. I, pp. 67–77.

Bergamasco, A., Piola, S. and Deligios, M. (1996) Model Oriented GIS for Marine and Coastal Environmental Applications, in Rumor, M. *et al.* (eds) *op. cit.*, Vol. 1, pp. 418–26.

Bernert, J.A., Eilers, J.M., Sullivan, T.J., Freemark, K.E. and Ribic, C. (1997) A Quantitative Method for Delineating Regions: An Example for the Western Corn Belt Plains Ecoregion of the USA, *Environmental Management*, Vol. 21, No. 3 (May/June), pp. 405–20.

Berry, J.K. (1998) Who's Minding the Farm?, *GIS World* (February), pp. 46–8.

Bertozzi, R., Buscaroli, A., Garde, C., Sequi, P. and Vianello, G. (1994) A GIS Application for the Evaluation of the Soil's Vulnerability Map, *Proceedings of the EGIS/MARI '94 Conference*, Paris (March 29–April 1), Vol. 1, pp. 1016–25.

Bettinetti, A., Pypaert, P. and Sweerts, J.-P. (1996) Application of an Integrated Management Approach to the Restoration Project of the Lagoon of Venice, *Journal of Environmental Management*, Vol. 46, No. 3 (March), pp. 207–27.

Bian, L. (2000) Component Modeling for the Spatial Representation of Wildlife Movements, *Journal of Environmental Management*, Vol. 59, No. 4 (August – Special issue), pp. 235–45.

Bishop, I.D. and Gimblett, H.R. (2000) Management of Recreational Areas: GIS, Autonomous Agents, and Virtual Reality, *Environment and Planning B: Planning and Design*, Vol. 27, No. 3 (May), pp. 423–35.

Bojorquez-Tapia, L.A., Diaz-Mondragon, S. and Gomez-Priego, P. (1999) GIS-approach for Land Suitability Assessment in Developing Countries: A Case Study of Forest Development Project in Mexico, in Thill, J.-C. (ed.) *op. cit.*, Ch. 14, pp. 335–52.

Bowker, G.C. (2000) Mapping Biodiversity, *International Journal of Geographical Information Science*, Vol. 14, No. 8 (December), pp. 739–54.

Brainard, J., Lovett, A. and Bateman, I. (1999) Integrating Geographical Information Systems Into Travel Cost Analysis and Benefit Transfer, *International Journal of Geographical Information Science*, Vol. 13, No. 3 (April–May), pp. 227–46.

Brimicombe, A.J. and Barlett, J.M. (1996) Linking GIS with Hydraulic Modeling for Flood Risk Assessment: The Hong Kong Approach, in Goodchild, M.F., Steyaert, L.T., Parks, B.O., Johnston, C., Maidment, D., Crare, M. and Glendinning, S. (eds) *op. cit.*, Ch. 30, pp. 165–8.

Brown, D.G., Pijanowski, B.C. and Duh, J.D. (2000) Modeling the Relationships Between Land Use and Land Cover on Private Lands in the Upper Midwest, USA, *Journal of Environmental Management*, Vol. 59, No. 4 (August – Special issue), pp. 247–63.

Brown, S. and Shrestha, B. (2000) Market-driven Land-use Dynamics in the Middle Mountains of Nepal, *Journal of Environmental Management*, Vol. 59, No. 3 (July), pp. 217–25.

Burlando, P., Mancini, M. and Rosso, R. (1994) FLORA: A Distributed Flood Risk Analyser, in Guariso, G. and Page, B. (eds) *Computer Support for Environmental IA*, The IFTP TC5/WG5.11 Working Conference on Computer Support for Environmental IA – CSEIA 93, Como, Italy (6–8 October 1993), IFIP Transactions B: Applications in Technology (B-16), North Holland: Elsevier Science BV, pp. 91–102.

Burley, J.B. and Brown, T.J. (1995) Constructing Interpretable Environments for Multidimensional Data: GIS Suitability Overlays and Principal Component Analysis, *Journal of Environmental Planning and Management*, Vol. 38, No. 3 (September), pp. 537–50.

Cambruzzi, T., Fiduccia, A. and Novelli, L. (1999) A Dynamic Geomonitoring System for Venetian Coastal Ecosystem: WATERS project, *Computers, Environment and Urban Systems*, Vol. 23, No. 6 (November), pp. 469–84.

Campbell, G. and Hastie, R. (1993) Management of First Nations Lands Using GIS Technology, *Proceedings of the Canadian Conference on GIS '93*, Ottawa, Canada (March 23–5), pp. 141–53.

Canter, L.W., Chowdhury, A.K.M.M. and Vieux, B.E. (1994) Geographic Information Systems: A Tool for Strategic Ground Water Quality Management, *Journal of Environmental Planning and Management*, Vol. 37, No. 3, pp. 251–66.

Carriquiry, A., Breidt, F.J. and Lakshminarayan, P.G. (1998) Sampling Schemes for Policy Analyses Using Computer Simulation Experiments, *Environmental Management*, Vol. 22, No. 4 (July/August), pp. 505–15.

Carver, S. (1991) Spatial Decision Support Systems for Facility Location: A Combined GIS and Multicriteria Evaluation Approach, in Klosterman, R.E. (ed.) *op. cit.*, Vol. 1, pp. 75–90.

Carver, S. (1999) Developing Web-based GIS/MCE: Improving Access to Data and Spatial Decision Support Tools, in Thill, J.-C. (ed.) *op. cit.*, Ch. 3, pp. 49–75.

Carver, S., Heywood, I., Cornelius, S. and Sear, D. (1995) Evaluating Field-Based GIS for Environmental Characterization, Modelling and Decision Support, *International Journal of Geographical Information Systems*, Vol. 9, No. 4, pp. 475–86.

Cassel-Gintz, M. and Petschel-Held, G. (2000) GIS-based Assessment of the Threat to World Forests by Patterns of Non-sustainable Civilisation Nature Interaction, *Journal of Environmental Management*, Vol. 59, No. 4 (August – Special issue), pp. 279–98.

Cedfeldt, P.T., Watzin, M.C. and Richardson, B.D. (2000) Using GIS to Identify Functionally Significant Wetlands in the Northeastern United States, *Environmental Management*, Vol. 26, No. 1 (July), pp. 13–24.

Chang, K.-T. and Li, Z. (2000) Modelling Snow Accumulation With a Geographical Information System, *International Journal of Geographical Information Science*, Vol. 14, No. 7 (October–November), pp. 693–707.

Chang, K.-T., Verbyla, D.L. and Yeo, J.J. (1995) Spatial Analysis of Habitat Selection by Sitka Black-Tailed Dear in Southeast Alaska, USA, *Environmental Management*, Vol. 19, No. 4 (July/August), pp. 579–89.

Chang, N.-B. and Wei, Y.L. (1999) Strategic Planning of Recycling Drop-Off Stations and Collection Network by Multiobjective Programming, *Environmental Management*, Vol. 24, No. 2 (August), pp. 247–63.

Charvat, K. (2001) Next steps in Precision Farming and GIS, *GeoEurope*, Issue 8 (August), pp. 44–6.

Chien, P. (2000) Endeavour Maps the World in Three Dimensions, *GeoWorld*, No. 4 (April), pp. 32–8.

Chou, Y.-H. (1992) Management of Wildfires With a Geographical Information System, *International Journal of Geographical Information Systems*, Vol. 6, No. 2, pp. 123–40.

Chou, Y.-H. and Soret, S. (1996) Neighborhood Effects in Bird Distributions, Navarre, Spain, *Environmental Management*, Vol. 20, No. 5 (September/October), pp. 675–87.

Church, R., Stoms, D., Davis, F. and Okin, B.J. (1996) Planning Management Activities to Protect Biodiversity with a GIS and an Integrated Optimization Model, in Goodchild, M.F. *et al.* (1996b) *op. cit.*

Chuvieco, E. and Salas, J. (1996) Mapping the Spatial Distribution of Forest Fire Danger Using GIS, *International Journal of Geographical Information Systems*, Vol. 10, No. 3, pp. 333–45.

Cihlar, J., D'Iorio, M., Mullins, D. and St-Laurent, L. (1989) Use of Satellite Data and GIS for Environmental Change Studies, *Proceedings of the National Canadian Conference on GIS "Challenge for the 1990s"*, Ottawa, Canada (February 27–March 3), pp. 933–43.

Clarke, C.J., Hobbs, R.J. and George, R.J. (1999) Incorporating Geological Effects in Modeling of Revegetation of Strategies for Salt-Affected Landscapes, *Environmental Management*, Vol. 24, No. 1 (July), pp. 99–109.

Clayson, J. (1996) On a Wing and a Prayer, *Mapping Awareness* (November), pp. 24–6.

Collins, R.P., Jenkins, A. and Sloan, W.T. (1998) A GIS Framework for Modelling Nitrogen Leaching from Agricultural Areas in the Middle Hills, Nepal, *International Journal of Geographical Information Science*, Vol. 12, No. 5 (July–August), pp. 479–90.

Condes, S., Martos, J. and Martinez-Millan, J. (1996) Simulation of the Propagation of Forest Fires, Integrated Within a GIS, in Rumor, M. *et al.* (eds) *op. cit.*, Vol. 2, pp. 1306–15.

Cordonnery, L. (1999) Implementing the Protocol on Environmental Protection to the Antarctic Treaty: Future Applications of Geographic Information Systems within the Committee for Environmental Protection, *Journal of Environmental Management*, Vol. 56, No. 4 (August), pp. 285–98.

Craig, W.J., Harris, T.M. and Weiner, D. (2002) (eds) *Community Participation and Geographic Information Systems*, Taylor & Francis, London and New York.

Cronshey, R.G., Theurer, F.D. and Glenn, R.L. (1996) GIS Water-Quality Model Interface: A Prototype, in Goodchild, M.F., Steyaert, L.T., Parks, B.O., Johnston, C., Maidment, D., Crane, M. and Glendinning, S. (eds) *op. cit.*, Ch. 54, pp. 287–91.

Cruickshank, M.M., Tomlinson, R.W. and Trew, S. (2000) Application of CORINE Land-cover Mapping to Estimate Carbon Stored in the Vegetation of Ireland, *Journal of Environmental Management*, Vol. 58, No. 4 (April), pp. 269–87.

Cserny, T., Hidvegi, M. and Tullner, T. (1997) From Degradation to Conservation, *GIS Europe* (October), pp. 37–41.

D'Agnese, F.A., Turner, A.K. and Faunt, C.C. (1996) Using Geoscientific Information Systems for Three-Dimensional Regional Groundwater Flow Modeling, in Goodchild, M.F., Steyaert, L.T., Parks, B.O., Johnston, C., Maidment, D., Crane, M. and Glendinning, S. (eds) *op. cit.*, Ch. 50, pp. 265–70.

Davidson, D.A. (1991) Forestry and GIS, *Mapping Awareness*, Vol. 5, No. 5 (June), pp. 43–5.

Davidson, D.A., Theocharopoulos, S.P. and Bloksma, R.J. (1994) A Land Evaluation Project in Greece Using GIS and Based on Boolean and Fuzzy Set Methodologies, *International Journal of Geographical Information Systems*, Vol. 8, No. 4, pp. 369–84.

Davis, F.W., Quattrochi, D.A. and Ridd, M.K. (1991) Environmental Analysis Using Integrated GIS and Remotely Sensed Data: Some Research Needs and Priorities, *Photogrametric Engineering and Remote Sensing*, Vol. 57 (June), pp. 689–97.

DePinto, J.V., Atkinson, J.F., Calkins, H.W., Densham, P.J., Guan, W., Lin, H., Xia, F., Rodgers, P.W. and Slawecki, T. (1996) Development of GEO-WAMS: A Modeling Support System to Integrate GIS with Watershed Analysis Models, in Goodchild, M.F., Steyaert, L.T., Parks, B.O., Johnston, C., Maidment, D., Crane, M. and Glendinning, S. (eds) *op. cit.*, Ch. 51, pp. 271–6.

De Roo, A.P.J., Wesseling, C.G., Cremers, N.H.D.T., Offermans, R.J.E., Ritsema, C.J. and Van Oostindie, K. (1994) LISEM: A Physically-Based Hydrological and Soil Erosion Model Incorporated in a GIS, *Proceedings of the EGIS/MARI '94 Conference*, Paris (March 29–April 1), Vol. 1, pp. 207–16.

Dev, D.S., Venkatachalam, P. and Natarajan, C. (1993) Geographic Information Systems for Environmental IA (EIS) – A Case Study, *International Journal of Environmental Studies*, Vol. 43, pp. 115–22.

Dippon, D.R., Green, P. and Pearson, D. (1989) Building the Western Oregon Database for the 1990's Resource Management Planning Effort, *Proceedings of the GIS/LIS '89 Conference*, Orlando, Florida (November 26–30), Vol. 1, pp. 276–85.

Douglas, W.J. (1995) *Environmental GIS Applications to Industrial Facilities*, Lewis Publishers, Boca Raton (Florida).

Duguay, C.R. and Walker, D.A. (1996) Environmental Modeling and Monitoring with GIS: Niwot Ridge Long-Term Ecological Research Site, in Goodchild, M.F., Steyaert, L.T., Parks, B.O., Johnston, C., Maidment, D., Crane, M. and Glendinning, S. (eds) *op. cit.*, Ch. 41, pp. 219–23.

Dusart, J., Toure, A. and Diop, S. (1994) Cartes du Couvert Vegetal, d'Utilisation et d'Occpation des Sols par Teledetection Dans le Cadre de l'Amenagement des Forets Naturelles du Gonakie Dans la Vallee du Fleuve Senegal, *Proceedings of the EGIS/MARI '94 Conference*, Paris (March 29–April 1), Vol. 1, pp. 579–88.

Eade, J.D.O. and Moran, D. (1996) Spatial Economic Valuation: Benefits Transfer using Geographical Information Systems, *Journal of Environmental Management*, Vol. 48, No. 1 (September), pp. 97–110.

Eaton, L. (1995) Statistical Models Mesh with Maps to Improve Land-Use Planning, *GIS World* (July), pp. 66–9.

Emani, S., Ratick, S.J., Clark, G.E., Dow, K., Kasperson, J.X., Kaspersons, R.E., Moser, S. and Schwarz, H. (1993) Assessing Vulnerability to Extreme Storm Events and Sea-Level Rise Using Geographical Information Systems (GIS), *Proceedings of the GIS/LIS '93 Conference*, Minneapolis, Minnesota (November 2–4), Vol. 1, pp. 201–9.

Ernst, T.L., Leibowitz, N.C., Roose, D., Stehman, S. and Urquhart, N.S. (1995) Evaluation of US EPA Environmental Monitoring and Assessment Program's (EMAP) – Wetlands Sampling Design and Classification, *Environmental Management*, Vol. 19, No. 1 (January/February), pp. 99–113.

Etzelmuller, B. and Bjornsson, H. (2000) Map Analysis Techniques for Glaciological Applications, *International Journal of Geographical Information Science*, Vol. 14, No. 6 (September), pp. 567–81.

Fedra, K. (1993) Clean Air, *Mapping Awareness*, Vol. 7, No. 6 (July/August), pp. 24–7.

Fedra, K. (1999) Urban Environmental Management: Monitoring, *Computers, Environment and Urban Systems*, Vol. 23, No. 6 (November), pp. 443–57.

Ford, S. and Lahage, B. (1996) Massachusetts Water Resource Authority: A Water-Quality GIS Showcase, *Journal of the Urban and Regional Information Systems Association*, Vol. 8, No. 1 (Spring), pp. 87–90.

Francek, M., Klopcic, J. and Klopcic, R. (1999) You Can't Put That Here!, *GeoWorld* (April), pp. 52–4.

Francek, M., Klopcic, J. and Klopcic, R. (2001) You Can't Put That Here, *GeoWorld*, Vol. 14, No. 4 (April), pp. 52–4.

Freeman, W. and Fox, J. (1995) ALAWAT: A Spatially Allocated Watershed Model for Approximating Stream, Sediment, and Pollutant Flows in Hawaii, USA, *Environmental Management*, Vol. 19, No. 4 (July/August), pp. 567–77.

Fritch, T.G., Mcknight, C.L., Yelderman Jr., J.C. and Arnold, J.G. (2000) Aquifer Vulnerability Assessment of the Paluxy Aquifer, Central Texas, USA, Using GIS and a Modified DRASTIC Approach, *Environmental Management*, Vol. 25, No. 3 (March), pp. 337–45.

Fry, C. (1995) The MIDAS Touch – GIS Joins Europe's Hunt for Gold, *GIS Europe* (December), pp. 25–7.

Fuller, R. and Groom, G. (1993a) The Land Cover Map of Great Britain, *GIS Europe* (October), pp. 25–8.

Fuller, R. and Groom, G. (1993b) The Land Cover Map of Great Britain, *Mapping Awareness*, Vol. 7, No. 9 (November), pp. 18–20.

Gao, X., Sorooshian, S. and Goodrich, D.C. (1993) Linkage of a GIS to a Distributed Rainfall-Runoff Model, in Goodchild, M.F., Parks, B.O. and Steyaert, L.T. (eds) *op. cit.*, Ch. 17, pp. 182–7.

Garnier, M., Lo Porto, A., Marini, R. and Leone, A. (1998) Integrated Use of GLEAMS and GIS to Prevent Groundwater Pollution Caused by Agricultural Disposal of Animal Waste, *Environmental Management*, Vol. 22, No. 5 (September/October), pp. 747–56.

Geist, D.R. and Dauble, D.D. (1998) Redd Site Selection and Spawning Habitat Use by Fall Chinook Salmon: The Importance of Geomorphic Features in Large Rivers, *Environmental Management*, Vol. 22, No. 5 (September/October), pp. 655–69.

Giavelli, G. and Rossi, O. (1999) The AEOLIAN Project: A MAB-UNESCO Investigation to Promote Sustainable Tourism in the Mediterranean Area, *International Journal of Environmental Studies*, Vol. 56, No. 6 (Section A: Environmental Studies), pp. 833–47.

Giles, D. (1995) The Integration of GIS and Geostatistical Modelling for a Tunelling Geohazard Study, *Proceedings of the Joint European Conference and Exhibition on Geographic Information JEC-GI '95*, The Hague (March 26–31), Vol. 1, pp. 421–6.

Girard, C.M., Le Bas, C., Szujecka, W. and Girard, M.C. (1994) Remote Sensing and Fallow Land, *Journal of Environmental Management*, Vol. 41, No. 1 (May), pp. 27–38.

GIS Europe (1992) CORINE Land Cover Inventory Progress, *GIS Europe* (December), pp. 27–34.

Goncalves Henriques, R. (2000) GEOCID: Portugal's GIS for Every Citizen, *GeoEurope*, Issue 10 (October), pp. 47–9.

Goodchild, M.F., Parks, B.O. and Steyaert, L.T. (1993) (eds) *Environmental Modelling with GIS*, Oxford University Press, Oxford.

Goodchild, M.F., Steyaert, L.T., Parks, B.O., Johnston, C., Maidment, D., Crane, M. and Glendinning, S. (1996a) (eds) *GIS and Environmental Modeling: Progress and Research Issues*, GIS World Books.

Goodchild, M.F., Steyaert, L.T., Parks, B.O., Johnston, C., Maidment, D., Crane, M. and Glendinning, S. (1996b) *Proceedings of the Third International Conference/*

Workshop on Integrating GIS and Environmental Modelling, National Centre for Geographic Information and Analysis, Santa Fe (New Mexico) (in CD format).

Graf, W.L. (2000) Locational Probability for a Dammed, Urbanizing Stream: Salt River, Arizona, USA, *Environmental Management*, Vol. 25, No. 3 (March), pp. 321–35.

Gumbricht, T. (1996) Application of GIS in Training for Environmental Management, *Journal of Environmental Management*, Vol. 46, No. 1 (January), pp. 17–30.

Gupta, M.C. and Sahai, B. (1993) Use of GIS for Brackish Water Aquaculture Site Selection in West Bengal, India, *Proceedings of the Canadian Conference on GIS '93*, Ottawa, Canada (March 23–5), pp. 586–98.

Gustafson, E.J. and Crow, T.R. (1996) Simulating the Effects of Alternative Forest Management Strategies on Landscape Structure, *Journal of Environmental Management*, Vol. 46, No. 1 (January), pp. 77–94.

Gustafson, E.J., Murphy, N.L. and Crow, T.R. (2001) Using a GIS Model to Assess Terrestrial Salamander Response to Alternative Forest Management Plans, *Journal of Environmental Management*, Vol. 63, No. 3 (November), pp. 281–92.

Guzzetti, F., Cardinali, M., Reichenbach, P. and Carrara, A. (2000) Comparing Landslide Maps: A Case Study in the Upper Tiber Basin, Central Italy, *Environmental Management*, Vol. 25, No. 3 (March), pp. 247–63.

Haak, B. and Bechdol, M. (1999) Multisensor Remote Sensing Data for Land use/Cover mapping, *Computers, Environment and Urban Systems*, Vol. 23, No. 1 (January), pp. 53–69.

Haack, B. (1996) Monitoring Wetland Changes with Remote Sensing: An East African Example, *Environmental Management*, Vol. 20, No. 3 (May/June), pp. 411–19.

Hallett, S.H., Jones, R.J.A. and Keay, C.A. (1996) Environmental Information Systems Developments for Planning Sustainable Land Use, *International Journal of Geographical Information Systems*, Vol. 10, No. 1, pp. 47–64.

Hao, Q.-W. and Chugh, Y.P. (1993) Spatial Predictive Modelling of Mine Subsidence Risk With GIS, *Proceedings of the GIS/LIS '93 Conference*, Minneapolis, Minnesota (November 2–4), Vol. 1, pp. 282–91.

Harding, J.S. and Winterbourn, M.J. (1997) An Ecoregion Classification of the South Island, New Zealand, *Journal of Environmental Management*, Vol. 51, No. 3 (November), pp. 275–87.

Harris, J., Gupta, S., Woodside, G. and Ziemba, N. (1993) Integrated Use of a GIS and a Three-Dimensional, Finite-Element Model: San Gabriel Basin Groundwater Flow Analysis, in Goodchild, M.F., Parks, B.O. and Steyaert, L.T. (eds) *op. cit.*, Ch. 15, pp. 168–72.

Harrison, C. and Haklay, M. (2002) The Potential of Public Participation Geographic Information Systems in UK Environmental Planning Appraisals by Active Publics, *Journal of Environmental Planning and Management*, Vol. 45, No. 6, pp. 841–63.

Hart, D. and Zilkoski, D. (1994) Mapping a Moving Target: The Use of GIS to Support Development of a Subsidence Model in the New Orleans Region, *Proceedings of the Urban and Regional Information Systems Association (URISA) Conference*, Milwaukee, Wisconsin (August 7–11), Vol. I, pp. 555–69.

Hay, L.E., Battaglin, W., Parker, R.S. and Leavesley, G.H. (1993) Modeling the Effects of Climate Change on Water Resources in the Gunnison River Basin,

Colorado, in Goodchild, M.F., Parks, B.O. and Steyaert, L.T. (eds) *op. cit.*, Ch. 39, pp. 392–9.

Hay, L., Knapp, L. and Bromberg, J. (1996) Integrating GIS, Scientific Visualization Systems, Statistics, and an Orographic Precipitation Model for a Hydroclimatic Study of the Gunnison River Basin, in Goodchild, M.F., Steyaert, L.T., Parks, B.O., Johnston, C., Maidment, D., Crane, M. and Glendinning, S. (eds) *op. cit.*, Ch. 44, pp. 235–8.

Havens, K.J., Priest III, W.I. and Berquist, H. (1997) Investigation and Long-Term Monitoring of *Pragmites australis* within Virginia's Constructed Wetland Sites, *Environmental Management*, Vol. 21, No. 4 (July/August), pp. 599–605.

Hawks, M.M., Stanovick, J.S. and Caldwell, M.L. (2000) Demonstration of GIS Capabilities for Fisheries Management Decisions: Analysis of Acquisition Potential Within the Meramec River Basin, *Environmental Management*, Vol. 26, No. 1 (July), pp. 25–34.

Healey, R.G., Robertson, S.G., Magor, J.I., Pender, J. and Cressman, K. (1996) A GIS for Desert Locust Forecasting and Monitoring, *International Journal of Geographical Information Systems*, Vol. 10, No. 1 (January–February), pp. 117–36.

Heywood, I., Cornelius, S., Openshaw, S. and Cross, S. (1989) Using a Spatial Database for Environmental Radiation Monitoring and Analysis, *Proceedings of the National Canadian Conference on GIS "Challenge for the 1990s"*, Ottawa, Canada, (February 27–March 3), pp. 1257–73.

Hickey, R.J., Bush, D.M. and Boulay, R.S. (1997) GIS Supports Coastal Risk Assessment, *GIS World* (June), pp. 54–8.

Higgs, G., Aitchison, B.I., Crosweller, H. and Jones, P. (1994) The National GIS Demonstrator of Common Lands for England and Wales, *Journal of Environmental Planning and Management*, Vol. 37, No. 1, pp. 33–51.

Hogsett, W.E., Weber, J.E., Tingey, D., Herstrom, A., Lee, E.H. and Laurence, J.A. (1997) An Approach for Characterizing Tropospheric Ozone Risk to Forests, *Environmental Management*, Vol. 21, No. 1 (January/February), pp. 105–20.

Hunter, J.C., Willett, K.B., McCoy, M.C., Quinn, J.F. and Keller, K.E. (1999) Prospects for Preservation and Restoration of Riparian Forests in the Sacramento Valley, California, USA, *Environmental Management*, Vol. 24, No. 1 (July), pp. 65–75.

Hussin, Y.A., de Gier, A. and Hargyono (1994) Forest Cover Charge Detection Analysis Using Remote Sensing – A Test for the Spatially Resolved Area Prediction Model, *Proceedings of the EGIS/MARI '94 Conference*, Paris (March 29–April 1), Vol. 2, pp. 1825–34.

Hutchinson, M. (1993) Development of a Continent-Wide DEM With Applications to Terrain and Climate Analysis, in Goodchild, M.F., Parks, B.O. and Steyaert, L.T. (eds) *op. cit.*, Ch. 16, pp. 172–81.

Isachenko, G.A. and Reznikov, A.I. (1994) Natural Anthropogenic Dynamics of Landscape: Information Systems of Simulation for Ladoga Region, *Proceedings of the EGIS/MARI '94 Conference*, Paris (March 29–April 1), Vol. 1, pp. 661–7.

Iverson, L.R., Szafoni, D.L., Baum, S.E. and Cook, E.A. (2001) Riparian Wildlife Habitat Evaluation Scheme Developed Using GIS, *Environmental Management*, Vol. 28, No. 5 (November), pp. 639–54.

Jankowski, P. and Nyerges, T. (2001) *Geographic Information Systems for Group Decision Making*, Taylor & Francis, London and New York.

Jaagus, J. (1996) Estimation of the Impact of Climate Change on Snow Cover and River Runoff in Estonia Using GIS, in Rumor, M. *et al.* (eds) *op. cit.*, Vol. 1, pp. 517–26.

Jang, C.J., Nishigami, Y. and Yanagisawa, Y. (1996) Assessment of Global Forest Change Between 1986 and 1993 Using Satellite-derived Terrestrial Net Primary Productivity, *Environmental Conservation*, Vol. 23, No. 4 (December), pp. 315–21.

Johnson, A.K.L., Ebert, S.P. and Murray, A.E. (1999) Distribution of Coastal Freshwater Wetlands and Riparian Forests in the Herbert River Catchment and Implications for Management of Catchments Adjacent the Great Barrier Reef Marine Park, *Environmental Conservation*, Vol. 26, No. 3 (September), pp. 229–35.

Johnson, L.B., Host, G.E., Jordan, J.K. and Rogers, L.L. (1991) Use of GIS for Landscape Design in Natural Resource Management: Habitat Assessment and Management for the Female Black Bear, *Proceedings of the GIS/LIS '91 Conference*, Atlanta, Georgia (October 28–November 1), Vol. 2, pp. 507–17.

Johnson, L.E., O'Donnell, S. and Tibi, R. (1990) Interactive Hydromet Information System for Real-Time Flood Forecasting and Warning, *Proceedings of the Urban and Regional Information Systems Association (URISA) Conference*, Edmonton, Alberta, Canada (August 12–16), Vol. I, pp. 140–50.

Johnston, C.A. (1993) Introduction to Quantitative Methods and Modelling in Community, Population, and Landscape Ecology, in Goodchild, M.F., Parks, B.O. and Steyaert, L.T. (eds) *op. cit.*, Ch. 25, pp. 276–83.

Johnston, C.A. and Sales, J. (1994) Using GIS to Predict Erosion Hazard Along Lake Superior, *Journal of the Urban and Regional Information Systems Association*, Vol. 6, No. 1 (Spring), pp. 57–62.

Johnston, C.A., Cohen, Y. and Pastor, J. (1996) Modeling of Spatially Static and Dynamic Ecological Process, in Goodchild, M.F., Steyaert, L.T., Parks, B.O., Johnston, C., Maidment, D., Crane, M. and Glendinning, S. (eds) *op. cit.*, Ch. 27, pp. 149–54.

Jolly, I.D., Walker, G.R., Dowling, T.I., Christen, E.W. and Murray, E. (2001) Regional Planning for the Siting of Local Evaporation Basins for the Disposal of Saline Irrigation Drainage: Development and Testing of a GIS-based Suitability Approach, *Journal of Environmental Management*, Vol. 63, No. 1 (September), pp. 51–70.

Jones, R.J.A., Bradley, R.I. and Siddons, P.A. (1994) A Land Information System for Environmental Risk Assessment, *Mapping Awareness* (October), pp. 20–3.

Kato, Y. (1987) A Computerized Soil Information System for Arable Land in Japan (JAPSIS): Present Structure and Some Applications, *Japan Agricultural Research Quarterly*, Vol. 21, No. 1, pp. 14–21.

Karnohan, B.J., Millspaugh, J.J., Jenks, J.A. and Naugle, D.E. (1998) Use of an Adaptive Kernel Home-based Estimator in a GIS Environment to Calculate Habitat Use, *Journal of Environmental Management*, Vol. 3, No. 1 (May), pp. 83–9.

Khaemba, W.M. and Stein, A. (2000) Use of GIS for a Spatial and Temporal Analysis of Kenyan Wildlife with Generalised Linear Modelling, *International Journal of Geographical Information Science*, Vol. 14, No. 8 (December), pp. 833–53.

Kingston, R., Carver, S., Evans, A. and Turton, I. (2000) Web-based Public Participation Geographical Information Systems: An Aid to Local Environmental

Decision-making, *International Journal of Geographical Information Science*, Vol. 14, No. 2 (March), pp. 109–25.

Kinsley, A.D. (1995) The Role and Functionality of GIS as a Planning Tool in Natural-Resource Management, *Computers, Environment and Urban Systems*, Vol. 19, No. 1, pp. 15–22.

Kirkman, H. (1996) Baseline and Monitoring Methods for Seagrass Meadows, *Journal of Environmental Management*, Vol. 47, No. 2 (June), pp. 191–201.

Kittel, T.G.F., Ojima, D.S., Schimel, D.S., McKeown, R., Bromberg, J.G., Painter, T.H., Rosenbloom, N.A., Parton, W.J. and Giorgi, F. (1996) Model GIS Integration and Data Set Development to Assess Terrestrial Ecosystem Vulnerability to Climate Change, in Goodchild, M.F., Steyaert, L.T., Parks, B.O., Johnston, C., Maidment, D., Crane, M. and Glendinning, S. (eds) *op. cit.*, Ch. 55, pp. 293–7.

Klijn, F., De Waal, R.W. and Voshaar, J.H.O. (1995) Ecoregions and Ecodistricts: Ecological Regionalizations for the Netherlands' Environmental Policy, *Environmental Management*, Vol. 19, No. 6 (November/December), pp. 797–813.

Kliskey, A.D. (1998) Linking the Wilderness Perception Mapping Concept to the Recreation Opportunity Spectrum, *Environmental Management*, Vol. 22, No. 1 (January/February), pp. 79–88.

Kliskey, A.D., Hoogsteden, C.C. and Morgan, R.K. (1994) The Application of Spatial-Perceptual Wilderness Mapping to Protected Areas Management in New Zealand, *Journal of Environmental Planning and Management*, Vol. 37, No. 4, pp. 431–45.

Knight, J., McCarron, S.G., McCabe, A.M. and Sutton, B. (1999) Sand and Gravel Aggregate Resource Management and Conservation in Northern Ireland, *Journal of Environmental Management*, Vol. 546, No. 3 (July), pp. 195–207.

Knox, J.W. and Weatherfield, E.K. (1999) The Application of GIS to Irrigation Water Resource Management in England and Wales, *The Geographical Journal*, Vol. 165, No. 1 (March), pp. 90–8.

Knox, J.W., Morris, J., Weatherhead, E.K. and Turner, A.P. (2000) Mapping the Financial Benefits of Sprinkler Irrigation and Potential Financial Impact of Restrictions on Abstraction: A Case-study in Anglian Region, *Journal of Environmental Management*, Vol. 58, No. 1 (January), pp. 45–59.

Kovacs, J.M., Wang, J. and Blanco-Correa, M. (2001) Mapping Disturbances in a Mangrove Forest Using Multi-Date Landsat TM Imagery, *Environmental Management*, Vol. 27, No. 5 (May), pp. 763–76.

Krummel, J.R., Dunn, C.P., Eckert, T.C. and Ayers, A.J. (1996) A Technology to Analyze Spatiotemporal Landscape Dynamics: Application to Cadiz Township (Wisconsin), in Goodchild, M.F., Steyaert, L.T., Parks, B.O., Johnston, C., Maidment, D., Crane, M. and Glendinning, S. (eds) *op. cit.*, Ch. 31, pp. 169–74.

Lajeunesse, D., Domon, G., Drapeau, P., Cogliastro, A. and Bouchard, A. (1995) Development and Application of an Ecosystem Management Approach for Protected Natural Areas, *Environmental Management*, Vol. 19, No. 4 (July/August), pp. 481–95.

Lakhatakia, M.N., Miller, D.A., White, R.A. and Smith, C.B. (1996) GIS as an Integrative Tool in Climate and Hydrology Modeling, in Goodchild, M.F., Steyaert, L.T., Parks, B.O., Johnston, C., Maidment, D., Crane, M. and Glendinning, S. (eds) *op. cit.*, Ch. 58, pp. 309–14.

Lammert, M. and Allan, J.D. (1999) Assessing Biotic Integrity of Streams: Effects of Scale in Measuring the Influence of Land Use/Cover and Habitat Structure on Fish and Macroinvertebrates, *Environmental Management*, Vol. 23, No. 2 (February), pp. 257–70.

Lankhorst, J.R.-K. (1992) Linking Landscape Ecological Knowledge With GIS, *Proceedings of the EGIS '92 Conference*, Munich (March 23–6), Vol. 2, pp. 1467–74.

Lanza, L. and Conti, M. (1994) Remote Sensing and GIS: Potential Application for Flood Hazard Forecasting, *Proceedings of the EGIS/MARI '94 Conference*, Paris (March 29–April 1), Vol. 2, pp. 1835–44.

Lardon, S., Triboulet, P., Duvernoy, I. and Albadalejo, C. (1994) Analyse Spatiale et Diagnostic de l'Activite Agricole Sur la Frontiere Agraire en Anglaterre, *Proceedings of the EGIS/MARI '94 Conference*, Paris (March 29–April 1), Vol. 2, pp. 1751–60.

Lee, T.J., Pielke, R.A., Kittel, T.G.F. and Weaver, J.F. (1993) Atmospheric Modeling and its Spatial Representation of Land Surface Characteristics, in Goodchild, M.F., Parks, B.O. and Steyaert, L.T. (eds) *op. cit.*, Ch. 10, pp. 108–22.

Lee, T.J. and Pielke, R.A. (1996) GIS and Atmospheric Modeling: A Case Study, in Goodchild, M.F., Steyaert, L.T., Parks, B.O., Johnston, C., Maidment, D., Crane, M. and Glendinning, S. (eds) *op. cit.*, Ch. 45, pp. 239–42.

Leggett, D.J. and Jones, A. (1996) The Application of GIS for Flood Defence in the Anglian Region: Developing for the Future, *International Journal of Geographical Information Systems*, Vol. 10, No. 1, pp. 103–16.

Lehman, D.R. (1994) NPDES Stormwater Discharge Program for Los Angeles County Department of Public Works, *Proceedings of the Urban and Regional Information Systems Association (URISA) Conference*, Milwaukee, Wisconsin (August 7–11), Vol. I, pp. 297–309.

Liang, C. and Mackay, D.S. (2000) A General Model of Watershed Extraction and Representation Using Globally Optimal Flow Paths and Up-slope Contributing Areas, *International Journal of Geographical Information Science*, Vol. 14, No. 4 (June), pp. 337–58.

Liao, H.-H. and Tim, S. (1994a) Interactive Water Quality Modeling Within a GIS Environment, *Computers, Environment and Urban Systems*, Vol. 18, No. 5, pp. 343–63.

Liao, H.-H. and Tim, S. (1994b) Interactive Modeling of Soil Erosion Within a GIS Environment, *Proceedings of the Urban and Regional Information Systems Association (URISA) Conference*, Milwaukee, Wisconsin (August 7–11), Vol. I, pp. 608–20.

Liebig, W. (1996) Mathematical Models and GIS, in Rumor, M. *et al.* (eds) *op. cit.*, Vol. 1, pp. 527–36.

Ljesevic, M. and Filipovic, D. (1995) Environmental Information System as the Basis of Environmental Protection in Serbia, *Computers, Environment and Urban Systems*, Vol. 19, No. 3, pp. 123–30.

Loomis, J. and Echohawk, C. (1999) Using GIS to Identify Under-represented Ecosystems in the National Wilderness Preservation System in the USA, *Environmental Conservation*, Vol. 26, No. 1 (March), pp. 53–8.

Lowell, K. (1991) Utilizing Discriminant Function Analysis With a Geographical Information System to Model Ecological Succession Spatially, *International Journal of Geographical Information Systems*, Vol. 5, No. 2, pp. 175–91.

Lyon, J.G. and Adkins, K.F. (1995) Use of a GIS for Wetland Identification, the St Clair Flats, Michigan, in Lyon, J.G. and McCarthy, J. (eds) *op. cit.*, Ch. 5, pp. 49–60.

Macfarlane, R. (1998) Implementing Agri-environmental Policy: A Landscape Ecology Perspective, *Journal of Environmental Planning and Management*, Vol. 41, No. 5 (September), pp. 575–96.

Mackey, B.G., Sims, R.A., Baldwin, K.A. and Moore, I.D. (1996) Spatial Analysis of Boreal Forest Ecosystems: Results from the Rinker Lake Case Study, in Goodchild, M.F., Steyaert, L.T., Parks, B.O., Johnston, C., Maidment, D., Crane, M. and Glendinning, S. (eds) *op. cit.*, Ch. 34, pp. 187–90.

Mackey, B.G. (1996) The Role of GIS and Environmental Modelling in the Conservation of Biodiversity, in Goodchild, M.F. *et al.* (1996b) *op. cit.*

Maidment, D.R. (1993) GIS and Hydrologic Modelling, in Goodchild, M.F., Parks, B.O. and Steyaert, L.T. (eds) *op. cit.*, Ch. 14, pp. 147–67.

Maidment, D.R. (1996a) Environmental Modeling within GIS, in Goodchild, M.F., Steyaert, L.T., Parks, B.O., Johnston, C., Maidment, D., Crane, M. and Glendinning, S. (eds) *op. cit.*, Ch. 59, pp. 315–23.

Maidment, D.R. (1996b) GIS and Hydrologic Modeling – an Assessment of Progress, in Goodchild, M.F. *et al.* (1996b) *op. cit.*

Malanson, G.P., Armstrong, M.P. and Bennett, D.A. (1996) Fragmented Forest Response to Climatic Warming and Disturbance, in Goodchild, M.F., Steyaert, L.T., Parks, B.O., Johnston, C., Maidment, D., Crane, M. and Glendinning, S. (eds) *op. cit.*, Ch. 46, pp. 243–7.

Malczewski, J. (1999) Spatial Multicriteria Decision Analysis, in Thill, J.-C. (ed.) *op. cit.*, Ch. 2, pp. 11–48.

Martin, K., Robl, W. and Pitzer, F. (1999) Avalanche!, *GeoEurope* (October), pp. 32–5.

Mayaux, P., Achard, F. and Malingreau, J.-P. (1998) Global Tropical Forest Area Measurements Derived from Coarse Resolution Satellite Imagery, *Environmental Conservation*, Vol. 25, No. 1 (March), pp. 37–52.

McKenney, D.W., Mackey, B.G. and Zavitz, B.L. (1999) Calibration and Sensitivity Analysis of a Spatially Distributed Solar Radiation Model, *International Journal of Geographical Information Science*, Vol. 13, No. 1 (January/February), pp. 49–65.

McKinney, D.C. and Tsai, H.-L. (1996) Solving Groundwater Problems Using Multigrid Methods in a Grid-Based GIS, in Goodchild, M.F., Steyaert, L.T., Parks, B.O., Johnston, C., Maidment, D., Crane, M. and Glendinning, S. (eds) *op. cit.*, Ch. 47, pp. 249–53.

McMahon, G. and Harned, D.A. (1998) Effect of Environmental Setting on Sediment, Nitrogen, and Phosphorus Concentrations in Albemarle-Pamlico Drainage Basin, North Carolina and Virginia, USA, *Environmental Management*, Vol. 22, No. 6 (November/December), pp. 887–903.

McWilliam, F. (1999) Blinded with Vision, *GeoEurope* (May), pp. 26–9.

Meiner, A., Saare, L., Rosaare, J. and Roose, A. (1990) Geographic Information Systems for Nature Management in the Industrial Region of North Estonia (USSR), *Proceedings of the EGIS '90 Conference*, Amsterdam (April 10–13), Vol. 2, pp. 756–61.

Melli, P. and Zannetti, P. (1992) (eds) *Environmental Modelling*, Computational Mechanics Publications, Southampton and Boston.

Memmi, J.M. (1995) Expertise and GIS Converge for Diamond Exploration, *GIS World* (February), pp. 54–7.

Mendonca-Santos, M.L. and Claramunt, C. (2001) An Integrated Landscape and Local Analysis of Land Cover Evolution in an Alluvial Zone, *Computers, Environment and Urban Systems*, Vol. 25, No. 6 (November), pp. 557–77.

Merrill, T., Wright, R.G. and Scott, J.M. (1995) Using Ecological Criteria to Evaluate Wilderness Planning Options in Idaho, *Environmental Management*, Vol. 19, No. 6 (November/December), pp. 815–25.

Mertens, B., Forni, E. and Lambin, E.F. (2001) Prediction of the Impact of Logging Activities on Forest Cover: A Case-study in the East Province of Cameroon, *Journal of Environmental Management*, Vol. 62, No. 1 (May), pp. 21–36.

Mitasova, H., Hofierka, J., Zlocha, M. and Iverson, L.R. (1996) Modelling Topographic Potential for Erosion and Deposition Using GIS, *International Journal of Geographical Information Systems*, Vol. 10, No. 5 (July–August), pp. 629–41.

Mladenoff, D.J., Host, G.E., Boeder, J. and Crow, T.R. (1996) LANDIS: A Spatial Model of Forest Landscape Disturbance, Succession, and Management, in Goodchild, M.F., Steyaert, L.T., Parks, B.O., Johnston, C., Maidment, D., Crane, M. and Glendinning, S. (eds) *op. cit.*, Ch. 32, pp. 175–9.

Montagu, S. (2000) GIS and Natural Resource Planning in Papua New Guinea: A Contextual Analysis. *Environment and Planning B: Planning and Design*, Vol. 27, No. 2, 187–96.

Moore, H.M., Fox, H.R., Harrouni, M.C. and El Alami, A. (1998) Environmental Challenges in the Rif Mountains, Northern Morocco, *Environmental Conservation*, Vol. 25, No. 4 (December), pp. 354–65.

Moore, I.D. (1996) Hydrologic Modeling and GIS, in Goodchild, M.F., Steyaert, L.T., Parks, B.O., Johnston, C., Maidment, D., Crane, M. and Glendinning, S. (eds) *op. cit.*, Ch. 26, pp. 143–8.

Moore, I.D., Turner, A.K., Wilson, J.P., Jenson, S.K. and Band, L.E. (1993) GIS and Land-Surface-Subsurface Process Modeling, in Goodchild, M.F., Parks, B.O. and Steyaert, L.T. (eds) *op. cit.*, Ch. 19, pp. 196–230.

Moreira, J.M., Gimenez-Azcarate, F. and Gould, M. (1994) Evolution of an Environmental Information System, *GIS World* (November), pp. 46–9.

Murray, K.S. and Rogers, D.T. (1999) Groundwater Vulnerability, Brownfield Redevelopment and Land Use Planning, *Journal of Environmental Planning and Management*, Vol. 42, No. 6 (November), pp. 801–10.

Namken, J.C. and Stuth, J.W. (1997a) A Prototype Graphic Landscape Analysis System: Part 1. Predicting Spatial Patterns of Grazing Pressure Using GIS, *International Journal of Geographical Information Science*, Vol. 11, No. 8 (December), pp. 785–98.

Namken, J.C. and Stuth, J.W. (1997b) A Prototype Graphic Landscape Analysis System: Part 2. A Bioeconomic Analysis Model for Grazing Land Development, *International Journal of Geographical Information Science*, Vol. 11, No. 8 (December), pp. 799–812.

Nelson, D. (1995) GIS/CAD Solutions: An Alaskan Case Study, *GIS World* (May), p. 45.

Nkambwe, M. (1991) Resource Utilization and Regional Planning Information Systems (RURPIS) in Botswana, *International Journal of Geographical Information Systems*, Vol. 5, No. 1, pp. 111–27.

Noguchi, T., Sugiyama, T., Setojima, M. and Mori, M. (1991) Prediction of Disaster Occurence Point on the Slope Land Along the Railway Using Raster GIS, *Proceedings of the GIS/LIS '91 Conference*, Atlanta, Georgia (October 28–November 1), Vol. 1, pp. 257–66.

Novak, J.H. and Dennis, R.L. (1993) Regional Air Quality and Acid Deposition Modeling and the Role of Visualization, in Goodchild, M.F., Parks, B.O. and Steyaert, L.T. (eds) *op. cit.*, Ch. 13, pp. 142–6.

Ochoa-Gaona, S. (2001) Traditional Land-Use Systems and Patterns of Forest Fragmentation in the Highlands of Chiapas, Mexico, *Environmental Management*, Vol. 27, No. 4 (April), pp. 571–86.

Ortega-Huerta, M.A. and Medley, K.E. (1999) Landscape Analysis of Jaguar (*Panthera onca*) Habitat Using Sighting Records in the Sierra de Tamaulipas, Mexico, *Environmental Conservation*, Vol. 26, No. 4 (December), pp. 257–69.

Osborn, S. and Cook, H.F. (1997) Nitrate Vulnerable Zones and Nitrate Sensitive Areas: A Policy and Technical Analysis of Groundwater Source Protection in England and Wales, *Journal of Environmental Planning and Management*, Vol. 40, No. 2 (March), pp. 217–33.

O'Sullivan, D. and Turner, A. (2001) Visibility Graphs and Landscape Visibility Analysis, *International Journal of Geographical Information Science*, Vol. 15, No. 3 (April–May), pp. 221–37.

Parrish, D.A., Townsend, L., Saunders, J., Carney, G. and Langston, C. (1993) US EPA Region 6 Comparative Risk Project: Evaluating Ecological Risk, in Goodchild, M.F., Parks, B.O. and Steyaert, L.T. (eds) *op. cit.*, Ch. 33, pp. 348–52.

Peccol, E., Bird, A.C. and Brewer, T.R. (1996) GIS as a Tool for Assessing the Influence Of Countryside Designations and Planning Policies On Landscape Change, *Journal of Environmental Management*, Vol. 47, No. 4 (August), pp. 355–67.

Pereira, J.M.C. and Duckstein, L. (1993) A Multiple Criteria Decision-Making Approach to GIS-Based Land Suitability Evaluation, *International Journal of Geographical Information Systems*, Vol. 7, No. 5, pp. 407–24.

Petit, C.C. and Lambin, E.F. (2001) Integration of Multi-source Remote Sensing Data for Land Cover Change Detection, *International Journal of Geographical Information Science*, Vol. 15, No. 8 (December), pp. 785–803.

Phinn, S., Franklin, J., Hope, A., Stow, D. and Huenneke, L. (1996) Biomass Distribution Mapping Using Airborne Digital Video Imagery and Spatial Statistics in a Semi-Arid Environment, *Journal of Environmental Management*, Vol. 47, No. 2 (June), pp. 139–64.

Pierce, G.J., Wang, J., Zheng, X., Bellido, J.M., Boyle, P.R., Denis, V. and Robin, J.P. (2001) A Cephalopod Fishery GIS for the Northeast Atlantic, *International Journal of Geographical Information Science*, Vol. 15, No. 8 (December), pp. 763–84.

Power, S. and Barnes, J.L. (1993) GIS Model for Wildlife Habitat Suitability Based on the New Brunswick Forest Inventory Database, *Proceedings of the Canadian Conference on GIS '93*, Ottawa, Canada (March 23–5), pp. 799–800.

Pratt, N.D., Bird, A.C., Taylor, J.C. and Carter, R.C. (1997) Estimating Areas of Land Under Small-Scale Irrigation Using Satellite Imagery and Ground Data for a Study Area in N.E. Nigeria, *The Geographical Journal*, Vol. 163, No. 1 (March), pp. 65–77.

Price, L. (1993) GIS Technology to Provide "Easy Access" for USDA Customers, *Journal of the Urban and Regional Information Systems Association*, Vol. 5, No. 2 (Fall), pp. 85–7.

Priya, S. and Shibasaki, R. (1997) Application of Geographic Information Systems (GIS) and Remote Sensing (RS) for Land Cover Mapping – A Case Study, *Proceedings of the 5th International Conference on Computers in Urban Planning and Urban Management*, Bombay (India), Vol. 2, pp. 542–51.

Quiel, F. (1995) Modelling and Optimizing Water Usage in Irrigated Areas in a GIS, *Proceedings of the Joint European Conference and Exhibition on Geographic Information JEC-GI '95*, The Hague (March 26–31), Vol. 1, pp. 415–20.

Ramirez-Sanz, L., Alcaide, T., Cuevas, J.A., Guillen, D.F. and Sastre, P. (2000) A Methodology for Environmental Planning in Protected Natural Areas, *Journal of Environmental Planning and Management*, Vol. 43, No. 6 (November), pp. 785–98.

Rhind, D. (2000) Current Shortcomings of Global Mapping and the Creation of a New Geographical Framework for the World, *The Geographical Journal*, Vol. 166, No. 2 (June), pp. 295–305.

Riezebos, H.Th., Jetten, V.G. and de Jong, S.M. (1990) The Use of GIS for Land Use Evaluation Studies, *Proceedings of the EGIS '90 Conference*, Amsterdam (April 10–13), Vol. 2, pp. 915–19.

Ringrose, S., Chanda, R., Nkambwe, M. and Sefe, F. (1996) Environmental Change in the Mid-Boteti Area of North-Central Botswana, *Environmental Management*, Vol. 20, No. 3 (May/June), pp. 397–410.

Rodriguez-Bachiller, A. (2000) Geographical Information Systems and Expert Systems for Impact Assessment. Part I: GIS, *Journal of Environmental Assessment Policy and Management*, Vol. 2, No. 3 (September), pp. 369–414.

Rogowski, A.S. (1996) Conditional Simulation of Percolate Flux Below a Root Zone, in Goodchild, M.F., Steyaert, L.T., Parks, B.O., Johnston, C., Maidment, D., Crane, M. and Glendinning, S. (eds) *op. cit.*, Ch. 48, pp. 255–9.

Rybaczuk, K.Y. (2001) GIS as an Aid to Environmental Management and Community Participation in the Negril Watershed, Jamaica, *Computers, Environment and Urban Systems*, Vol. 25, No. 2 (March), pp. 141–65.

Saghafian, B. (1996) Implementation of a Distributed Hydrology Model within GRASS, in Goodchild, M.F., Steyaert, L.T., Parks, B.O., Johnston, C., Maidment, D., Crane, M. and Glendinning, S. (eds) *op. cit.*, Ch. 38, pp. 205–8.

Samulski, J.C. (1991) Developing a GIS Based Storm Water Analysis Model, *Proceedings of the EGIS '91 Conference*, Brussels (April 2–5), Vol. I, pp. 965–74.

Sarch, M.-T. and Birkett, C. (2000) Fishing and Farming at Lake Chad: Responses to Lake-level Fluctuations, *The Geographical Journal*, Vol. 166, No. 2 (June), pp. 156–72.

Sato, S., Arima, T., Hsiao, N., Hagashima, S. and Sugahara, T. (1995) Using GIS Topographic Data for Quantitative Landscape Analysis of City Hall Views in 76 Kyushu Cities, *Proceedings of the 4th International Conference on Computers in Urban Planning and Urban Management*, Melbourne, Australia (July 11–14), Vol. 1, pp. 463–76.

Scharek, P., Tullner, T. and Turczi, G. (1995) Digging Deeper: Hungary's Geological Survey Increases its GIS Activities, *GIS Europe* (May), pp. 28–30.

Schmidt, M.G., Schreier, H.E. and Shah, P.B. (1995) A GIS Evaluation of Land Use Dynamics and Forest Soil Fertility in a Watershed in Nepal, *International Journal of Geographical Information Systems*, Vol. 9, No. 3, pp. 317–27.

Scott, H.D. and Udouj, T.H. (1999) Spatial and Temporal Characterization of Land-use in the Buffalo National River Watershed, *Environmental Conservation*, Vol. 26, No. 2 (June), pp. 94–101.

Scott, T.A. and Sullivan, J.E. (2000) The Selection and Design of Multiple-Species Habitat Preserves, *Environmental Management*, Vol. 26 (Supplement), pp. S37–S53.

Sham, C.H., Brawley, J.W. and Moritz, M.A. (1995) Quantifying Septic Nitrogen Loadings to Receiving Waters: Waquoit Bay, Massachusetts, *International Journal of Geographical Information Systems*, Vol. 9, No. 4, pp. 527–42.

Sham, C.H., Brawley, J.W. and Moritz, M.A. (1996) Analizing Septic Nitrogen Loading to Receiving Waters: Waquoit Bay, Massachusetts, in Goodchild, M.F., Steyaert, L.T., Parks, B.O., Johnston, C., Maidment, D., Crane, M. and Glendinning, S. (eds) *op. cit.*, Ch. 49, pp. 261–4.

Shivlani, M.P. and Suman, D.O. (2000) Dive Operator Use Patterns in the Designated No-Take Zones of the Florida Keys National Marine Sanctuary (FKNMS), *Environmental Management*, Vol. 25, No. 6 (June), pp. 647–59.

Sifakis, N., Kontoes, H. and Elias, P. (1999) Tracking the Toxic Time Bomb, *GeoEurope* (October), pp. 39–41.

Skidmore, A.K., Gauld, A. and Walker, P. (1996) Classification of Kangaroo Habitat Distribution Using Three GIS Models, *International Journal of Geographical Information Systems*, Vol. 10, No. 4, pp. 441–54.

Smallwood, K.S., Wilcox, B., Leidy, R. and Yarris, K. (1998) Indicators Assessment for Habitat Conservation Plan of Yolo County, California, USA, *Environmental Management*, Vol. 22, No. 6 (November/December), pp. 947–58.

Smith, C.S., McDonald, G.T. and Thwaites, R.W. (2000) TIM: Assessing the Sustainability of Agricultural Land Management, *Journal of Environmental Management*, Vol. 60, No. 4 (December), pp. 267–88.

Srinivasan, R., Arnold, J., Rosenthal, W. and Muttiah, R.S. (1996) Hydrologic Modeling of Texas Gulf Basin Using GIS, in Goodchild, M.F., Steyaert, L.T., Parks, B.O., Johnston, C., Maidment, D., Crane, M. and Glendinning, S. (eds) *op. cit.*, Ch. 40, pp. 213–17.

Stein, A., Staritsky, I., Bouma, J. and van Groenigen, J.W. (1995) Interactive GIS for Environmental Risk Assessment, *International Journal of Geographical Information Systems*, Vol. 9, No. 5, pp. 509–25.

Strobol, J. (1992) Comprehensive GIS Support for Managing Forest Resources, *Proceedings of the EGIS '92 Conference*, Munich (March 23–6), Vol. 2, pp. 968–77.

Sui, D.Z. and Maggio, R.C. (1999) Integrating GIS with Hydrological Modeling: Practices, Problems, and Prospects, *Computers, Environment and Urban Systems*, Vol. 23, No. 1 (January), pp. 33–51.

Swetnam, R.D., Mountford, J.O., Armstrong, A.C., Gowing, D.J.G., Brown, N.J., Manchester, S.J. and Treweek, J.R. (1998) Spatial Relationships Between Site Hydrology and the Occurrence of Grassland of Conservation Importance: A Risk Assessment with GIS, *Journal of Environmental Management*, Vol. 54, No. 3 (November), pp. 189–203.

Tang, S.M. and Montgomery, D.R. (1995) Riparian Buffers and Potentially Unstable Ground, *Environmental Management*, Vol. 19, No. 5 (September/October), pp. 741–9.

Tappan, G.G., Moore, D.G. and Knausenberger, W.I. (1991) Monitoring Grasshopper and Locust Habitats in Sahelian Africa Using GIS and Remote Sensing Technology, *International Journal of Geographical Information Systems*, Vol. 5, No. 1, pp. 123–35.

Taupier, R. and Terner, M. (1991) MassGIS: A Case Study of the Massachusetts Environmental Geographic Information System, *Proceedings of the Urban and Regional Information Systems Association (URISA) Conference*, San Francisco (August 11–15), Vol. II, pp. 87–96.

Taylor, J.R. (1994) Application of GIS to Cultural Landscape Assessment Within the Niagara Escarpment Planning Area, *Proceedings of the Canadian Conference on GIS '94*, Ottawa, Canada (June 6–10), Vol. 1, pp. 742–50.

Taylor, P.J., Walker, G.R., Hodgson, G., Hatton, T.J. and Correll, R.L. (1996) Testing of a GIS Model of *Eucalyptus largiflorens* Health on a Semiarid Saline Floodplain, *Environmental Management*, Vol. 20, No. 4 (July/August), pp. 553–64.

Thapa, G.B. and Weber, K.E. (1995) Status and Management of Watersheds in the Upper Pokhara Valley, Nepal, *Environmental Management*, Vol. 19, No. 4 (July/August), pp. 497–513.

Thivierge, A.R. (1994) Automatic ARC/INFO To Facilitate Forest Management in British Colombia, *Proceedings of the Canadian Conference on GIS '94*, Ottawa, Canada (June 6–10), Vol. 2, pp. 1101–12.

Thomas, G., Sannier, C.A.D. and Taylor, J.C. (2000) Mapping Systems and GIS: A Case Study using the Ghana National Grid, *The Geographical Journal*, Vol. 166, No. 2 (June), pp. 306–11.

Thumerer, T., Jones, A.P. and Brown, D. (2000) A GIS Based Coastal Management System for Climate Change Associated Flood Risk Assessment on the East Coast of England, *International Journal of Geographical Information Science*, Vol. 14, No. 3 (April–May), pp. 265–81.

Thuresson, T., Nasholm, B., Holm, S. and Hagner, O. (1996) Using Digital Image Projections to Visualize Forest Landscape Changes Due to Management Activities and Forest Growth, *Environmental Management*, Vol. 20, No. 1 (January/February), pp. 35–40.

Tortosa, D. and Beach, P. (1993) Application of Portable GPS/Desktop Mapping-GIS for Fire Management Support, *Proceedings of the Canadian Conference on GIS '93*, Ottawa, Canada (March 23–5), pp. 802–19.

Trozzi, C. and Vaccaro, R. (1993) Air Pollution Emissions Inventory and GIS, *Proceedings of the EGIS '93 Conference*, Genoa (Italy), Vol. 1, pp. 47–56.

Uchida, A., Satani, N., Nakano, H., Deguchi, A. and Hagishima, S. (1997) Study on Method for Visual Evaluation of Sloped Wooded Area, *Proceedings of the 5th International Conference on Computers in Urban Planning and Urban Management*, Bombay (India), Vol. 1, pp. 146–56.

Van der Heijde, P.K.M. (1992) Developments in Computer Technology Enhancing the Application of Groundwater Models, in Melli, P. and Zannetti, P. (eds) *op. cit.*, Ch. 2, pp. 23–33.

Van Horssen, P. (1996) Ecological Modelling in GIS, in Goodchild, M.F. *et al.* (1996b) *op. cit.*

Ventura, S.J. (1988) Combining Automated Data to Create Decision-Making Information; The Dane County Land Conservation Department Example, *Proceedings of the Urban and Regional Information Systems Association (URISA) Conference*, Los Angeles (August 7–11), Vol. I, pp. 100–8.

Vieux, B.E., Farajalla, N.S. and Gaur, N. (1996) Integrated GIS and Distributed Storm Water Runoff Modeling, in Goodchild, M.F., Steyaert, L.T., Parks, B.O.,

Johnston, C., Maidment, D., Crane, M. and Glendinning, S. (eds) *op. cit.*, Ch. 37, pp. 199–204.

Vieux, B.E., Mubaraki, M.A. and Brown, D. (1998) Wellhead Protection Area Delineation Using a Coupled GIS and Groundwater Model, *Journal of Environmental Management*, Vol. 54, No. 3 (November), pp. 205–14.

Wagner, J. (1995) Environmental Planning for a World Heritage Site: Case Study of Angkor, Cambodia, *Journal of Environmental Planning and Management*, Vol. 38, No. 3 (September), pp. 419–34.

Walker, P.A. and Young, M.D. (1997) Using Integrated Economic and Ecological Information to Improve Government Policy, *International Journal of Geographical Information Science*, Vol. 11, No. 7 (October–November), pp. 619–32.

Wang, S.-Q. and Unwin, D.S. (1992) Modelling Landslide Distribution on Loess Soils in China: An Investigation, *International Journal of Geographical Information Systems*, Vol. 6, No. 5, pp. 391–405.

Wang, X., White-Hull, C., Dyer, S. and Yang, Y. (2000) GIS-ROUT: A River Model for Watershed Planning, *Environment and Planning B: Planning and Design*, Vol. 27, No. 2 (March), pp. 231–46.

Wang, X. (2001) Integrating Water-quality Management and Land-use Planning in a Watershed Context, *Journal of Environmental Management*, Vol. 61, No. 1 (January – Special issue), pp. 25–36.

Watkins, R.L., Cocklin, C. and Laituri, M. (1997) The Use of Geographic Information Systems for Resource Evaluation: A New Zealand Example, *Journal of Environmental Planning and Management*, Vol. 40, No. 1 (January), pp. 37–57.

Webb, A.D. and Bacon, P.J. (1999) Using GIS for Catchment Management and Freshwater Salmon Fisheries in Scotland: The DeeCAMP Project, *Journal of Environmental Management*, Vol. 55, No. 2 (February), pp. 127–43.

Webber, C.M., Watzin, M.C. and Wang, D. (1996) Role of Wetlands in Reducing Phosphorus Loading to Surface Water in Eight Watersheds in the Lake Champlain Basin, *Environmental Management*, Vol. 20, No. 5 (September/October), pp. 731–9.

Weber, R.S. (1990) Effective Use of Geographic Information System Technology in Municipal Scale Environmental Management, *Proceedings of the Urban and Regional Information Systems Association (URISA) Conference*, Edmonton, Alberta, Canada (August 12–16), Vol. I, pp. 124–39.

Wickham, J.D., Jones, K.B., Riitters, K.H., O'Neill, R.V., Tankersley, R.D., Smith, E.R., Neale, A.C. and Chaloud, D.J. (1999) An Integrated Environmental Assessment of the US Mid-Atlantic Region, *Environmental Management*, Vol. 24, No. 4 (November), pp. 553–60.

Wilson, J. (1996) GIS-Based Land Surface/Subsurface Modeling: New Potential for New Models?, in Goodchild, M.F. *et al.* (1996b) *op. cit.*

Wing, M.G. and Johnson, R. (2001) Quantifying Forest Visibility with Spatial Data, *Environmental Management*, Vol. 27, No. 3 (March), pp. 411–20.

Wolf, D. (1996) Development and Use of the Soil Protection Administration in Baden-Wurthemberg, in Rumor, M. *et al.* (eds) *op. cit.*, Vol. 2, pp. 915–22.

Woodhouse, S., Lovett, A., Dolman, P. and Fuller, R. (2000) Using a GIS to Select Priority Areas for Conservation, *Computers, Environment and Urban Systems*, Vol. 24, No. 2 (March), pp. 79–93.

Xia, L. (1997) A Sustainable Land Allocation Model with the Integration of Remote Sensing and GIS – A Case Study in Dongguan, *International Journal of Environmental Studies*, Vol. 53, No. 4 (Section A: Environmental Studies), pp. 325–48.

Xiang, W.-N. (1993) A GIS Method for Riparian Water Quality Buffer Generation, *International Journal of Geographical Information Systems*, Vol. 7, No. 1, pp. 57–70.

Yarie, J. (1996) A Forest Ecosystem Dynamics Model Integrated within a GIS, in Goodchild, M.F. *et al.* (1996b) *op. cit.*

Zalidis, G.C. and Gerakis, A. (1999) Evaluating Sustainability of Watershed Resources Management through Wetland Functional Analysis, *Environmental Management*, Vol. 24, No. 2 (August), pp. 193–207.

5 GIS and expert systems for impact assessment

5.1 INTRODUCTION

Chapters 3 and 4 reviewed a wide range of GIS applications to environmental matters, also showing how limited their capabilities are when used on their own and without pre-programming. This chapter discusses the use of expert systems (ES) technology, in particular in combination with GIS, arguing that "partial" technologies like GIS maximise their contribution within the framework of decision-support tools. The chapter first discusses the use of ES without GIS, and then with GIS, in Impact Assessment and environmental management, following the same distinction used when reviewing GIS applications in the previous chapters. Decision support systems are discussed afterwards.[13]

In contrast to the previous review of GIS applications, ES and decision-support technologies are more novel and the proportion of references appearing in research journals and books – as opposed to magazines and conference papers without follow-up publications – is much greater, a reflection of the greater research interest these types of GIS applications still have. Another consequence of this is that the proportion of publications discussing *methodological issues* is far greater than that in more established types of GIS use.

5.2 EXPERT SYSTEMS WITHOUT GIS FOR ENVIRONMENTAL ASSESSMENT

It is interesting that, in parallel to ES not making inroads in areas like town planning – as already mentioned – such systems seem to be attracting fresh interest in new areas like IA and environmental management. The process appears to be starting all over again in this new field, with articles highlighting

13 Rodriguez-Bachiller (2000) includes an earlier version of this bibliographical review.

the potential of ES appearing in the environmental literature, and prototypes starting to be developed and used.

5.2.1 Expert systems without GIS for impact assessment

Looking first at IA as such, most of the early articles performed what can be called an *eye-opener* function, and at the same time some were monitoring what was happening (like Spooner, 1985, in the US Environmental Protection Agency), some were pointing out the potential of ES for IA in general (Chalmers, 1989; Lein, 1989), and some were pointing at particular areas of IA:

- For project *screening* (determining if a project requires an impact assessment study), Geraghty (1992) reviewed briefly some systems in Japan, Italy and Canada and proposed the GAIA system, an ES for guidance to help assess the significance of likely impacts from a project in order to see if an Environmental Statement is needed. Later, Brown *et al.* (1996) developed it into the HyperGAIA system (which they labelled as decision support system) to diffuse IA expertise, and they used project screening as an example. This group of researchers have made the issue of expertise and its diffusion, central to ES, their main focus of interest, even if their discussions are not always linked to any computerised system in particular: Geraghty *et al.* (1996) are interested in the future use of guidance manuals for EIA (which can be seen as "paper" ES), and Geraghty (1999) undertakes a comparative study of guidance documents to support practice.
- For the *scoping* of project impacts (identifying the impacts to be studied and how "key" they are), Fedra *et al.* (1991) provide an early example for the Lower Mekong Basin in South-East Asia, and Edward-Jones and Gough (1994) developed the ECOZONE system to scope the impacts on agriculture of projects of any kind.
- For *impact prediction* as such, Huang (1989) developed the early system MIN-CYANIDE for the minimisation of cyanide waste in electroplating plants, and Kobayashi *et al.* (1997) incorporate environmental considerations in an ES to help with the location of industrial land uses.
- For the *review* of Environmental Statements, Schibuola and Byer (1991) proposed the REVIEW system (written in Prolog) to overcome the problem of Environmental Statements being reviewed in an *ad hoc* way, and he illustrated the system concentrating on only one aspect of ES: the consideration of alternatives for a project.
- Echoing similar developments in other areas (like GIS), Hughes and Schirmer (1994) point out the potential of expert systems for *public participation* in IA as part of an interactive multimedia approach.

5.2.2 Expert systems without GIS for environmental management

In the more general area of *environmental management*, a few "eye-opener" articles on the potential of ES have been appearing since the 1980s (Hushon, 1987; Borman, 1989; Lein, 1990), while some early prototypes were already being developed mainly to help with two types of tasks:

- Environmental *analysis*, where *geology* is quite prominent: Krystinik (1985) proposed a system for the interpretation of depositional environments, Fang and Schultz (1986) and Schultz *et al.* (1988) discuss the XEOD system for the geological interpretation of sedimentary environments, and Liang (1988) developed a system for environmental analysis of sedimentation; Miller (1991) applies a system to sedimentary basin analysis, while Besio *et al.* (1991) apply a non-geological ES to classify and analyse the landscape in an area.
- *Management* as such: Coulson *et al.* (1989) designed a system for pest management in forests, Greathouse *et al.* (1989) applied to environmental control a system for land management developed earlier (Davis *et al.*, 1988) and, more recently, Clayton and Waters (1999) also developed a land management system, for the Northwest Territories in Canada.

These are just a few examples. Fedra *et al.* (1991) review a number of early projects from the 1980s combining ES and hydrologic *modelling*, and a comprehensive review of environmental management expert systems in the 1980s can be found in Warwick *et al.* (1993).

5.3 EXPERT SYSTEMS WITH GIS

Turning now to ES in combination with GIS, the notion of linking GIS technology to other advanced tools like expert systems was already emerging in the early 1990s, as calls for so-called "intelligent" GIS were frequent and in wide-ranging arenas (Laurini and Milleret-Raffort, 1990; Burrough, 1992; Openshaw, 1993a). *Eye-opener* articles were starting to suggest the types of structures that such combined systems would have, and also starting to show examples of ES–GIS combinations (Smith *et al.*, 1987; Bouille, 1989; Heikkila *et al.*, 1990; Fedra *et al.*, 1991; Lam and Swayne, 1991; Evans *et al.*, 1993; Leung and Leung, 1993a; Vessel, 1993), not forgetting the considerable difficulties involved in linking these two technologies, which were identified at quite an early stage (Navinchandra, 1989).

Because of the greater novelty of this technology in the early 1990s (at least in this field), there was a greater emphasis on methodological issues than for GIS alone (see previous chapters), which had undergone similar

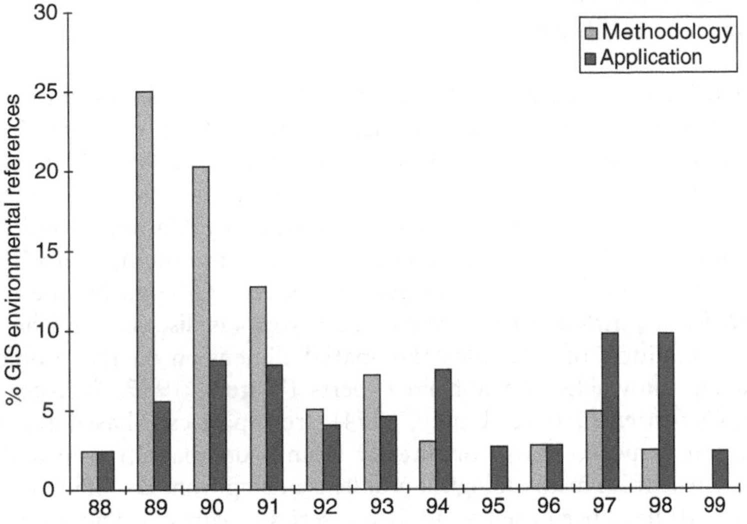

Figure 5.1 The change of emphasis from methodology to application.

methodological discussions a decade earlier but are now raising more issues about their diffusion than about their methodology. Figure 5.1 shows the frequency in GIS–ES usage of methodological and application references during the 1990s expressed as percentages of all environmental GIS references reviewed each year (see Rodriguez-Bachiller, 2000), and we can see how the methodological emphasis in the early 1990s gradually fades away and is replaced by discussions of practical applications.

5.3.1 GIS and expert systems: methodological issues

What dominated the methodological discussion in those years was undoubtedly the question of *how* to integrate ES and GIS, and many authors contributed to that debate in the early 1990s (Webster, 1990a; Fedra *et al.*, 1991; Smith and Yiang, 1991; Zhu and Healey, 1992; Fischer, 1994), mapping out the possible *forms of integration* between the two technologies – in a way similar to earlier discussions about linking GIS with models:

- ES logic can be used simply to enhance the GIS database with rules.
- An ES (the same as a model) can be "loosely coupled" with an external GIS, calling its database through an interface.
- Using "tight coupling", one of the two technologies can be a "shell" for the other and run it: the ES can be running the GIS or the GIS can run the ES.

- In full integration, ES operations can be built into GIS functionality (or spatial information handling can be built into the ES, although that is much more difficult).

Related to the problem of GIS–ES integration, the development of suitable *interface tools* for the connection, usually in the form of "shells" which could talk to both technologies (Buehler and Wright, 1989; Maidment and Djokic, 1991; Leung and Leung, 1993b) also attracted considerable attention, a prominent example being the interface written by Maidment and Djokic to connect the NEXPERT expert-systems "shell" and the Arc-Info GIS.

Apart from the form of the integration between GIS and ES, the issue of *knowledge acquisition* is ever-present in ES work (as discussed in Chapter 2) and the addition of GIS adds the spatial dimension to the problem of extracting knowledge, be it from experts (Waters, 1989; Webster *et al.*, 1989; Cowen *et al.*, 1990; Linsey, 1994), from past case-based experience (Holt and Benwell, 1996), or directly from a database (Deren and Tao, 1994). Apart from methodological problems arising from ES–GIS integration, ES (and AI) have been used to address a series of *cartographic* problems in GIS work, mainly in areas having to do with visualisation presentation of maps, and with the interpretation of certain type of data.

5.3.1.1 Methodological issues: visualisation

The visualisation problem that has probably attracted most attention in connection with the use of AI techniques with GIS has been that of *map generalisation*, central to any cartographic system where a decision has to be made each time a map is produced, at a given scale, about how much detail to use at that scale. Such decisions can be about *what* to include (what sizes of settlements to leave out, for instance), or in terms of *how* to represent lines (line generalisation) on the map.[14] To deal with this problem, two different types of AI approaches have been explored, with unequal interest:

- Using *neural networks* (see Chapter 2) started to attract interest in the late 1980s, to generalise settlements (Powitz and Meyer, 1989) or for general-purpose line generalisation (Pariente, 1994; Werschlein and Weibel, 1994).
- But, by far, the most researched approach to "intelligent" map generalisation is *rule-based* – similar to how ES work – sometimes involving "knowledge acquisition" (Muller and Mouwes, 1990) to determine the

14 The issue of how much detail to use when representing a line at a particular scale leads directly to the perplexing realisation that at different scales, lines appear to change in length as their scale of representation changes, and the concept that links these two variables (scale and size) is that of *fractal dimension*, which opens the door into the field of fractal analysis, fascinating in itself and with wide-ranging ramifications (an easy introduction to the subject can be found in Lauwerier, 1987).

rules, which are used to replicate how a cartographer would do it (Richardson, 1989; Armstrong and Bennett, 1990; Mackaness and Beard, 1990), or to select the best generalisation algorithm from a range produced over the years by research into automatic generalisation (Joao *et al.*, 1990, 1991; Herbert, 1991; Herbert and Joao, 1991; Herbert *et al.*, 1992; Offermann, 1993).

Other examples in the ES–GIS literature covering issues of *visualisation/interfacing* include a variety of problem areas:

- automatic *name-placement* on maps (Freeman and Ahn, 1984; Doerschler, 1987; Doerschler and Freeman, 1989; Jones, 1989);
- *symbolisation*, the automatic selection of symbols for map features (Mackaness and Fisher, 1987; Siekierska, 1989; Greven, 1995; Zhan and Buttenfield, 1995);
- dealing with map *projections* automatically (Jankowski and Nyerges, 1989) and making earth, aerial and satellite pictures compatible (Logan *et al.*, 1988);
- general human–computer *interfacing* (Morse, 1987; Tzafestas and Hatzivasilou, 1990);
- automatic map *error-correction*, like for example the removal of so-called "sliver polygons"[15] in GIS maps (Rybaczuk, 1993).

5.3.1.2 Methodological issues: classification

The use of "intelligent" methods together with GIS for the *classification* of satellite data attracted interest since the early years of satellite data becoming widely available (Estes *et al.*, 1986; Mckeown, 1986), and the same two main approaches were researched as for map generalisation: neural networks and rule-based systems.

Neural networks are particularly suited to pattern recognition, and they have been used to *classify* a wide range of data, including:

- *land cover*, which is probably the most common problem addressed this way since the early 1990s (Fisher and Pathirana, 1990; Buch *et al.*, 1994a,b; Maruchi *et al.*, 1994; Foody, 1995; Atkinson and Cutler, 1996; Dai and Khorram, 1999);
- *features*, be it the recognition of lines (Mower, 1988) or the identification of architectural types (Maiellaro and Barbanente, 1993);
- *multi-factor data-sets*, for example to perform "cluster analysis" (Openshaw and Wymer, 1990) or to assess land suitability (Wang, 1994).

15 Such polygons usually result from double-digitising or from the superposition of several maps of the same features.

Rule-based classification systems attracted attention earlier than neural-network systems (which only came to the forefront of research in the 1990s), and *what* they have covered has varied:

- Classification of *land cover* has been a favourite theme since the 1980s (Ying *et al.*, 1987; Wharton, 1987, 1989; De Jong and Riezebos, 1991; Hong, 1991; Leung and Leung, 1993b), with early applications to forestry (Goldberg *et al.*, 1984), and also applications to agricultural land use in particular (Kontoes *et al.*, 1993; Van der Laan, 1994; Hassani *et al.*, 1996). Srinivasan and Richards (1993) apply these methods to classify "mixed" data from radar, satellite and other sources. An interesting variation to the theme is to reverse the logic of these methods and use a training set of "ground-truthed" data in comparison with satellite data in order to *derive the rules* (by "rule induction") for the knowledge base of the future ES (Barbanente *et al.*, 1991; Dymond and Luckman, 1993).
- Identification of *roads* ("road extraction") from satellite data has also attracted considerable attention (Goodenough *et al.*, 1987; Wang and Newkirk, 1987a,b; Newkirk and Wang, 1989; Newkirk, 1991; Van Cleynenbreugel *et al.*, 1991; Goodenough and Fung, 1991), to overcome the difficulty of identifying linear features in data sources which only give *areal* information. In a variation on the theme, O'Neill and Grenney (1991) built a rule-based prototype for road identification *not* using satellite data, but data from the Road Inventory Files and the digital address files (TIGER) in the US.
- Identification of *geographical features* from satellite data using decision rules: Hartnett *et al.* (1994) used this approach in Antarctica to identify clouds, topographical edges, ice, etc., and Cambridge *et al.* (1996) use a similar approach to model acid rain.

To finish this section, it is worth mentioning the approach of Shaefer (1992), who proposed long ago combining the two approaches discussed above (ES and neural networks) so that the ES rules could improve the performance of neural networks by taking their output and making choices among the different probability options suggested, and then feeding back these suggestions into the network's operation.

5.3.2 GIS and expert systems in the Regional Research Laboratories

At the time this review started – the early 1990s – GIS technology itself was relatively new outside America. In the diffusion process that was taking place, the setting up of the Regional Research Laboratories (the RRLs already referred to in Chapter 1) in the UK was a crucial step and provided the front-line in that process. An examination of the work carried out as part of the Regional Research Laboratory Initiative of 1988–91 in the UK provides an insight into the issues dominating GIS research at the time, and acts as a "pilot survey" of the issues and prospects concerning the combination of these two technologies. Given the emphasis of GIS work at the time on "diffusion and acceptance", the scope of the RRL survey was widened from the outset to include links between GIS and not just expert systems, but general artificial intelligence (AI) on the one hand and, on the other, a wider range of decision-support tools leading to the so-called decision support systems (DSS), already discussed in Chapter 2.

The Regional Research Laboratory Initiative (Masser, 1990) was launched by the UK Economic and Social Research Council (ESRC) in 1987 in a trial phase, with its main phase starting in 1988, and with over two million pounds invested up to its conclusion at the end of 1991. It polarised GIS research in the UK into 8 Regional Research Laboratories (RRLs), some of them with more than one site, so that in total there were a dozen research sites linked to this programme spread evenly throughout the country, mostly academic departments of geography, sometimes other social science or environment-related departments, sometimes computer centres. These departments had different degrees of involvement in the programme, and tended to support research carried out mainly by "resident" researchers at those sites, having the additional practical aim of stimulating and helping local (private and public) decision-makers in the use of the new GIS technology. This contrasts, for instance, with the parallel experience of the US National Centre for Geographic Information and Analysis – funded with a comparable budget by the National Science Foundation – concentrated in only three centres for the whole country (Santa Barbara, Buffalo and Maine), and financing research projects done both inside and outside those centres, with mainly theoretical aims (Openshaw *et al.*, 1987; Openshaw, 1990).

5.3.2.1 The RRLs research agenda

Taking the technical research profile of the different RRLs, as summarised by Plummer (1990) and also in a series of articles in the Mapping Awareness magazine during 1989 and 1990, a short-list of technical research topics can be extracted which set out the extent to which the AI–GIS connection was expected to be explored "on paper":

Midlands RRL (Geography, Leicester and Loughborough Universities), see also Maguire *et al.* (1989):

- spatial databases and data transfer;
- data integration and de-referencing of multi-referenced spatial data;
- human–computer interfaces.

North East RRL (Geography and Town Planning, Newcastle University), see also Openshaw *et al.* (1989):

- tools for spatial analysis of vector data;
- fuzzy geodemographics, locational errors, homogeneity of catchment;
- areas, design of zone aggregation methods;
- automated data-clustering pattern detection and map overlay;
- spatial error propagation when integrating multi-source data.

Northern Ireland RRL (Geoscience, Belfast University; Environmental Studies, Ulster University in Coleraine), see also Stringer and Bond (1990):

- spatial resolution of aggregated spatial data;
- multi-model database structures;
- human–computer interfaces.

North West RRL (Geography, Lancaster University), see also Flowerdew (1989):

- comparison of sets of data;
- area interpolation;
- fast digitisation techniques;
- environmental "plume" models.

RRL for Scotland (Geography, Edinburgh University), see also Healey *et al.* (1990):

- parallel processing;
- a system-independent cartographic "browser".

South East RRL (Geography, Birkbeck College in London and London School of Economics), see also Rhind and Shepherd (1989):

- efficient data storage;
- intelligent front-ends for Arc-Info;
- data encoding and integration;
- remote sensing for land use change;
- data exchange and integration.

Wales and South West RRL (Town Planning, Cardiff University), see also Green *et al.* (1989):

- information systems;
- GIS and expert systems;
- Artificial Intelligence and remote sensing;
- fractal geometries;
- error structures and propagation.

Manchester and Liverpool RRL (Geography, Manchester University; Civic Design, Liverpool University), see also Hirschfield *et al.* (1989):

- address-referencing systems;
- geodemographics and cluster analysis methods.

The first impression from this listing already shows how limited the interest in expert systems or related approaches seemed to be in general, with only indirect reference to such methods in the North East RRL, the RRL for

Scotland, and the South East RRL. The only notable exception was the Wales and South West RRL where explicit interest was expressed in artificial intelligence methods from the beginning.

5.3.2.2 RRL-related work and publications

A literature review of the material produced by the researchers in these laboratories, and informal interview surveys by telephone or in person tended to confirm the preliminary views of the RRL work:

Midlands RRL: At Leicester University, no RRL-linked research was directed to artificial intelligence techniques as such, but Peter Fisher (personal communication) extended his personal interests in this direction. He considers "search" techniques to be central to all artificial intelligence methods (Fisher, 1990a,b), and he sees AI and ES' worth in relation to GIS to be in two related areas: the handling of spatially distributed errors, and uncertainty linked to the data explosion of today and compounded through cartographic manipulation, having illustrated his ideas with applications in soil taxonomy (Fisher and Balachandran, 1990) and in fuzzy land classification from satellite data (Fisher and Pathirana, 1990). In Fisher's view, ES should be able to do non-trivial GIS tasks like telling what an object is, mapping out the history of how the object was created and its values derived, and should also be capable of explaining its reasoning.

Related research at Loughborough University did not focus on expert systems as such but was directed at the issue of intelligent information retrieval from databases (David Walker, personal communication) involving natural language processing and understanding, linked to the general issue of "meta-data" (Medyckyj-Scott *et al.*, 1991) and user-oriented interfaces, from the simple menu-based type (Robson and Adlam, 1991) to more "intelligent" approaches (Medyckyj-Scott, 1991).

North East RRL: Most of the work at this RRL concentrated on the use of "zoned" data of the Census type (Mike Coombes, personal communication), on issues related to the "ecological fallacy", and on questions linked to the regionalisation of such zones using large matrices of data. Stan Openshaw[16] tended to prefer approaches based on "patterns", while Mike Coombes tended to prefer more "craft-based" approaches and, as an automated alternative to the latter, the potential of ES was explored, but there was some disillusionment with them because it was not felt they really produced the flexibility required. There was some work on AI, linked to Stan Openshaw's own personal interests listing AI as one of the most important research topics for the introduction of spatial analysis functionality into GIS (Openshaw, 1990, 1993a), although he found it difficult (personal communication) even

16 In Newcastle University at the time.

to define the field covered by AI. Expert systems as such were not used because, as Stan Openshaw put it, "they don't work"; instead, the interest in AI at the laboratory concentrated on the use of neural nets to help with the regionalisation problem, applied to the 1991 Census (Openshaw and Wymer, 1990), and Openshaw (1993b) explored the use of neural nets to model spatial interaction.

North West RRL: There is no explicit research on expert systems at this RRL, but some limited reference to the related question of so-called spatial decision support systems, focusing on a possible application for evacuation planning (De Silva, 1991), an area of "disaster planning", one of the growth areas in the application of GIS technology.

RRL for Scotland: No research at this RRL was focused on ES (Richard Healey, personal communication), the only area of work remotely related to artificial intelligence was that of parallel processing of large geographical databases; the only work focused on this ES–GIS relationship was that carried out by a doctoral student working on a system for geographical analysis interfacing with "loose coupling" the Arc-Info GIS and the NASA expert systems shell CLIPS (Zhu and Healey, 1992).

South East RRL: At the *London School of Economics* site, both Craig Whitehead (Geographical Information Research Laboratory manager) and Derek Diamond had not been in favour of exploring the route of GIS–expert system links, because "Expert Systems are tainted with the failures of Artificial Intelligence" (Derek Diamond, personal communication); on the other hand, considering how the GIS industry was dominated by technology and by software companies, the interest went in the direction of the idea of a "federal" GIS: proprietary packages inter-linked into an evolutionary system moving from the simple to the complex, a database linked to a mapping system for purposes of both spatial analysis and decision support, towards a spatial decision support system for Landuse Planning (Hershey, 1991a,b).

At the *University College* site, Rhind (1990) suggested "the role of Expert Systems" as one of the main foci for the GIS research agenda, particularly in the following areas of GIS: pattern recognition and "object extraction", integration of diverse data, data search, cartographic generalisation, "idiot-proofing" of systems, GIS teaching, and elicitation of "soft" knowledge from humans. The actual research at this site (Graeme Herbert, personal communication) concentrated on issues of map generalisation and name-placement (the location of labels on maps). Artificial intelligence techniques were incorporated to choose and apply the best map design or generalisation algorithms depending on the characteristics of the map and the feature being generalised (Joao *et al.*, 1990, 1991; Herbert, 1991; Herbert and Joao, 1991).

Wales and South West RRL: As in other RRLs, the programme at this RRL evolved incrementally (Chris Webster, personal communication), reflecting on the one hand existing strengths and interests and, on the other, new opportunities that emerged in the process. Artificial intelligence was not on the original agenda for the RRL in 1986, but was introduced by Chris Webster and Mike Batty, and developed in several phases:

In the *first phase*, starting even before the RRL contract, there were some experiments with expert systems with *no connection to GIS*: the first was linked to MPhil work (De Souza, 1988) focusing on the use of expert systems for Development Control (following a line not too different from that followed at Oxford Polytechnic at the time, see Rodriguez-Bachiller, 1991), using the ESDA expert-system "shell"; the second looked at text animation as a possible alternative approach to knowledge acquisition, using as test-bed the comparison between a system that extracted knowledge from a manual on planning standards for Malaysia, and a system based on standard knowledge acquisition from an expert to deal with possible hazards in the South Wales valleys (Webster *et al.*, 1989). In addition, an expert system for Permitted Development (the issue of whether a development requires planning permission) was developed using the Prolog shell PEXPERT, combining rules and case-law to answer the basic question, seen as "testing the hypothesis" that a development IS permitted.

In the *second phase*, a former research student[17] tried alternative approaches to risk assessment using GIS, and explored with Chris Webster the integration of spatial data and expert systems and how to automate the process of spatial search within the framework of general decision-making by building spatial knowledge into the knowledge base. After a first exploration of logic programming using PROLOG (Webster, 1989a), the same author wrote an experimental system to express spatial databases in "predicate calculus" form using PROLOG, and then built the spatial and topological knowledge as well as the generic search-algorithm using the ESDA shell, to increase the functionality of the predicate-calculus system (Webster, 1989b); a data-set consisting of a few polygons was digitised in Arc-Info format, then exported as unstructured segments, and then converted into Prolog-readable form as "predicates". The functionality of the system was quite trivial (Chris Webster, personal communication), all it did was to go into a map and decide if a search was inside or outside an area, but it showed how the functionality of an expert system could be embellished by bringing spatial data into it.

In what can be called the *third phase*, Ian Bracken and Chris Webster collaborated with an organisation linked to remote sensing, participating in a working group on GIS in Utrecht (with Peter Burrough), and this prompted interest in looking at GIS from the point of view of decision

17 Anthony Wislocki.

support systems (Bracken and Webster, 1989; Webster 1990a). Chris Webster reviewed the field of object-oriented approaches (Webster, 1990b), and he went on to investigate the design of an object-oriented urban and regional planning database using the artificial intelligence tool known as "semantic net" (Webster and Omare, 1990, 1991; Webster, 1991), already mentioned in Chapter 2.

In the *fourth phase*, inspired by the previous work and by the discussions within the Netherlands workshop, interest developed in the areas of vision and pattern recognition, and the possibility of building some intelligence into GIS and their capacity to recognise objects: in conventional or object-oriented systems, objects are explicitly defined and classified as groups of pixels, by adding a label to a classified series of vectors; the question became how to ask the system to find objects that "look like...", and store them. This was speared on by the work done at Utrecht about small-scale buildings, and further work at the RRL explored the methodology and techniques to answer the above question (Webster *et al.*, 1991, 1992). Using SPOT images of Harare (Zimbabwe), some prototypical morphological areas were extracted exploring several pattern-recognition methodologies, which were then tested by predicting housing and population densities and comparing the predictions with the actual values. More recent SPOT data for Cardiff and Bristol was then being obtained (a much better data-set to link up to), and the next phase was intended to be to incorporate the "population surfaces" of Ian Bracken and Dave Martin (Bracken and Martin, 1989; Martin, 1988, 1990). This whole area of work introduced another angle: the possibility of linking Remote Sensing and GIS into a single framework. Also, as Chris Webster explored the combination of AI techniques with a remote sensing process of data capture for GIS, Ian Bracken was exploring a similar combination applied to more conventional data-capture techniques like digitising (Bracken, 1989). The next stage, according to Chris Webster, was probably going to move in the direction of "intelligent" retrieval and spatial search, using non-Euclidean spatial reasoning using "fuzzy" concepts like "near", "far", etc.

Manchester and Liverpool RRL: The initial impression was confirmed by Peter Brown (personal communication), that the client-oriented work at this RRL led to relatively little interest in geographical information systems, and certainly not in the direction of expert systems or artificial intelligence, probably a reflection of the relative state of infancy of those technologies at the time when the RRLs were in operation.

5.3.2.3 GIS and AI in the RRLs: conclusions

The more detailed survey of RRL work confirmed to a large extent the indications from the first impressions:

- Only in a very limited number of RRLs was the possible connection of GIS and AI considered worthy of exploration, partly due to the relative novelty of the GIS technology itself (in fact, in one of the laboratories even GIS technology was almost deliberately ignored), and also partly due to some degree of distrust of the so-called "intelligent" approaches, which were considered too new and unproven.
- Where the connection between AI and GIS was explored, it tended to be applied to the solution of *cartographic problems* present in GIS (error propagation, map generalisation, name-placement, object recognition, classification of satellite information) or to the improvement of database interrogation, but with little or no reference to taking advantage of the combination of these two types of technologies to improve *decision-making*, which one would expect to be potentially the most important reason for using expert systems technology.
- When it concerned decision-making, the emphasis of RRL work seemed to have moved beyond the relatively simple and "inflexible" expert systems in favour of more general systems which were becoming increasingly popular in the literature under the generic name of decision support systems (DSS), and their natural extensions into the spatial dimension (SDSS).

5.3.3 ES and GIS for impact assessment

The potential of combining ES and GIS for impact assessment was pointed out from the early 1990s (Fedra *et al.*, 1991). Fedra (1993) discusses this potential and illustrates it for air pollution *impact analysis* related to climatic change, along similar lines as other authors did later for impact monitoring/analysis, like Kondratiev *et al.* (1996) who combine a GIS (IDRISI) with remote-sensing data to model environmental pollution.

For *impact prediction*, Lundgard *et al.* (1992) use ES to predict noise impacts, and many authors apply similar approaches to the prediction of *pollution* impacts: Appelman *et al.* (1993) develop a system to forecast the effects of sand pits on underground water, Cuddy *et al.* (1996) predict the environmental damage from army training exercises using the ES to handle qualitative information, and Burde *et al.* (1994) develop the SAFRAN system to evaluate the impact of atmospheric acid on soil and ground water, combining Arc-Info and an ES shell using rules (instead of "map algebra", as in GIS) to combine impact maps into overall results. On a slightly different approach, Calori *et al.* (1994) use an ES to select the right air pollution model depending on the scenario, articulated with different models by a "semantic net". For *areawide* impact prediction, Ciancarella *et al.* (1994) describe the SIBILLA system which combines an ES of legal knowledge, prediction models, and GIS – all with "hypertext" to facilitate zooming in and out of each – to analyse and compare the prescriptive contents of land use plans as well as the design of new ones; the aim for the future was to develop it into a proper decision support system to estimate how human activities can affect environmental resources under different legal constraints, and the authors illustrate its potential with an application to the Comacchio wetlands in Italy.

Some systems are designed to serve mainly one purpose within IA, like the EIA system for the *screening* of projects in the basin of the Mekong river (Fedra *et al.*, 1991) using GIS and satellite data. Others try to serve several purposes within IA, sometimes in an evolutionary process, like the case reported by Daniel *et al.* (1994): ESSA Technologies developed the SCREENER system for project *screening*, and then started developing another one, SPEARS (Spatial Environmental Assessment and Review System), to assess (*scope*) possible impacts and select a range of possible *mitigation* measures.

The most common way of coupling ES and GIS is by the latter being "run" from the former. Fully integrated coupling (so-called "tight coupling") is very rare because of the limitations of commercial GIS packages. Of the "loose-coupling" alternatives, only occasionally do we see ES being called – treated as GIS subroutines – by the GIS to perform pre-processing operations on some of the GIS data, for example when they are used to interpret and classify satellite data. The most common approach is for ES to act as "managers" of the problem-solving procedure, and GIS are called to: (i) provide geographical information; (ii) perform certain forms of spatial analysis on it. This also applies to the systems to be discussed in the next section.

5.3.4 ES and GIS for environmental management

Fedra *et al.* (1991) review early examples of GIS–ES integration for environmental management. Maidment (1993) reviews and discusses extensively the integration of GIS, models, ES and other AI techniques like semantic nets (see Chapter 2) in *water* modelling and management, which has attracted considerable attention since the 1980s (Heatwole *et al.*, 1987; Crossland, 1990; Roberts and Ricketts, 1990; Robillard, 1990). For coastal management in particular, Roberts and Ricketts (1990) describe the ASPENEX model combining the NEXPERT shell with Arc-Info, and Lee *et al.* (1991) use a knowledge-based approach to predict wetland conversion and shoreline reconfigurations during long-term sea-rise. On a different note, Wang (1997) discusses an expert system for the selection of groundwater models for protection programmes.

In *ecology*, the potential of adding ES to GIS is also pointed out by Hanson and Baker (1993) in the field of rangeland modelling, using ES to pre-compute data for models and to select the links between the right parts of the right model (acting almost as a decision support system).

Along similar lines, Lam (1993) discusses the RAISON system that uses ES to select the right model (in this case for acid-rain simulation) according to the data and the geographical regime. Miller and Morrice (1991, 1993) give examples related to the prediction of vegetation change, and Miller (1996) deals with the same issue combining *two* knowledge bases for the ES, one to deal with the spatial data and one specific to the subject matter.

Mapping *land-slide* and *erosion risks* has made extensive use of ES technology linked to GIS: Pearson *et al.* (1991, 1992) applied it to Cyprus, combining NEXPERT (an object-oriented ES "shell") and Arc-Info using the interface designed by Maidment and Djokic (1991). Ferrier *et al.* (1993) and Ferrier and Wadge (1997) applied the same set of tools to the Cheshire Basin in the UK. Adinarayana *et al.* (1994) used a raster GIS and rules to define the probabilities of soil erosion in an area of the Western Ghats (India), and Kolejka and Pokorny (1994) used an ES to identify the characteristics of areas of land-slide hazards, which were then mapped with a GIS in Southern Moravia (Czech Republic).

In *geology*, Miller (1994) describes a system integrating geologic knowledge for the San Juan Basin (New Mexico); ES–GIS combinations have been suggested in this field since the 1980s (Katz, 1988; Usery *et al.*, 1988, 1989; Vogel, 1989), and Cheng *et al.* (1994) discuss a system for the estimation of mineral potential in different areas.

Applied to *rural* management, Archambault (1990) used an ES to diagnose pest-risks in Quebec. In *forestry*, Skidmore *et al.* (1991) used ES to classify satellite data in New South Wales (Australia) and decide with production rules the type of forest soil landscape in each area.[18] Gouldstone Gronlund and Xiang (1993) and Gouldstone Gronlund *et al.* (1994) combine ES and GIS to define priority management areas to combat forest fires. In the general area of *environmental monitoring*, Lam and Pupp (1996) introduce ES to integrate several databases and models and produce environmental reports.

A typical model of ES–GIS combination emerges again (Yazdani, 1993): the role ES play when linked to GIS is often that of "managers" of the operation of the GIS – which provide data and some modelling – guiding the correct use of GIS functions or data and helping with their interpretation, in a way similar to what "decision support systems" (DSS) do, although proper DSS do it in their own distinctive way (see Chapter 2). But the similarities are apparent, and point us in the direction of some applications of these technologies which can be said to represent practically a *borderline* between ES and DSS, where the former is used very directly to help with management practices: Radwan and Bishr (1994) deal with several kinds of non-point pollution and erosion models – in a multi-model

18 This issue of land classification relates directly to the methodological problem of classification of data already discussed in Section 5.3.1.2.

system looking very much like a DSS – where the ES is used to pre-process data for them and to analyse their results, linking once more the NEXPERT shell and Arc-Info; Xiang (1997) describes the system CRITIC, which uses rules to identify deficiencies in fire-control plans (for North Carolina State Park authorities) in the form of undesirable relationships between planned decisions.

5.4 DECISION SUPPORT SYSTEMS (AND ES) WITH GIS

We have seen how expert systems are often programmed as "managers" of the problem-solving logic where GIS makes an efficient contribution when applied to small problems with relatively straightforward aims and well-defined solution methods. However, as problems become bigger and more complex, the simple rule-based logic of ES (see Chapter 2) can prove inadequate and needs a more open-ended framework within which to "explore" and perform its problem-solving. Decision Support Systems (DSS) were developed to respond to such needs in more complex situations and, accordingly, GIS technology has also become involved with these new-style systems. The potential – the need even – for integration of these different types of tools is now deep-rooted in the GIS user-community, as already identified in a survey amongst planners (Baumewerd-Ahlmann *et al.*, 1994). As discussed in Chapter 2, the call for DSS originated mainly from the tradition of "Management Information Systems", but within the GIS-related literature we could see similar pressures towards a wide-ranging framework – within which GIS, ES, and models are constituent parts – coming from several directions:

- interest in multi-criteria decision-making with GIS, where – it was argued – a DSS framework is essential (Heywood *et al.*, 1994; Peckham, 1997);
- fields such as urban and regional planning – also interested in multi-criteria decision-making – where the pioneering idea of "desktop planning" (Newton *et al.*, 1988) and calls for improved information systems (Han and Kim, 1989; Clarke, 1990; Nijkamp and Scholten, 1991, 1993) could be seen as antecedents to spatial DSS;
- spatial analysis and modelling, where the flexibility of DSS was seen as having the potential to resolve some of the "bottlenecks" in this field (Copas and Medyckyj-Scott, 1991; Fischer and Nijkamp, 1992, 1993);
- the GIS field itself, where DSS were seen as the logical framework for GIS (and ES) to achieve their potential as decision-making tools (Abel *et al.*, 1992; Richer and Chevalier, 1992; Caron and Buogo, 1993; Chevallier, 1993; Laaribi *et al.*, 1993; Chevalier, 1994; Holmberg, 1996).

Another aspect of this complementarity between DSS and ES is the fact that, over time, the interest in the latter in the GIS-related literature seems to be declining as the interest in the former increases. Based on an updated version of the bibliography in Rodriguez-Bachiller (2000), Figure 5.2 shows the relative frequency of DSS and ES references each year (expressed as percentages of all environmental GIS references reviewed) gradually changing during the 1990s. The relative decline of expert systems is not because DSS are replacing ES, but because they provide an *envelope* for them. DSS references are often *also* about ES, which they mention as "components" of DSS, but ES are not any more the central focus of interest.

Even more than when dealing with ES, the literature on GIS-related DSS focuses heavily on *methodological* issues, undoubtedly reflecting how new this technology still is. Concentrating only on references dealing with environmental issues, important work by Fedra (1993b, 1994, 1995) discusses the basic structures to integrate GIS, models, and ES in pairs or into an environmental DSS combining all three, illustrating the discussion with examples on air and water quality management, technological risk assessment, and general environmental management. Abel *et al.* (1992) discuss the SISKIT system suggesting architectures for GIS which are suitable for DSS, Van Voris *et al.* (1993) emphasise the importance of the visualisation of the information while it is being processed in the DSS, Frysinger *et al.* (1996) propose an open architecture to integrate models and GIS into

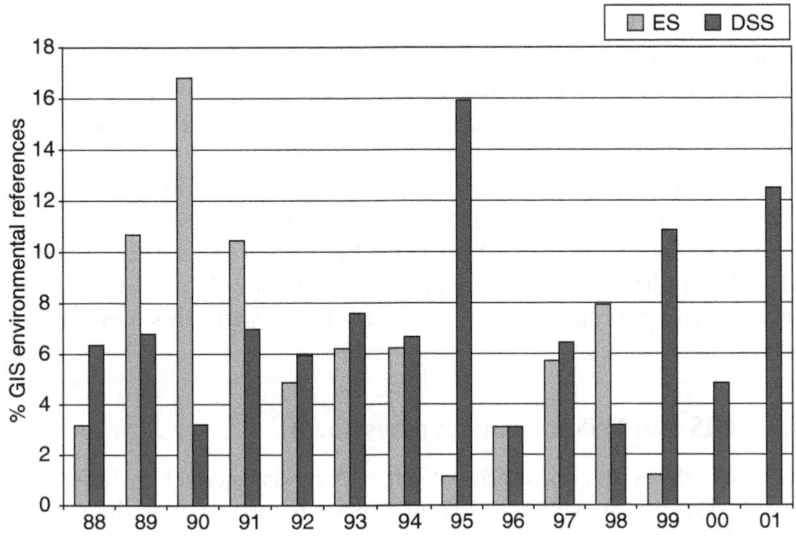

Figure 5.2 GIS with expert systems and decision support systems.

an environmental DSS, and Romao *et al.* (1996) propose the COASTMAP system for coastal zone management using *hypermedia* techniques to integrate the various modules in the DSS.

A thorough discussion of methodological issues can be found in Leung and Leung (1993a) and Leung (1993) related to the development of one particular example of "intelligent" DSS, and an accepted structure for these DSS with GIS is now widespread in the literature (Arbeit, 1993; Grothe and Scholten, 1993; Enache, 1994; Birkin *et al.*, 1996). Also, a growing literature on so-called "spatial" DSS – or SDSS – (Densham and Goodchild, 1989; Armstrong and Densham, 1990; Ryan, 1992; Densham, 1991, 1993, 1994; Densham and Rushton, 1996; Ayeni, 1997) and extensions like "group" DSS (Jankowski *et al.*, 1997; Jones, *et al.* 1997) also reinforced this discussion in the 1990s, although these systems do not always involve GIS, but other technologies for spatial referencing and mapping.

As before, one of the dominant methodological issues is the question of *how* to integrate the different modules in the DSS – similar to the question of integrating GIS and models or ES. As Badji and Mallans (1991) considered quite early on: (i) it can be *ad hoc*, with each module being developed separately; (ii) using *partial* linkage, either a GIS can be developed around a model or a model around a GIS; (iii) with *full* linkage, the respective data of the two systems are tailored to each other's needs. In their example, Badji and Mallans apply a "partial" approach to the development of a DSS for irrigation-water management. In terms of the actual programming of the modules (including GIS) that make up a DSS, Peckham (1997) provides a similar list of how it can be done: (i) programming all the elements from scratch; (ii) using a commercial GIS and its macro language; (iii) with a "federated" approach, using different packages for the different modules, all operating on the same "windowing" environment, although he recognises that not many commercial GIS can do this. A good example of integration can be found in Djokic (1996), describing a general purpose "shell" for Spatial DSS, based on the already mentioned link between Arc-Info and the ES shell NEXPERT (Maidment and Djokic, 1991).

Let us now look at GIS applications integrated within a DSS (which, strictly speaking, constitute a Spatial DSS) for the purposes of IA, often also involving ES in the armoury of the DSS. Because of the nature of DSS, they tend to be applied to tasks more complex than simple models or even ES, especially in later applications, as confidence with this new approach grows.

5.4.1 GIS and DSS for impact assessment

The use of DSS with GIS, specifically for IA tends to cover various "stages" in the IA process as well as different types of impacts. For the *scoping* of impacts (identifying which impacts to study and how "key" they are) and the *review* of Environmental Statements, Haklay *et al.* (1998) discuss an interesting system for Israel – where Statements are prepared centrally and not by the developer – which compares project characteristics with an environmental database to suggest the impacts to investigate, and evaluate Environmental Statements accordingly.

For *impact prediction*, systems of this kind have been designed to cover virtually all types of impacts:

- *Waste*: Peckham (1993) describes a system to do scenario generation and evaluation for industrial waste management in the Lombardy region.
- *Water*: Booty *et al.* (1994) discuss the RAISON system (to run on PCs) developed for the Water Research Institute of Canada to do EIA of discharges into water streams, combining GIS, ES, models and statistics; the ES shell is used to construct rules in dialogue with the expert, and those rules are used to run models and are also extended into spreadsheet "IF" formulae to manipulate the data; Rushton *et al.* (1995) discuss the Northeast-ESRC Land Use Programme (NELUP) to predict the consequences of land use changes in water catchments in Northeast England, and Wadsworth and Callaghan (1995) show examples of use of the same system.
- *Air pollution*: Briassoulis and Papazoglou (1994) develop a multi-criteria DSS to evaluate land suitability in terms of accident risk based on proximity to major hazard facilities using dispersion models to estimate the risks, and Chang *et al.* (1997) develop a system for disaster planning for chemical emergencies, combining Arc-Info (programmed in AML), air diffusion models to simulate impacts, and a knowledge base to evaluate the rescue actions needed.
- *Noise*: Altenhoff and Lee (1993) use Arc-Info for the simulation of noise emissions and abatement measures.
- *Traffic* impacts: Appelman and Piepers (1993) discuss a system for the Ministry of Public Works in the Netherlands to apply a landscape-ecological approach in IA for the planning of highways, and Miyamoto *et al.* (1995) design a location/land-use model integrated with a traffic/transport model and a model to simulate traffic impacts for specific projects or land use plans in Bangkok; the models are written in Fortran, the interfaces in Visual Basic, and the rest are "off-the-shelf" packages.
- *Landscape*: Goncalves *et al.* (1995) combine a cellular-automata model of landscape change with a multi-criteria impact-evaluation methodology (using IDRISI) and illustrate it with an example about a new freeway being planned in central Portugal.
- *Multiple impacts*: Although it is not presented as a DSS but as a modelling application, Biagi and Pozzana (1994) present a structure which has in fact the ingredients of a DSS: various models are used to predict the geographical distribution of impacts derived from land-use changes, and to assess their effect on the environmental situation.

For impact *mitigation*, Kusse and Wentholt (1992) discuss the RIM system which combines an ES and a GIS to simulate emission levels into ground water before and after mitigation measures, and their SENSE system extends this capacity into *suggesting* such measures; Salt and Culligan Dunsmore discuss SDSS for post-emergency management of radioactively contaminated land, using examples from Scotland. On a different note, Fedra (1999) discusses the *monitoring* of urban environmental impacts using DSS (including ES).

A rare example of DSS application to help with land reclamation at the *decommissioning* stage of a project can be found in Hickey and Jankowski (1997) for a smelter project, including the production of re-vegetation priority maps using Arc-Info's GRID and programming it in AML.

5.4.2 GIS and DSS for environmental management

As we would expect, environmental management tasks can reach considerable levels of complication and "open-endedness", and it is for tasks of this kind that DSS are ideally suited. Let us look at some typical areas of application for DSS with GIS to deal with environmental matters.

The use of these systems to help with various aspects of *agriculture* and rural management is quite wide-ranging:

- General *management* and *policy making* include a wide variety of uses from the early 1990s:

  (i) For general *land-use management* and planning, Yang and Sharpe (1991) describe a prototype system to help design "buffer zones" around environmental conservation areas, De Sede *et al.* (1992) describe the GERMINAL project developed at the Swiss Federal Institute of Technology to aid decision-making in rural planning at regional level, and Sharifi (1992) discusses a system for agricultural land-use planning. Shvebs *et al.* (1994) propose a system for the optimisation of rural land resources for Ukraine, and McClean *et al.* (1995) discuss a similar system for land-use planning applicable to both rural and urban environments. Keller and Strapp (1996) use the "Application Programming Interface" to interact with a GIS, and apply it to the management of land consolidation, MacDonald and Faber (1999) propose a system for sustainable land-use planning, Zeng and Chou (2001) propose the REGIS system for "optimal" spatial decision-making for Southern Sydney (Australia), and Recatala *et al.* (2000) use the LUPIS model (Newton *et al.*, 1988) for land-use planning for the Valencia Region in Spain.

  (ii) On *water-related* issues, Ye *et al.* (1992) describe a DSS (including ES and GIS) to support irrigation scheduling in Belgium, and Watson and Wadsworth (1996) integrate economic, ecological and hydrologic models to investigate the effects of different rural policies in the UK.

Negahban *et al.* (1996) describe an agricultural DSS for the Lake Okeechobee (LOADSS), Martin *et al.* (1999) develop a Spatial DSS for watershed management in the Saint-Charles river (Quebec), and Qureshi and Harrison (2001) discuss DSS for riparian "revegetation" in North Queensland (Australia).

(iii) On *environmental* management, Zhu *et al.* (1998) propose a knowledge-based approach to designing environmental DSS, and Seder *et al.* (2000) discuss "intelligent" DSS for environmental management in urban systems; with a specific focus, Douven *et al.* (1993), Douven and Scholten (1994) and Beinat (1996) discuss the development of a DSS to decide the admissibility of new pesticides in an area (for use in the Netherlands).

- *Classification* of environmental situations: Leung and Leung (1993b) illustrate the use of an "intelligent" DSS to classify land types from Landsat data, and climate types from database information on rain, etc.
- *Site selection* tasks are not so complicated, and DSS are less frequent: for a rare example, Jain *et al.* (1995) design a Spatial DSS to evaluate alternatives, applied to an example for livestock site selection in Lake Icaria (Iowa).
- Management of *urban development* – which can be seen as a general case of "site-selection" – in potential conflict with the natural environment, as in the suggested approach by Despotakis *et al.* (1992) to help design sustainable development strategies for the Greek islands.
- *Pollution* management: Van Tiel *et al.* (1991) describe the BOBIS system to deal with the whole cycle of pollution management (detection, analysis, forecasting, clean-up). On the other hand, pollution forecasting is a quite specific and a relatively simple task for a DSS, and Engel *et al.* (1993) discuss the integration of the AGNPSS pollution model and the USLE rainfall erosion model with a raster GIS (GRASS), for an environmental field station near West Lafayette (Indiana).

To manage *landscape*, Liu *et al.* (1993) developed an early DSS at Virginia Polytechnic to manage landscape resources in national parks, and Cudlip *et al.* (1999) suggest a system (PLAINS) combining GIS and expert systems for landscape assessment.

We can see examples of applications to *forestry* management from the 1980s: Sieg and McCollum (1988) discuss the LAMPS system for general management, Reisinger *et al.* (1990) present a system to analyse the effects of forest harvesting, and Dubois and Gold (1994) develop an interactive system with the emphasis on the instant visualisation of the effects of actions/policies to help decision-makers. Bishop and Karadaglis (1996) combine Arc-Info and a linear-programming model of optimisation of forestry resources with a visualisation toolkit IRIS (written in "C") which, because of the speed of C, becomes the "base" of the system. Ideally – the authors argue – these different units should be "tightly coupled", but proprietary GIS like Arc-Info make it impossible, as already mentioned. Cocks and Ive (1996) propose the SIRO-MED system (an adaptation of the LUPIS system, see Newton *et al.*, 1988) for forest land allocation. On the related topic of *forest fires*, Casale *et al.* (1993) design a system combining GIS and ES into a DSS to help choose the resources needed to combat a fire and to give advice on the best way to fight it, and Kessell (1996) combines dynamic modelling with visualisation to help with bushfires in Australia.

Related to *water* management, Grenney *et al.* (1994) use a graphical interface (like a GIS but interactive) to construct maps of stream networks, with a rule-based SPECS to discuss data about the streams and provide advice on their environmental situation and actions to be taken, and Taylor *et al.* (1999) discuss a similar multi-model system for water-resources planning in Sydney. Epstein *et al.* (1993) discuss the CLAIR system, an all-inclusive DSS with multi-agency integration – touching on another key area of development in DSS work, that of *group* decision-support – for air and water quality control around urban and industrial developments. Along similar lines, Westmacott (2001) discusses the possibilities of DSS for integrated coastal management in the tropics. For general *pollution-risk* analysis, Franco *et al.* (1996) discuss the SIGRI system combining Arc-Info, simulation models and ES to assess industrial pollution risk at regional and sub-regional scale and produce risk-index maps.

5.5 CONCLUSIONS

Expert systems have been around since the 1960s. After some fading of initial enthusiasm in some traditional areas of decision-making, they now seem to be attracting fresh interest in new areas like IA and environmental management. Eye-opener articles highlighting the potential of ES have appeared in the environmental literature, and prototypes have been developed and used. The connection of GIS to expert systems has been suggested from quite early in the history of GIS applications, as links between GIS and models became more ambitious, and as the aims of GIS applications moved closer to decision support and away from pure analysis and modelling. Calls for "intelligent" GIS are frequent, but their development has been slow to get started, particularly in the UK, where a survey of the Regional Research Laboratory experience (which lasted into the early 1990s) confirmed the impression that only a very limited number of RRLs explored the possible connection of GIS and AI – let alone ES – partly due to the relative novelty of the GIS technology itself, partly due to some degree of distrust of the so-called "intelligent" approaches, which were considered too new and unproven. Where the connection between AI and GIS was explored, it was in the solution to cartographic problems, but with little or no reference to improving *decision-making*, potentially the most important reason for using ES. When decision-making was investigated, the emphasis of RRL work moved beyond the relatively simple expert systems in favour of more general decision support systems (DSS).

Because of the relative novelty of this technology, greater emphasis has been put on methodological issues compared with the discussion of GIS alone (see previous chapters). In particular the question of *how* to integrate the two technologies – loosely coupled, tightly coupled, fully integrated – was paramount, replicating to some extent previous discussions relating to the linking of GIS with models. The most common way of coupling ES and GIS is by the latter being "run" from the former, as fully integrated coupling is very rare because of the limitations of commercial GIS. Expert Systems act as "managers" of the problem-solving procedure, and GIS are called to provide geographical information or to perform certain forms of spatial analysis. Then, as problems become bigger and more complex, the simple logic of ES starts to prove inadequate and needs a wider framework within which to operate. While IA as such often involves relatively simple operations of a technical nature, where models and/or GIS can be used to achieve a solution, environmental management tasks can reach considerable levels of complication and "open-endedness". For tasks of this kind DSS are ideally suited and, within such systems, ES play an important role together with GIS, models, and other procedures.

Previous chapters have reviewed examples of specialised computer technologies, such as GIS and models, which could be used quite fruitfully to help with IA. They also show how the potential of such technologies diminishes as the problems facing the environmental professionals increase in complexity. Tasks in IA more akin to environmental management, like environmental monitoring, modelling and forecasting, are much more open-ended and require much more varied expertise. Some degree of automation could be used in these areas, more as an "aid" than a substitute, for which the more flexible DSS seem more appropriate. ES can be used within them to help decide the approach to use, GIS can help with the data and perform some of the tasks involved, and modelling can be called on to do the more intricate and specific simulations, but the overall management of the process must be left open-ended and variable. Full automation of these tasks may never be possible or advisable: experts are more difficult (impossible?) to replace at this level, and computer tools like ES, GIS or models are more likely to be a complement to the expert rather than a replacement.

REFERENCES

Abel, D.J., Yap, S.K., Ackland, R., Cameron, M.A., Smith, D.F. and Walker, G. (1992) Environmental Decision Support System Project: An Exploration of Alternative Architectures for Geographical Information Systems, *International Journal of Geographical Information Systems*, Vol. 6, No. 3, pp. 193–204.

Adinarayana, J., Krishna, N.R., Suri, J.K., Rao, K.G. and Venktachalam, P. (1994) Assessment of Soil Erosion of a Watershed Environment Using GIS, *Proceedings of the Canadian Conference on GIS '94*, Ottawa, Canada (June 6–10), Vol. 1, pp. 609–20.

Altenhoff, D. and Lee, S. (1993) A Decision Support System for the Simulation of Noise Emissions and Noise Abatement Measures Using ARC/INFO, *Proceedings of the 3rd International Conference on Computers in Urban Planning and Urban Management*, Atlanta (Georgia), Vol. 2, p. 3 (Abstract).

Appelman, K. and Piepers, A.A.G. (1993) Implementation of a Decision Support System for Environmental Impact Assessment, *Proceedings of the EGIS '93 Conference*, Genoa, Italy (March 29–April 1), Vol. 2, pp. 1492–6.

Appelman, K., Van der Vegt, H.J.W. and Allewijn, R. (1993) Environmental Impact Assessment of a Sand Pit Using a Water Management Information System, *Proceedings of the EGIS '93 Conference*, Genoa, Italy (March 29–April 1), Vol. 2, pp. 1061–6.

Arbeit, D. (1993) Resolving the Data Problem: A Spatial Information Infostructure for Planning Support, *Proceedings of the 3rd International Conference on Computers in Urban Planning and Urban Management*, Atlanta (Georgia), Vol. 1, pp. 3–26.

Archambault, L. (1990) Des Nouveaux Outils Pour L'Amenagement des Plantations en Relation Avec les Ravageurs, *Proceedings of the "GIS for the 1990s" National Canadian Conference*, Ottawa, Canada (March 5–8), pp. 406–13.

Armstrong, M.P. and Densham, P.J. (1990) Database Organisation Strategies for Spatial Decision Support Systems, *International Journal of GIS*, Vol. 4, No. 1, pp. 3–20.

Armstrong, M.P. and Bennett, D.A. (1990) A Knowledge Based Object-Oriented Approach to Cartographic Generalisation, *Proceedings of the GIS/LIS '90 Conference*, Anaheim, California (November 7–10), Vol. 1, pp. 48–57.

Atkinson, P.M. and Cutler, M.E.J. (1996) Unmixing Mixed Pixels, *GIS Europe* (June), pp. 18–19.

Ayeni, B. (1997) The Design of Spatial Decision Support Systems in Urban and Regional Planning, in Timmermans, H. (ed.) *Decision Support Systems in Urban Planning*, E & FN Spon, London, Ch. 1, pp. 3–22.

Badji, M. and Mallans, D. (1991) Integrating GIS and Process Models: A New Perspective for Decision Support Systems (DSS) in Irrigation Water Management, *Proceedings of the EGIS '91 Conference*, Brussels (April 2–5), Vol. 1, pp. 48–54.

Barbanente, A., Borri, D., Esposito, F., Maciocco, G. and Selicato, F. (1991) Automatically Acquiring Knowledge by Digital Maps in Artificial Intelligence Planning Techniques, *Proceedings of the 2nd International Conference on Computers in Urban Planning and Urban Management*, Oxford Polytechnic (July 6–8), Vol. 1.

Baumewerd-Ahlmann, A., Scolles, F., Achwabl, A., Simon, K.-H. and Waschkowski, R. (1994) Integrated Computer Support for Environmental Impact Assessment, in Guariso, G. and Page, B. (eds) *op. cit.*, pp. 289–99.

Beinat, E. (1996) Decision Support and Spatial Analysis for Risk Assessment of New Pesticides, in Rumor, M. *et al.* (eds) *op. cit.*, Vol. 1, pp. 757–66.

Besio, M., Frixione, M. and Lavaggi, R. (1991) Knowledge Representation in Environmental and Landscape Systems – Part II, *Proceedings of the 14th Urban Data Management Symposium (UDMS)*, Vol. 1, pp. 265–80.

Biagi, B. and Pozzana, G. (1994) A GIS-Based Information System for the Assessment of the Variations in the Environment Load, Owed to Changes in Socio-Economic Factors, *Proceedings of the EGIS/MARI '94 Conference*, Paris (March 29–April 1), Vol. 1, pp. 642–51.

Birkin, M., Clarke, G., Clarke, M. and Wilson, A. (1996) *Intelligent GIS. Location decisions and Strategic planning*, Geo Information International, London.

Bishop, I.D. and Karadaglis, C. (1997) Linking Modelling and Visualisation for Natural Resources Management, *Environment and Planning B: Planning and Design*, Vol. 24, No. 3 (May), pp. 345–58.

Booty, W.G., Wong, I.W.S., Lam, D.C.L., Kerby, J.P., Ruddock, R. and Kay, D.F. (1994) Application of an Expert System for Point Source Water Quality Modelling, in Guariso, G. and Page, B. (eds) *op. cit.*, pp. 233–44.

Borman, S. (1989) Use of Environmental Expert Systems Growing, *Chemical and Engineering News*, Vol. 67 (October), pp. 25–7.

Bouille, F. (1989) Applying an Integrated Expert System to Urban Management and Planning, *Proceedings of the 1st International Conference on Computers in Urban Planning and Urban Management*, Hong Kong (August 22–5).

Bracken, I. (1989) *"Intelligent" Digitising for Complex Map Bases*, Technical Reports in Geo-Information Systems, Computing and Cartography No. 20, Wales and South West Regional Research Laboratory, Department of Town Planning, University of Wales College of Cardiff, Cardiff.

Bracken, I. and Martin, D. (1989) The Generation of Spatial Population Distribution from Census Centroid Data, *Environment and Planning A*, No. 21, pp. 537–44.

Bracken, I. and Webster, C. (1989) *A Systems Architecture View of GIS Functionality*, Technical Reports in Geo-Information Systems, Computing and Cartography No. 8, Wales and South West Regional Research Laboratory, Department of Town Planning, University of Wales College of Cardiff, Cardiff.

Briassoulis, H. and Papazoglou (1994) Determining the Uses of Land in the Vicinity of Major Hazard Facilities: An Optimization Algorithm, in Guariso, G. and Page, B. (eds) *op. cit.*, pp. 177–86.

Brown, A.G.P., Geraghty, P.J. and Horton, F.F. (1996) A Decision Support System for Environmental Impact Assessment Procedures Legislation, in *Improving Environmental Assessment Effectiveness: Research, Practice and Training*, Proceedings of IAIA '96, 16th Annual Meeting (June 17–23), Centro Escolar Turistico e Hoteleiro, Estoril (Portugal), Vol. III, pp. 31–6.

Buch, A.M., Narain, A. and Pandey, P.C. (1994a) Classification of Glacier Features Using Artifical Neural Networks, *Proceedings of the Canadian Conference on GIS '94*, Ottawa, Canada (June 6–10), Vol. 2, pp. 1250–8.

Buch, A.M., Narain, A. and Pandey, P.C. (1994b) Application of Artificial Neural Networks in Hydrological Modelling: A Case Study of Run-off Modelling of the Himalayan Glacier Basin, *Proceedings of the Canadian Conference on GIS '94*, Ottawa, Canada (June 6–10), Vol. 1, pp. 408–10.

Buehler, K.A. and Wright, J.R. (1989) *A Bayesian Expert System Shell for Spatial Modelling*, First International Conference on Expert Systems in Environmental Planning and Engineering, Lincoln Institute – Massachusetts Institute of Technology, Boston (September).

Burde, M., Jackel, T., Dieckmann, R. and Hemker, H. (1994) Environmental Impact Assessment for Regional Planning with SAFRaN, in Guariso, G. and Page, B. (eds) *op. cit.*, pp. 245–56.

Burrough, P.A. (1992) Development of Intelligent Geographical Information Systems, *International Journal of Geographical Information Systems*, Vol. 6, No. 1, pp. 1–11.

Calori, G., Colombo, F. and Finzi, G. (1994) FRAME: A Knowledge-based Tool to Support the Choice of the Right Air Pollution Model, in Guariso, G. and Page, B. (eds) *op. cit.*, pp. 211–21.

Caron, C. and Buogo, A. (1993) L'Integration des Nouvelles Fonctionalites des les SIG Pour la Prise de Decision: L'Example du Projet Synopse, *Proceedings of the Canadian Conference on GIS '93*, Ottawa, Canada (March 23–5), pp. 950–61.

Cambridge, H., Cinderby, S., Kuylenstierna, J. and Chadwick, M.J. (1996) 'A hard rain's gonna fall…' environmental modelling of acid rain, *GIS Europe* (February), pp. 20–2.

Casale, G., Cipressini, G., Catello, G. and Longobardi, A. (1993) AFFRESCO: Allocation of Fire-Fighting Resources in Emergency Situation Control, *Proceedings of the EGIS '93 Conference*, Genoa, Italy (March 29–April 1), Vol. 2, pp. 1149–67.

Chalmers, L. (1989) Prospects of the Use of Expert Systems in Environmental Impact Assessment, *Geography in Action*, New Zealand Geographical Society Conference Series, Vol. 15, pp. 175–9.

Chang, N.-B., Wei, Y.L., Tseng, C.C. and Kao, C.-Y.J. (1997) The Design of a GIS-based Decision Support System for Chemical Emergency Preparedness and Response in an Urban Environment, *Computers, Environment and Urban Systems*, Vol. 21, No. 1 (January), pp. 67–94.

Cheng, Q., Agterberg, F.P., Bonham-Carter, G.F. and Sun, J. (1994) Artificial Intelligence Modelling for Integrating Spatial Patterns in Mineral Potential Estimation With Incomplete Information, *Proceedings of the Canadian Conference on GIS '94*, Ottawa, Canada (June 6–10), Vol. 1, pp. 206–19.

Chevallier, J.-J. (1993) Systemes d'Aide a la Decision a Reference Spatiale (SADRAS): Methode de Conception et de Development, *Proceedings of the Canadian Conference on GIS '93*, Ottawa, Canada (March 23–5), pp. 561–72.

Chevalier, J.-J. (1994) De L'Information a l'Action: Vers des Systemes d'Aide a la Decision a Reference Spatiale (SADRAS), *Proceedings of the EGIS/MARI '94 Conference*, Paris (March 29–April 1), Vol. 1, pp. 9–21.

Ciancarella, L., De Sabbata, P., Litido, M., Secondini, P., Catellani, M. and Travasoni, S. (1994) SIBILLA: An Expert System for Analysis and Use of the Local Planning Acts, in Guariso, G. and Page, B. (eds) *op. cit.*, pp. 257–68.

Clarke, M. (1990) GIS and Model Based Analysis: Towards Effective DSS Making with GIS, in Scholten, H.J. and Stillwell, J.C.H. (eds) *Geographical Information Systems for Urban and Regional Planning*, Kluwer Academic Publishers, pp. 165–75.

Clayton, D. and Waters, N. (1999) Distributed Knowledge, Distributed Processing, Distributed Users: Integrating Case-based Reasoning and GIS for Multicriteria Decision Making, in Thill, J.-C. (ed.) *op. cit.*, Ch. 12, pp. 275–307.

Cocks, D. and Ive, J. (1996) Mediation Support for Forest Land Allocation: The SIRO-MED System, *Environmental Management*, Vol. 20, No. 1 (January/February), pp. 41–52.

Copas, C. and Medyckyj-Scott, D. (1991) Is There Life After Modelling? Putting the "Support" Back Into Decision Making with GIS, *Proceedings of the EGIS '91 Conference*, Brussels (April 2–5), Vol. 1, pp. 216–25.

Coulson, R.N., Saunders, M.C., Loh, D.K. and Oliveia, F.L. (1989) Knowledge System Environment for Integrated Pest Management in Forest Landscapes: The Southern Pine Beetle, *Bulletin of the Entomological Society of America*, Vol. 35, No. 2, pp. 26–32.

Cowen, D., Mitchell, L. and Meyer, W. (1990) Industrial Modelling Using a Geographical Information System: The First Step in Developing an Expert System for Industrial Site Selection, *Proceedings of the GIS/LIS '90 Conference*, Anaheim, California (November 7–10), Vol. 1, pp. 1–10.

Crossland, M.D. (1990) HydroLOGIC – A Prototype Geographical Information Expert System for Examining an Artificial Intelligence Application in a GIS Environment, *Proceedings of the GIS/LIS '90 Conference*, Anaheim, California (November 7–10), Vol. 1, pp. 225–33.

Cuddy, S.M., Davis, J.R. and Whigham, P.A. (1996) Integrating Time and Space in an Environmental Model to Predict Damage from Army Training Exercises, in Goodchild, M.F., Steyaert, L.T., Parks, B.O., Johnston, C., Maidment, D., Crane, M. and Glendinning, S. (eds) *op. cit.*, Ch. 56, pp. 299–303.

Cudlip, W., Lysons, C., Ley, R., Deane, G., Stroink, H. and Roli, F. (1999) A New Information System in Support of Landscape Assessment: PLAINS. *Computers, Environment and Urban Systems*, Vol. 23, No. 6, pp. 449–67.

Daniel, C., Webb, T. and Sully, L. (1994) A GIS-Based Expert System for Environmental Assessment, *Proceedings of the Canadian Conference on GIS '94*, Ottawa, Canada (June 6–10), Vol. 1, pp. 105–16.

Davis, J.R., Nanninga, P.M., Hoare, J.R.L. and Clarke, J.L. (1988) A Knowledge-based Approach to Land Management, in Newton, P.W., Taylor, M.A.P. and Sharpe, R. (eds) *op. cit.*, Ch. 23, pp. 213–21.

Dai, X. and Khorram, S. (1999) Data Fusion Using Artificial Neural Networks: A Case Study on Multitemporal Change Analysis, *Computers, Environment and Urban Systems*, Vol. 23, No. 1 (January), pp. 19–31.

De Jong, S.M. and Riezebos, H.Th. (1991) Use of a GIS-Database as "a priori" Knowledge in Multispectral Landcover Classification of Satellite Images, *Proceedings of the EGIS '91 Conference*, Brussels (April 2–5), Vol. 1, pp. 503–8.

Densham, P.J. (1991) Spatial Decision Support Systems, in Maguire *et al.* (eds) *op. cit.*, Ch. 26.

Densham, P.J. (1993) Integrating GIS and Parallel Processing to Provide Decision Support for Hierarchical Location Selection Problems, *Proceedings of the GIS/LIS '93 Conference*, Minneapolis, Minnesota (November 2–4), Vol. 1, pp. 170–9.

Densham, P.J. (1994) Integrating GIS and Spatial Modelling: Visual Interactive Modelling and Location Selection, *Geographic Systems*, Vol. 1, pp. 204–13.

Densham, P.J. and Goodchild, M.F. (1989) Spatial Decision Support Systems: A Research Agenda, *Proceedings of the GIS/LIS Conference* (Orlando, Florida), Vol. 2, pp. 707–16.

Densham, P.J. and Rushton, G. (1996) Providing Spatial Decision Support for Rural Public Service Facilities that Require A Minimum Workload, *Environment and Planning B*, Vol. 23 (September), pp. 553–74.

Department of the Environment (1996) *Changes in the Quality of Environmental Impact Statements for Planning Projects*, Report by the Impact Assessment Unit, School of Planning, Oxford Brookes University HMSO, London.

Deren, L. and Tao, C. (1994) KGD – Knowledge Discovery from GIS – Propositions in the Use of KDD in an Intelligent GIS, *Proceedings of the Canadian Conference on GIS '94*, Ottawa, Canada (June 6–10), Vol. 2, pp. 1001–12.

De Sede, M.H., Prelay-Droux, R., Claramunt, C. and Vidale, L. (1992) Development of a Decision Support Tool for Environmental Management: The GERMINAL Project, *Proceedings of the EGIS '92 Conference*, Munich (March 23–6), Vol. 2, pp. 1457–66.

De Silva, F.N. (1991) *Spatial Decision Support Systems for Evaluation Planning: Literature Review and Research Proposal*, Research Report No. 21, North West Regional Research Laboratory, Department of Geography, Lancaster University, Lancaster.

De Souza, A.V. (1988) *LUCTROL: A Land-Use Control Expert System*, unpublished M.Phil Thesis, Department of Town Planning, University of Wales College of Cardiff, Cardiff.

Despotakis, V., Giaoutzi, M. and Nijkamp, P. (1992) GIS as a Tool for Sustainable Development Strategies on Greek Islands, *Proceedings of the EGIS '92 Conference*, Munich (March 23–6), Vol. 1, pp. 173–85.

Djokic, D. (1996) Toward a General-Purpose Decision Support System Using Existing Technologies, in Goodchild, M.F., Steyaert, L.T., Parks, B.O., Johnston, C., Maidment, D., Crane, M. and Glendinning, S. (eds) *op. cit.*, Ch. 64, pp. 353–6.

Doerschler, J.S. (1987) *A Rule Based System for Dense-map Name Placement*, Technical Report SR-005, CAIP Centre, Rutgers University, New Brunswick, New Jersey.

Doerschler, J.S. and Freeman, H. (1989) An Expert System for Dense-map Name Placement, *Proceedings of the Autocarto9 ACSM/ASPRS Conference*, Falls Church (Virginia), Vol. 1, pp. 215–24.

Douven, W., Van Veldhuizen, H. and Scholten, H.J. (1993) The Development of Spatial Decision Support Systems for the Admission of Pesticides, *Proceedings of the EGIS '93 Conference*, Genoa, Italy (March 29–April 1), Vol. 1, pp. 597–605.

Douven, W. and Scholten, H.J. (1994) Towards a Spatial Decision Framework to Support Multi-Level Pesticide Regulation, *Proceedings of the EGIS/MARI '94 Conference*, Paris (March 29–April 1), Vol. 1, pp. 712–23.

Dubois, L. and Gold, C. (1994) Analyse Methodologique de Development d'Applications Interactives Pour les Systemes d'Aide a la Decision Forestiers, *Proceedings of the Canadian Conference on GIS '94*, Ottawa, Canada (June 6–10), Vol. 2, pp. 1406–12.

Dymond, J.R. and Luckman, P.G. (1993) Direct Induction of Compact Rule-Based Classifiers for Resource Mapping, *International Journal of Geographical Information Systems*, Vol. 8, No. 4, pp. 357–67.

Edwards-Jones, G. and Gough, M. (1994) ECOZONE: A Computerised Knowledge Management System for Sensitising Planners to the Environmental Impacts of Development Projects, *Project Appraisal*, Vol. 9, No. 1 (March), pp. 37–45.

Enache, M. (1994) Integrating GIS with DSS: A Research Agenda, *Proceedings of the Urban and Regional Information Systems Association (URISA) Conference*, Los Angeles (August 7–11), Vol. 1, pp. 154–66.

Engel, B.A., Srinivasan, R. and Rewerts, C. (1993) A Spatial Decision Support System for Modeling and Managing Agricultural Non-Point-Source Pollution, in Goodchild, M.F., Parks, B.O. and Steyaert, L.T. (eds) *op. cit.*, Ch. 20, pp. 230–7.

Epstein, A., Golovanov, S. and Litwin, U. (1993) Environmental Decision Support System "CLAIR" for Air and Surface Water Quality Control, Innovation Investment Strategy Optimization and Integrated Risk Assessment in Urban and Industrial Complex Developments, *Proceedings of the 16th Urban Data Management Symposium (UDMS) '93*, Vienna (September 6–10), pp. 567–79.

Estes, J.E., Sailer, C. and Tinney, L.R. (1986) Applications of Artificial Intelligence Techniques to Remote Sensing, *Professional Geographer*, Vol. 38, pp. 133–41.

Evans, T.A., Djokic, D. and Maidment, D.R. (1993) Development and Application of Expert Geographic Information System, *ASCE Journal of Computing and Civil Engineering*, Vol. 7, pp. 339–53.

Fang, J.H. and Schultz, A.W. (1986) XEOD: An Expert System for Interpreting Clastic Sedimentary Environments, *Abstracts of the Society of Economic Palaeontologists and Mineralogists*, Annual Midyear Meeting, Vol. 3, pp. 34–5.

Fedra, K., Winkelbauer, L. and Pantulu, V.R. (1991) *Expert Systems for Environmental Screening. An Application in the Lower Mekong Basin*, RR-91-19. International Institute for Applied Spatial Analysis. A-2361 Laxenburg, Austria.

Fedra, K. (1993) GIS and Environmental Modeling, in Goodchild, M.F., Parks, B.O. and Steyaert, L.T. (eds) *op. cit.*, Ch. 5, pp. 35–50.

Fedra, K. (1994) Integrated Environmental Information and Decision Support Systems, in Guariso, G. and Page, B. (eds) *op. cit.*, pp. 269–88.

Fedra, K. (1995) From Spatial Data to Spatial Information: GIS, Environmental Models, and Expert Systems, *Proceedings of the Joint European Conference and Exhibition on Geographic Information JEC-GI '95*, The Hague (March 26–31), Vol. 1, pp. 264–78.

Fedra, K. (1999) Urban Environmental Management: Monitoring, *Computers, Environment and Urban Systems*, Vol. 23, No. 6 (November), pp. 443–57.

Ferrier, G., Pan, P.S.Y., Wadge, G. and McDermott, C. (1993) An Integrated GIS and Expert System for Geological Interpretation of a Sedimentary Basin, *Proceedings of the EGIS '93 Conference*, Genoa, Italy, (March 29–April 1), Vol. 1, pp. 738–45.

Ferrier, G. and Wadge, G. (1997) An Integrated GIS and Knowledge-based System as an Aid for the Geological Analysis of Sedimentary Basins, *International Journal of Geographical Information Systems*, Vol. 11, No. 3 (April–May), pp. 291–7 .

Fisher, P.F. (1990a) Introduction, *Computers & Geosciences*, Vol. 16, No. 6, pp. 751–2.

Fisher, P.F. (1990b) A Primer of Geographic Search Using Artificial Intelligence, *Computers & Geosciences*, Vol. 16, No. 6, pp. 753–76.

Fisher, P.F. and Balachandran, C.S. (1990) STAX: A Turbo Prolog Rule-based System for Soil Taxonomy, *Computers & Geosciences*, Vol. 15, No. 3, pp. 295–324.

Fisher, P. and Pathirana, S. (1990) The Evaluation of Fuzzy Membership of Land Cover Classes in the Suburban Zone, *Remote Sensing Environment*, No. 34, pp. 121–32.

Fischer, M.M. (1994) From Conventional to Knowledge-Based Geographic Information Systems, *Computers, Environment and Urban Systems*, Vol. 18, No. 4, pp. 233–42.

Fischer, M.M. and Nijkamp, P. (1992) Geographical Information Systems and Spatial Modelling: Potentials and Bottlenecks, *Proceedings of the EGIS '92 Conference*, Munich (March 23–6), Vol. 1, pp. 214–25.

Fischer, M.M. and Nijkamp, P. (1993) Design and Use of GIS and Spatial Models, in Fischer, M.M. and Nijkamp, P. (eds) *op. cit.*, Ch. 1, pp. 1–13.

Fischer, M.M. and Nijkamp, P. (1993) (eds) *Geographical Information Systems, Spatial Modelling and Policy Evaluation*, Springer Verlag.

Flowerdew, R. (1989) The North West Regional Research Laboratory, *Mapping Awareness*, Vol. 3, No. 2 (May/June).

Foody, G.M. (1995) Land Cover Classification by an Artificial Neural Network With Ancillary Information, *International Journal of Geographical Information Systems*, Vol. 9, No. 5, pp. 527–42.

Franco, C., Tuffery, C., Venturini, E., Dusserre, G., Casal, J. and Morari, F. (1996) Integrated Information System for Industrial Risk Management (SIGRI) at Regional/Subregional Scale, in Rumor, M. *et al.* (eds) *op. cit.*, Vol. 1, pp. 583–91.

Freeman, H. and Ahn, J. (1984) AUTOMAP – An Expert System for Automatic Map Name Placement, *Proceedings of the 1st International Symposium on Spatial Data Handling*, International Geographical Union, Ohio, Vol. 1, pp. 544–71.

Frysinger, S.P., Copperman, D.A. and Levantino, J.P. (1996) Environmental Decision Support Systems (EDSS): An Open Architecture Integrating Modeling and GIS, in Goodchild, M.F., Steyaert, L.T., Parks, B.O., Johnston, C., Maidment, D., Crane, M. and Glendinning, S. (eds) *op. cit.*, Ch. 65, pp. 357–61.

Geraghty, P.J. (1992) Environmental Assessment and the Application of an Expert System Approach, *Town Planning Review*, Vol. 63, No. 2, pp. 123–141.

Geraghty, P.J., Brown, A.G.C. and Horton, F.F. (1996) The Use of Guidance Manuals for the Preparation of Environmental Impact Statements, in *Improving Environmental Assessment Effectiveness: Research, Practice and Training*, Proceedings of IAIA '96, 16th Annual Meeting (June 17–23), Centro Escolar Turistico e Hoteleiro, Estoril (Portugal), Vol. I, pp. 243–8.

Geraghty, P.J. (1999) A Comparative Study of Guidance Documents for EIA and their Potential for Supporting Practice, Paper presented at the conference *Forecasting the Future: Impact Assessment for a New Century*, IAIA '99, 19th Annual Meeting (June 15–19), The University of Strathclyde (John Anderson Campus), Glasgow.

Goldberg, M., Alvo, M. and Karam, G. (1984) The Analysis of Satellite Imagery Using an Expert System: Forestry Applications, *Proceedings of the Autocarto6 ACSM/ASPRS Conference*, Falls Church (Virginia), Vol. 1, pp. 493–553.

Goncalves, P., Antunes, P., Santos, R., Jordao, R., Alves, H. and Videira, N. (1995) Georeferenced Decision Support System for Environmental Impact Assessment: A Case Study, *Proceedings of the Joint European Conference and Exhibition on Geographic Information JEC-GI '95*, The Hague (March 26–31), Vol. 1, pp. 571–8.

Goodchild, M.F., Parks, B.O. and Steyaert, L.T. (1993) (eds) *Environmental Modelling with GIS*, Oxford University Press, Oxford.

Goodchild, M.F., Steyaert, L.T., Parks, B.O., Johnston, C., Maidment, D., Crane, M. and Glendinning, S. (1996) (eds) *GIS and Environmental Modeling: Progress and Research Issues*, GIS World Books.

Goodenough, D.G., Goldberg, H. and Zelek, J. (1987) An Expert System for Remote Sensing, *IEEE Transactions on Geoscience and Remote Sensing*, GE-25, pp. 349–59.

Goodenough, D.G. and Fung, K. (1991) Knowledge-Based Methods for Temporal and Spatial Data Fussion, *Proceedings of the Canadian Conference on GIS '91*, Ottawa, Canada (March 18–22), pp. 994–5.

Gouldstone Gronlund, A. and Xiang, W.-N. (1993) A Knowledge Based GIS Approach to Priority Area Identification for Forest Fire Management, *Proceedings of the Urban and Regional Information Systems Association (URISA) Conference*, Atlanta, Georgia (July 25–9), Vol. II, pp. 102–10.

Gouldstone Gronlund, A., Xiang, W.-N. and Sox, J. (1994) GIS Expert System Technologies Improve Forest Fire Management Techniques, *GIS World*, Vol. 7, No. 2, pp. 32–6.

Greathouse, D., Clements, J. and Mordis, K. (1989) The Use of Expert Systems to Assist in Decisions Concerning Environmental Control, *Critical Reviews in Environmental Control*, Vol. 19, No. 4, pp. 341–57.

Green, A., Higgs, G., Mathews, S. and Webster, C. (1989) The Wales and South West Regional Research Laboratory, *Mapping Awareness*, Vol. 3, No. 3 (July/August).

Grenney, W., Senti, T. and Bovee, K. (1994) Knowledge-Based System for Evaluating In-Stream Habitat, in Guariso, G. and Page, B. (eds) *op. cit.*, pp. 223–32.

Greven, E.G. (1995) Greaking the Data Barrier: From Aeronautical Charting to Total Information Management, *GIS Europe* (November), pp. 28–30.

Grothe, M. and Scholten, H. (1993) Modelling Catchment Areas: Towards the Development of Spatial Decision Support Systems for Facility Location Problems, in Fischer, M.M. and Nijkamp, P. (eds) *op. cit.*, Ch. 16, pp. 264–80.

Guariso, G. and Page, B. (eds) (1994) *Computer Support for Environmental Impact Assessment*, The IFTP TC5/WG5.11 Working Conference on Computer Support for Environmental Impact Assessment – CSEIA 93, Como, Italy (6–8 October 1993), IFIP Transactions B: Applications in Technology (B-16), North Holland: Elsevier Science B.V.

Haklay, M., Feitelson, E. and Doytsher, Y. (1998) The Potential of a GIS-based Scoping System: An Israeli Proposal and Case Study, *Environmental Impact Assessment Review*, Vol. 18, No. 5 (September), pp. 439–59.

Han, S.Y. and Kim, T.J. (1989) Intelligent Urban Information Systems: Review and Prospects, *Journal of the American Planners Association*, No. 3, pp. 296–308; also in Kim, T.J., Wiggins, L.L. and Wright, J.R. (eds) (1990) *Expert Systems: Applications to Urban Planning*, Springer Verlag.

Hanson, J.D. and Baker, B.B. (1993) Simulation of Rangeland Production: Future Applications in Systems Ecology, in Goodchild, M.F., Parks, B.O. and Steyaert, L.T. (eds) *op. cit.*, Ch. 30, pp. 305–13.

Hartnett, J., Williams, R. and Crowther, P. (1994) Per Pixel Reasoning Using a GIS Closely Coupled to an Expert System to Produce Surface Classifications Based on Remotely Sensed Data and Expert Knowledge, *Proceedings of the EGIS/MARI '94 Conference*, Paris (March 29–April 1), Vol. 1, pp. 677–83.

Hassani, M., Van der Laan, F. and Honig, M. (1996) Agricultural Fact-finding From Space, *GIS Europe* (January), pp. 19–21.

Healey, R., Burnhill, P. and Dowie, P. (1990) Regional Research Laboratory for Scotland, *Mapping Awareness*, Vol. 4, No. 2 (March).

Heatwole, C.D., Dillaha, T.A. and Mostaghinu, S. (1987) *Integrating Water Research Tools: Process Models, GIS and Expert Systems*, Paper of the American Society of Agricultural Engineers, No. 87–2043, p. 12.

Heikkila, E., Moore, J.E. and Kim, T.J. (1990) *Future Directions for EGIS: Applications to Land Use and Transportation Planning*, First International Conference on Expert Systems in Environmental Planning and Engineering, Lincoln Institute – Massachusetts Institute of Technology, Boston (September), also in Kim, T.J., Wiggins, L.L. and Wright, J.R. (eds) (1990) *Expert Systems: Applications to Urban Planning*, Springer Verlag.

Herbert, G. (1991) *The Use of Knowledge-Based Techniques for Map Design and Generalisation*, Working Report No. 25, South East Regional Research Laboratory, Department of Geography, Birkbeck College, London.

Herbert, G. and Joao, E.M. (1991) *Automating Map Design and Generalisation: A Review of Systems and Prospects for Future Progress in the 1990's*, Working Report No. 27, South East Regional Research Laboratory, Department of Geography, Birkbeck College, London.

Herbert, G., Joao, E. and Rhind, D. (1992) Use of an Artificial Intelligence Approach to Increase Use Control of Automatic Line Generalisation, *Proceedings of the EGIS '92 Conference*, Munich (March 23–6), Vol. 1, pp. 554–63.

Hershey, R.R. (1991a) *Towards an Effective Spatial Decision Support System for Landuse Planning*, Working Paper No. 4, London School of Economics Geographical Information Research Laboratory, Department of Geography, London School of Economics, London.

Hershey, R.R. (1991b) *An Spatial Decision Support System for Landuse Management and Planning: A Pilot Study for Thetford Forest District*, Working Paper No. 5, London School of Economics Geographical Information Research Laboratory, Department of Geography, London School of Economics, London.

Heywood, I., Oliver, J. and Tomlinson, S. (1994) Building an Exploratory Multicriteria Modelling Environment for Spatial Decision Support, *Proceedings of the EGIS/MARI '94 Conference*, Paris (March 29–April 1), Vol. 1, pp. 632–41.

Hickey, R. and Jankowski, P. (1997) GIS and Environmental Decision Making to Aid Smelter Reclamation Planning, *Environment and Planning A*, Vol. 29 (January), pp. 5–19.

Hirschfield, A., Barr, R., Batey, P. and Brown, P. (1989) The Urban Research and Policy Evaluation Regional Research Laboratory, *Mapping Awareness*, Vol. 3, No. 6 (December).

Holmberg, S.C. (1996) Problem Solving and Decision Support with GIS, in Rumor, M. *et al.* (eds) *op. cit.*, Vol. 1, pp. 566–9.

Holt, A. and Benwell, G.L. (1996) Case-Based Reasoning and Spatial Analysis, *URISA Journal* (September), pp. 27–36.

Hong, C.K. (1991) An Intelligent Geographic Information System, *Proceedings of the Canadian Conference on GIS '91*, Ottawa, Canada (March 18–22), pp. 796–808.

Huang, Y.L. (1989) MIN-CYANIDE: An Expert System for Cyanide Waste Minimization in Electroplating Plants, *Environmental Progress*, Vol. 10 (May), pp. 89–95.

Hughes, G. and Schirmer, D. (1994) Interactive Multimedia, Public Participation and Environmental Assessment, *Town Planning Review*, Vol. 65, No. 4, pp. 399–414.

Hushon, J.M. (1987) Expert Systems for Environmental Problems, *Environmental Science and Technology*, Vol. 21 (September), pp. 838–41.

Jain, D.K., Tim, U.S. and Jolly, R. (1995) Spatial Decision Support System for Planning Sustainable Livestock Production, *Computers, Environment and Urban Systems*, Vol. 19, No. 1, pp. 57–75.

Jankowski, P. and Nyerges, T. (1989) Design Considerations for MAPKBS – Map Projection Knowledge Based Systems, *The American Cartographer*, Vol. 16, pp. 85–95.

Jankowski, P., Nyerges, T.L., Smith, A., Moore, T.J. and Horvath, E. (1997) Spatial Group Choice: A SDSS Tool for Collaborative Spatial Decision Making,

International Journal of Geographical Information Science, Vol. 11, No. 6 (September), pp. 579–602.

Joao, E.M., Rhind, D., Openshaw, S. and Kelk, B. (1990) *Generalisation and GIS Databases*, Working Report No. 17, South East Regional Research Laboratory, Department of Geography, Birkbeck College, London.

Joao, E.M., Herbert, G. and Rhind, D. (1991) *The Measurement and Control of Generalisation Effects*, Working Report No. 19, South East Regional Research Laboratory, Department of Geography, Birkbeck College, London.

Jones, C. (1989) Cartographic Name Placement with PROLOG, *IEEE Computer Graphics and Applications*, Vol. 9, No. 5, pp. 36–47.

Jones, R.M., Copas, C.V. and Edmonds, E.A. (1997) GIS Support for Distributed Group-work in Regional Planing, *International Journal of Geographical Information Science*, Vol. 11, No. 1 (January–February), pp. 53–71.

Katz, S.S. (1988) Emulating the Prospector Expert System with a Raster GIS, in Thomas, H.G. (ed.) *GIS: Integrating Technology and Geoscience Applications*, National Academy of Sciences, Washington DC.

Keller, C.P. and Strapp, J.D. (1996) Multicriteria Decision Support for Land Reform Using GIS and API, in Goodchild, M.F., Steyaert, L.T., Parks, B.O., Johnston, C., Maidment, D., Crane, M. and Glendinning, S. (eds) *op. cit.*, Ch. 66, pp. 363–6.

Kessell (1996) The Integration of Empirical Modeling, Dynamic Process Modeling, Visualization, and GIS for Bushfire Decision Support in Australia, in Goodchild, M.F., Steyaert, L.T., Parks, B.O., Johnston, C., Maidment, D., Crane, M. and Glendinning, S. (eds) *op. cit.*, Ch. 67, pp. 367–71.

Kobayashi, Y., Sato, S., Arima, T., Hagishima, S., Oshima, K., Hitaka, K., Kim, N.-G., Lee, D.-b., Park, T.-c., Kim, Y.-h., Kim, I.-c. and Yudono, A. (1997) Industrial Location Plan Expert System – Expert System for Industrial Development Planning with Environmental Management, *Proceedings of the 5th International Conference on Computers in Urban Planning and Urban Management*, Bombay (India), Vol. 1, pp. 134–45.

Kolejka, J. and Pokorny, J. (1994) GEORISK – Local GIS for Hazardous Production/Storage Sites, *Proceedings of the Canadian Conference on GIS '94*, Ottawa, Canada (June 6–10), Vol. 1, pp. 395–407.

Kondratiev, K.Y., Bobylev, L.P., Donchenko, V.K., Rastoskuev, V.V. and Shalina, E.V. (1996) Finding a Solution for Pollution, *GIS Europe* (August), pp. 20–2.

Kontoes, C., Wilkinson, G.G., Burrill, A., Goffredo, S. and Megier, J. (1993) An Experimental System for the Integration of GIS Data in Knowledge-Based Image Analysis for Remote Sensing of Agriculture, *International Journal of Geographical Information Systems*, Vol. 7, No. 3, pp. 247–62.

Krystinik, K.B. (1985) An Example Expert System for the Interpretation of Depositional Environments, in *Open File Report of the US Geological Survey*, USGS, Reston (Virginia).

Kusse, B. and Wentholt, A. (1992) SENSE: An Active Approach to Environmental Decision Support, *Proceedings of the EGIS '92 Conference*, Munich (March 23–6), Vol. 2, pp. 1168–76.

Laaribi, A., Chevalier, J.-J. and Martel, J.-M. (1993) Integration des SIG et de l'Analyse Multicritere Pour l'Aide a la Decision a Reference Spatiale, *Proceedings of the Canadian Conference on GIS '93*, Ottawa, Canada (March 23–5), pp. 968–74.

Lam, D.C.L. (1993) Combining Ecological Modeling, GIS and Expert Systems: A Case Study of Regional Fish Species Richness Model, in Goodchild, M.F., Parks, B.O. and Steyaert, L.T. (eds) *op. cit.*, Ch. 24, pp. 270–5.

Lam, D. and Pupp, C. (1996) Integration of GIS, Expert Systems and Modeling for State-of-Environment Reporting, in Goodchild, M.F., Steyaert, L.T., Parks, B.O., Johnston, C., Maidment, D., Crane, M. and Glendinning, S. (eds) *op. cit.*, Ch. 75, pp. 419–22.

Lam, D.C.L. and Swayne, D.A. (1991) *Integrating Database, Spreadsheet, Graphics, GIS, Statistics, Simulation Models and Expert Systems: Experiences with the Raison System on Microcomputers*, NATO ASI Series, Vol. G26, Heidelberg, Springer.

Laurini, R. and Milleret-Raffort, F. (1990) Towards Intelligent GIS, in Polydorides, N. (ed.) *Computers in Planning* 6, Patras (June), pp. 127–56.

Lee, J.K., Park, R.A., Mausel, P.W. and Howe, R.C. (1991) GIS-Related Modeling of Impacts of Sea Level Rise on Coastal Areas, *Proceedings of the GIS/LIS '91 Conference*, Atlanta, Georgia (October 28–November 1), Vol. 1, pp. 356–67.

Lein, J. (1989) An Expert System Approach to Environmental Impact Assessment, *International Journal of Environmental Studies*, Vol. 33, pp. 13–27.

Lein, J. (1990) Explaining a Knowledge-Based Procedure for Developing Suitability Analysis, *Applied Geography*, Vol. 10, pp. 171–86.

Leung, Y. and Leung, K.S. (1993a) An Intelligent Expert System Shell for Knowledge-Based Geographical Information Systems: 1. The Tools, *International Journal of Geographical Information Systems*, Vol. 7, No. 3, pp. 189–99.

Leung, Y. and Leung, K.S. (1993b) An Intelligent Expert System Shell for Knowledge-Based Geographical Information Systems: 2. Some Applications, *International Journal of Geographical Information Systems*, Vol. 7, No. 3, pp. 201–13.

Leung, Y. (1993) Towards the Development of an Intelligent Spatial Decision Support System, in Fischer, M.M. and Nijkamp, P. (eds) *op. cit.*, Ch. 9, pp. 131–45.

Liang, B. (1988) An Expert System for Environmental Analysis of Sedimentation, *Geological Science and Technology Information*, Vol. 7, No. 3, pp. 125–30.

Linsey, T.K. (1994) A Geographer's Knowledge Base, *Proceedings of the EGIS/MARI '94 Conference*, Paris (March 29–April 1), Vol. 1, pp. 626–31.

Liu, B., Shanholtz, V.O., Fox, E. and Desai, C. (1993) Development of Landscape Resources Management Decision Support System, *Proceedings of the GIS/LIS '93 Conference*, Minneapolis, Minnesota (November 2–4), Vol. 1, p. 427 (Abstract).

Logan, T.L., Ritter, N.D., Parks, G.S., Nichols, S.D. and Bryant, N.A. (1988) Real-time Georeferencing with Stereo Imagery: Ruled Surface and Neural Network Approaches, *Proceedings of the GIS/LIS '88 Conference*, San Antonio, Texas (November 30–December 2), Vol. 1, pp. 122–31.

Lundgard, L.E., Tangen, G. and Skyberg, B. (1992) Acoustic Diagnoses of GIS, Field Experience and Development of Expert System, IEEE *Transactions on Power Delivery*, Vol. 7 (January), pp. 287–93.

MacDonald, M.L. and Faber, B.G. (1999) Exploring the Potential of Multi-criteria Spatial Decision Support Systems: A System for Sustainable Land-use Planning and Design, in Thill, J.-C. (ed.) *op. cit.*, Ch. 15, pp. 353–77.

Mackaness, W.A. and Fisher, P.F. (1987) Automatic Recognition and Resolution of Spatial Conflicts in Cartographic Symbolization, *Proceedings of the Autocarto8 ASPRS Conference*, Falls Church (Virginia), Vol. 1, pp. 709–18.

Mackaness, W. and Beard, K. (1990) Development of an Interface for User Interaction in Rule-Based Map Generalisation, *Proceedings of the GIS/LIS '90 Conference*, Anaheim, California (November 7–10), Vol. 1, pp. 106–16.

Maiellaro, N. and Barbanente, A. (1993) Use of Knowledge Management Techniques for an Urban Information System, *Proceedings of the EGIS '93 Conference*, Genoa, Italy (March 29–April 1), Vol. 2, pp. 891–900.

Maruchi, B.E., Sauchyn, D. and Kite, G. (1994) A Solution to the Mixed Pixel Problem, *Proceedings of the Canadian Conference on GIS '94*, Ottawa, Canada (June 6–10), Vol. 1, pp. 812–21.

Mckeown, D.M. (1986) *The Role of Artificial Intelligence in the Integration of Remotely Sensed Data with GIS*, Report CMU-CS-86-174, Department of Computer Science, Carnegie-Mellon University, Pittsburgh, Pennsylvania.

Maguire, D.J., Strachan, A.J. and Unwin, D.J. (1989) The Midlands Regional Research Laboratory, *Mapping Awareness*, Vol. 3, No. 1 (March–April).

Maguire, D.J., Goodchild, M.F. and Rhind, D.W. (1991) (eds) *Geographical Information Systems: Principles and Applications*, Longman, London (2 Vols).

Maidment, D.R. and Djokic, D. (1991) *Creating an Expert GIS: The Arc-Nexpert Interface*, paper given at "Planning Transatlantic", Joint ACSP/AESOP International Congress, School of Planning, Oxford Polytechnic (July 8–12).

Maidment, D.R. (1993) GIS and Hydrologic Modelling, in Goodchild, M.F., Parks, B.O. and Steyaert, L.T. (eds) *op. cit.*, Ch. 14, pp. 147–67.

Martin, D. (1988) *An Approach to Surface Generation from Centroid-Type Data*, Technical Reports in Geo-Information Systems, Computing and Cartography No. 5, Wales and South West Regional Research Laboratory, Department of Town Planning, University of Wales College of Cardiff, Cardiff.

Martin, D. (1990) *A Suite of Programs for Socioeconomic Surface Generation*, Technical Reports in Geo-Information Systems, Computing and Cartography No. 28, Wales and South West Regional Research Laboratory, Department of Town Planning, University of Wales College of Cardiff, Cardiff.

Martin, N., St-Onge, B. and Waaub, J.-P. (1999) Geographic Tools for Decision Making in Watershed Management, in Thill, J.-C. (ed.) *op. cit.*, Ch. 13, pp. 309–34.

Masser, I. (1990) *The Regional Research Laboratory Initiative: An Overview*, ESRC: Regional Research Laboratory Initiative, Discussion Paper No. 1.

McClean, C.J., Watson, P.M., Wadsworth, R.A., Blaiklock, J. and O'Callaghan, J.R. (1995) Land Use Planning: A Decision Support System, *Journal of Environmental Planning and Management*, Vol. 38, No. 1, pp. 77–92.

Medyckyj-Scott, D., Newman, I., Ruggles, C. and Walker, D. (1991) (eds) *Metadata in the Geosciences*, Group D Publications Ltd, Loughborough.

Medyckyj-Scott, D. (1991) User-oriented Enquiry Facilities, in Medyckyj-Scott, D., Newman, I., Ruggles, C. and Walker, D. (eds) *Metadata in the Geosciences*, Group D Publications Ltd, Loughborough.

Miller, B.M. (1990, 1991) Applications of EXPERT Systems and Geographic Information Systems to Basin Characterisation, in Carter, L.H.M. (ed.) *United States Geological Survey Research on Energy Resources*, Program of Abstracts, USGS Circular, pp. 53–5.

Miller, B. (1991) An Object-oriented Expert System for Sedimentary Basin Analysis, abstracts in *AAPG Bulletin*, Vol. 72, No. 2, pp. 223.

Miller, B.M. (1994) The Role of Geographical Information Systems and Expert Systems Technology in Natural Resource Management, *Proceedings of the Canadian Conference on GIS '94*, Ottawa, Canada (June 6–10), Vol. 1, pp. 723–32.

Miller, D.R. (1996) Knowledge-Based Systems for Coupling GIS and Process-Based Ecological Models, in Goodchild, M.F., Steyaert, L.T., Parks, B.O., Johnston, C., Maidment, D., Crane, M. and Glendinning, S. (eds) *op. cit.*, Ch. 43, pp. 231–4.

Miller, D. and Morrice, J. (1991) An Expert System and GIS Approach to Predicting Changes in the Upland Vegetation of Scotland, *Proceedings of the GIS/LIS '91 Conference*, Atlanta, Georgia (October 28–November 1), Vol. 1, pp. 11–20.

Miller, D. and Morrice, J. (1993) Coupling of Process-based Vegetation Models to GIS and Knowledge Based Systems for Analysis of Vegetation Change, *Proceedings of the EGIS '93 Conference*, Genoa, Italy (March 29–April 1), Vol. 1, pp. 334–43.

Miyamoto, K., Udomsri, R., Sathyaprasad, S. and Ren, F. (1995) A Decision Support System for Integrated Planning and Implementation Regarding Land-Use, Transport and the Environment in Developing Metropolises, *Proceedings of the 4th International Conference on Computers in Urban Planning and Urban Management*, Melbourne, Australia (July 11–14), Vol. 2, pp. 55–68.

Morse, B.W. (1987) Expert Interface to GIS, *Proceedings of the Autocarto7 ASPRS Conference*, Falls Church (Virginia), Vol. 1, pp. 535–41.

Mower, J.E. (1988) A Neural Network Approach to Feature Recognition along Cartographic Lines, *Proceedings of the GIS/LIS '88 Conference*, San Antonio, Texas (November 30–December 2), Vol. 1, pp. 250–5.

Muller, J.C. and Mouwes, P.J. (1990) Knowledge Acquisition and Representation for Rule Based Map Generalization: An Example from the Netherlands, *Proceedings of the GIS/LIS '90 Conference*, Anaheim, California (November 7–10), Vol. 1, pp. 58–67.

Navinchandra, D. (1989) *Observations on the Role of A.I. Techniques in Geographical Information Processing*, First International Conference on Expert Systems in Environmental Planning and Engineering, Lincoln Institute – Masachussetts Institute of Technology, Boston (September).

Negahban, B., Fonyo, C., Campbell, K.L., Jones, J.W., Boggess, W.G., Kiker, G., Hamouda, E., Flaig, E. and Lal, H. (1996) LOADSS: A GIS-Based Decision Support System for Regional Environmental Planning, in Goodchild, M.F., Steyaert, L.T., Parks, B.O., Johnston, C., Maidment, D., Crane, M. and Glendinning, S. (eds) *op. cit.*, Ch. 54, pp. 287–91.

Newkirk, R.T. and Wang, F. (1989) Integrating Remote Sensing and Geographical Information System Knowledge in an Expert System for Change Detection, *Proceedings of the 1st International Conference on Computers in Urban Planning and Urban Management*, Hong Kong (August 22–25), pp. 245–54.

Newkirk, R.T. (1991) Expert GIS for Spatial Decision Support, *Proceedings of the 2nd International Conference on Computers in Urban Planning and Urban Management*, Oxford Polytechnic (July 6–8), Vol. 1, pp. 91–105.

Newton, P.W., Taylor, M.A.P. and Sharpe, R. (1988) Introduction to Desktop Planning, in Newton, P.W., Taylor, M.A.P. and Sharpe, R. (eds) *op. cit.*, pp. 1–14.

Newton, P.W., Taylor, M.A.P. and Sharpe, R. (eds) (1988) *Desktop Planning*, Hargreen Publishing Company, Melbourne (Australia).

Nijkamp, P. and Scholten, H.J. (1991) Spatial Information Systems: Design, Modelling, and Use in Planning, in Polydorides, N.D. (ed.) *Computers in Planning 8*, Proceedings of the 3rd URSA-NET Seminar-Forum, Patras, Greece (June 7–9), pp. 13–24.

Nijkamp, P. and Scholten, H.J. (1993) Spatial Information Systems: Design, Modelling, and Use in Planning, *International Journal of Geographical Information Systems*, Vol. 7, No. 1, pp. 85–96.

Offermann, W. (1993) Cartographic Generalization with Amplified Intelligence, *Proceedings of the EGIS '93 Conference*, Genoa, Italy (March 29–April 1), Vol. 2, pp. 1019–24.

O'Neill, W.A. and Grenney, W.J. (1991) Prototype Knowledge-Based Model for Matching Non-Graphic Road Inventory Files With Existing Digital Cartographic Databases, *Proceedings of the Urban and Regional Information Systems Association (URISA) Conference*, San Francisco (August 11–15), Vol. 1I, pp. 159–68.

Openshaw, S. (1990) *A Spatial Analysis Research Strategy for the Regional Research Laboratory Initiative*, ESRC: Regional Research Laboratory Initiative, Discussion Paper No. 3.

Openshaw, S. (1993a) Some Suggestions Concerning the Development of Artificial Intelligence Tools for Spatial Modelling and Analysis in GIS, in Fischer, M.M. and Nijkamp, P. (eds) *op. cit.*, Ch. 2, pp. 17–33.

Openshaw, S. (1993b) Modelling Spatial Interaction Using a Neural Net, in Fischer, M.M. and Nijkamp, P. (eds) *op. cit.*, pp. 147–64.

Openshaw, S., Goddard, J. and Coombes, M. (1987) *Integrating Geographic Data for Policy Purposes: Some Recent UK Experience*, Research Report 87/1, North East Regional Research Laboratory, Department of Geography, University of Newcastle upon Tyne, Newcastle upon Tyne.

Openshaw, S., Gillard, A. and Charlton, M. (1989) The North East Regional Research Laboratory, *Mapping Awareness*, Vol. 3, No. 4 (September/October).

Openshaw, S. and Wymer, C. (1990) A Neural Net Classifier System for Handling Census Data, *Proceedings of the Workshop on Neural Networks for Statistical and Economic Data*, Dublin (December).

Pariente, D. (1994) Geographic Interpolation and Extrapolation by Means of Neural Networks, *Proceedings of the EGIS/MARI '94 Conference*, Paris (March 29–April 1), Vol. 1, pp. 684–93.

Powitz, B.M. and Meyer, U. (1989) Generalization of Settlements by Pattern Recognition Methods, *Proceedings of the ICA Conference* (Budapest), Vol. 1, p. 7.

Pearson, E.J., Wadge, G. and Wislocki, A.P. (1991) Mapping Natural Hazards With Spatial Modelling Systems, *Proceedings of the EGIS '91 Conference*, Brussels (April 2–5), Vol. 1I, pp. 847–55.

Pearson, E.J., Wadge, G. and Wislocki, A.P. (1992) An Integrated Expert System/GIS Approach to Modelling and Mapping Natural Hazards, *Proceedings of the EGIS '92 Conference*, Munich (March 23–6), Vol. 1, pp. 762–71.

Peckham, R.J. (1993) Application of a GIS in Decision Support for Industrial Waste Management, *Proceedings of the EGIS '93 Conference*, Genoa, Italy (March 29–April 1), Vol. 2, pp. 952–61.

Peckham, R. (1997) Geographical Information Systems and Decision Support for Environmental Management, in Timmermans, H. (ed.) *Decision Support Systems in Urban Planning*, E & FN Spon, London, Ch. 5, pp. 75–86.

Plummer, S.E. (1990) *The ESRC Regional Research Laboratory: A Technical Profile*, ESRC: Regional Research Laboratory Initiative, Discussion Paper No. 2.

Qureshi, M.E. and Harrison, S.R. (2001) A Decision Support Process to Compare Riparian Revegetation Options in Scheu Creek Catchment in North Queensland, *Journal of Environmental Management*, Vol. 62, No. 1 (May), pp. 101–12.

Radwan, M.M. and Bishr, Y.A. (1994) Integrating the Object Oriented Data Modelling and Knowledge System for the selection of the Best Management Practice in Watersheds, *Proceedings of The Canadian Conference on GIS*, Ottawa (June), Vol. 1, pp. 690–9.

Recatala, L., Ive, J.R., Baird, I.A., Hamilton, N. and Sanchez, I. (2000) Land-use Planning in the Valencian Mediterranean Region: Using LUPIS to Generate Issue Relevant Plans, *Journal of Environmental Management*, Vol. 59, No. 3 (July), pp. 169–84.

Reisinger, T.W., Kenney, D.P. and Goode, K.B. (1990) A Spatial Decision Support System for Opportunity Area Analysis on the Jefferson National Forest, *Proceedings of the GIS/LIS '90 Conference*, Anaheim, California (November 7–10), Vol. 2, pp. 733–40.

Rhind, D. (1990) *A GIS Research Agenda*, Working Report No. 6, South East Regional Research Laboratory, Department of Geography, Birkbeck College, London.

Rhind, D. and Shepherd, J. (1989) The South East Regional Research Laboratory, *Mapping Awareness*, Vol. 2, No. 6 (January–February).

Richardson, D.E. (1989) Rule Based Generalization for Base Map Production, *Proceedings of the National Canadian Conference on GIS "Challenge for the 1990s"*, Ottawa, Canada (February 27–March 3), pp. 718–39.

Richer, O. and Chevallier, J.-J. (1992) SIRS Pour la Recolte Forestiere: Outil d'Inventoire ou Moyen de Planification?, *Proceedings of the Canadian Conference on GIS '92*, Ottawa, Canada (March 24–6), pp. 291–302.

Roberts, S.I. and Ricketts, P.J. (1990) Developing Integrated Information Systems for Marine and Coastal Environments – Lessons for the FMG Project, *Proceedings of the "GIS for the 1990s" National Canadian Conference*, Ottawa, Canada (March 5–8), pp. 157–66.

Robillard, P.D. (1990) Linking GIS to Expert Systems for Water Resources Management, *Proceedings of the Applications of GIS, Simulation Models and Knowledge-based Systems for Landuse Management Conference* (Blacksburg, Virginia), pp. 1–10.

Robson, P.G. and Adlam, K.A.McL. (1991) A Menu-aided Retrieval System (MARS) for Use With a Relational Database Management System, in Medyckyj-Scott, D., Newman, I., Ruggles, C. and Walker, D. (eds) *Metadata in the Geosciences*, Group D Publications Ltd, Loughborough.

Rodriguez-Bachiller, A. (1991) Expert Systems in Planning: An Overview, *Planning Practice and Research*, Vol. 6, No. 3 (Winter), pp. 20–5.

Rodriguez-Bachiller, A. (1998) *GIS and Decision-Support : A Bibliography*, Working Paper No. 176, School of Planning, Oxford Brookes University.

Rodriguez-Bachiller, A. (2000) Geographical Information Systems and Expert Systems for Impact Assessment. Part II: Expert Systems and Decision Support Systems, *Journal of Environmental Assessment Policy and Management*, Vol. 2, No. 3 (September), pp. 415–48.

Romao, T., Molendijk, M. and Scholten, H. (1996) COASTMAP: A Collaborative Multimedia Environment for Coastal Zone Managers, in Rumor, M. *et al.* (eds) *op. cit.*, Vol. 1, pp. 137–46.

Rumor, M., McMillan, R. and Ottens, H.F.L. (1996) (eds) Geographical Information From Research to Application through Cooperation, *Proceedings of the Second Joint European Conference and Exhibition on Geographical Information*, Barcelona, Palacio de Congresos (March 27–9), 2 Vols.

Rushton, S.P., Cherrill, A.J., Tucker, K. and O'Callaghan, J.R. (1995) The Ecological Modelling System of NELUP, *Journal of Environmental Planning and Management*, Vol. 38, No. 1, pp. 35–52.

Ryan, T.C. (1992) Spatial Decision Support Systems, *Proceedings of the Urban and Regional Information Systems Association (URISA) Conference*, Washington, DC (July 12–16), Vol. 1II, pp. 49–59.

Rybaczuk, K. (1993) Using Information Based Rules for Sliver Polygon Removal in GIS, in Fischer, M.M. and Nijkamp, P. (eds) *op. cit.*

Salt, C.A. and Culligan Dunsmore, M. (2000) Development of a Spatial Decision Support System for Post-emergency Management of Radioactively Contaminated Land, *Journal of Environmental Management*, Vol. 58, No. 3 (March), pp. 169–78.

Schibuola, S. and Byer, P.H. (1991) Use of Knowledge-based Systems for the Review of Environmental Impact Statements, *Environmental Impact Assessment Review*, Vol. 11, pp. 11–27.

Schultz, A.W., Fang, J.H. and Chen, H.C. (1988) XEOD: An Expert System for Determining Depositional Environments, *Geobyte*, Vol. 3, No. 2, pp. 22–32.

Seder, I., Weinkauf, R. and Neumann, T. (2000) Knowledge-based Databases and Intelligent Decision Support for Environmental Management in Urban Systems, *Computers, Environment and Urban Systems*, Vol. 24, No. 3 (May), pp. 233–50.

Shaefer, P. (1992) Improved Neural Network Techniques for GIS Classification Applications, *Proceedings of the Canadian Conference on GIS '92*, Ottawa, Canada (March 24–6), pp. 1037–46.

Sharifi, M.A. (1992) Design of a Decision Support System for Land Use Planning, *Proceedings of the EGIS '92 Conference*, Munich (March 23–6), Vol. 1, pp. 118–26.

Shvebs, H.I., Svetlitchnyi, A.A. and Plotnitskiy, S.V. (1994) Elaboration of Decision Support Systems for Optimization of Land Resources Using, *Proceedings of the 15th Urban Data Management Symposium (UDMS)*, Lyon (November 16–20), Vol. 2, pp. 1876–83.

Sieg, G.E. and McCollum, M.P. (1988) Integrating Geographical Information Systems and Decision Support Systems, *Proceedings of the GIS/LIS '88 Conference*, San Antonio, Texas (November 30–December 2), Vol. 2, pp. 901–10.

Siekierska, E.M. (1989) Advantages of Expert Systems with Examples of the Use of "VP.EXPERT" System for Cartographic Applications, *Proceedings of the National Canadian Conference on GIS "Challenge for the 1990s"*, Ottawa, Canada (February 27–March 3), pp. 710–17.

Skidmore, A.K., Ryan, P.J., Dawes, W., Short, D. and O'Loughlin, E. (1991) Use of an Expert System to Map Forest Soils from a Geographical Information System, *International Journal of Geographical Information Systems*, Vol. 5, No. 4, pp. 431–45.

Smith, T.R., Peuquet, D.J., Menon, S. and Agarwal, P. (1987) KBGIS-II: A Knowledge Based GIS, *International Journal of GIS*, Vol. 1, No. 2, pp. 149–72.

Smith, T.R. and Yiang, J.E. (1991) Knowledge-based Approaches in GIS, in Maguire, D.J., Goodchild, M.F. and Rhind, D.W. (eds) *Geographical Information Systems: Principles and Applications*, Longman (Ch. 27).

Spooner, C.S. (1985) The Emerging Uses of Expert Systems at the US Environmental Protection Agency, *Expert Systems in Government Symposium*, pp. 73–7.

Srinivasan, A. and Richards, J.A. (1993) Analysis of GIS Spatial Data Using Knowledge-Based Methods, *International Journal of Geographical Information Systems*, Vol. 7, No. 6, pp. 479–500.

Stringer, P. and Bond, D. (1990) The Northern Ireland Regional Research Laboratory, *Mapping Awareness*, Vol. 4, No. 1 (January/February).

Taylor, K., Walker, G. and Abel, D. (1999) A Framework for Model Integration in Spatial Decision Support Systems. *International Journal of Geographical Information Science*, Vol. 13, No. 6, pp. 533–55.

Thill, J.-C. (1999) (ed.) *Spatial Multicriteria Decision Making and Analysis*, Ashgate, Aldershot, Brookfield USA.

Tzafestas, S.G. and Hatzivasiliou, F.V. (1990) Human-Computer Interface: Artificial Intelligence and Software Psychological Issues, *Proceedings of the EGIS '90 Conference*, Amsterdam (April 10–13), Vol. 2, pp. 1060–9.

Usery, E., Barr, D.J. and Deister, R.R. (1988) A Geological Engineering Application of a Knowledge-based GIS, Technical Papers, *American Congress for Surveying and Mapping Convention*, Vol. 2, pp. 176–85.

Usery, E., Altheide, P., Deister, R.R. and Barr, D.J. (1989) Knowledge-based GIS Techniques Applied to Geological Engineering, in Ripple, W.J. (ed.) *Fundamentals of GIS: A Compendium*, American Congress for Surveying and Mapping Convention, Bethseda, USA.

Van Cleynenbreugel, J., Fierens, F., Suetens, P. and Oosterlinck, A. (1991) A Strategy to Incorporate GIS Knowledge During Road Extraction from Satellite Imagery, *Proceedings of the EGIS '91 Conference*, Brussels (April 2–5), Vol. 1, pp. 197–206.

Van der Laan, F. (1994) Policing Europe's Common Agricultural Policy: GIS and Remote Sensing Take Up the Challenge, *GIS Europe* (July), pp. 32–5.

Van Tiel, R.M.E., Schalkwijk, J.H. and Henkes, E.J. (1991) BOBIS, An Environmental Decision Support System, *Proceedings of the EGIS '91 Conference*, Brussels (April 2–5), Vol. 1I, pp. 1078–84.

Van Voris, P., Millard, W.D. and Thomas, J. (1993) TERRA-Vision – The Integration of Scientific Analysis into the Decision-making Process, *International Journal of Geographical Information Systems*, Vol. 7, No. 2, pp. 143–64.

Vessel, D. (1993) Design and Implementation Issues of Cartographic Expert Systems in GIS Applications, *Proceedings of the GIS/LIS '93 Conference*, Minneapolis, Minnesota (November 2–4), Vol. 2, pp. 711–17.

Vogel, R. (1989) An Integrated GIS/Expert System, *Computer Oriented Geological Society*, Vol. 5, No. 3, p. 119.

Wadsworth, R.A. and O'Callahan, J.R. (1995) Empirical Searches of the NELUP Land Use Database, *Journal of Environmental Planning and Management*, Vol. 38, No. 1, pp. 107–16.

Wang, F. and Newkirk, R. (1987a) A Knowledge-Based System for Highway Network Extraction, *Proceedings of IGARSS '87*, pp. 384–7.

Wang, F. and Newkirk, R. (1987b) An Expert Systems for Remote Sensing Land Use Change Detection, *Journal of Imaging Technology*, Vol. 13, No. 3, pp. 116–22.

Wang, F. (1994) The Use of Artificial Neural Networks in a Geographical Information System for Agricultural Land-suitability Assessment, *Environment and Planning A*, Vol. 26, pp. 265–84.

Wang, X. (1997) Conceptual Design of a System for Selecting Appropriate Groundwater Models in Groundwater Protection Programs, *Environmental Management*, Vol. 21, No. 4 (July/August), pp. 607–15.

Warwick, C.J., Mumford, J.D. and Norton, G.A. (1993) Environmental Management Expert Systems, *Journal of Environmental Management*, Vol. 39, pp. 251–70.

Waters, N.M. (1989) Expert Systems with GIS: Knowledge-Acquisition for Spatial Decision Support Systems, *Proceedings of the National Canadian Conference on GIS "Challenge for the 1990s"*, Ottawa, Canada (February 27–March 3), pp. 740–59.

Watson, P.M. and Wadsworth, R.A. (1996) A Computerised Decision Support System for Rural Policy Formulation, *International Journal of Geographical Information Systems*, Vol. 10, No. 4, pp. 425–40.

Webster, C. (1989a) *An Introduction to the Use of Logic Programming in Spatial Analysis*, Technical Reports in Geo-Information Systems, Computing and Cartography No. 12, Wales and South West Regional Research Laboratory, Department of Town Planning, University of Wales College of Cardiff, Cardiff.

Webster, C. (1989b) *A Logic-Programming Formulation of the Point-in-Polygon Problem*, Technical Reports in Geo-Information Systems, Computing and Cartography No. 17, Wales and South West Regional Research Laboratory, Department of Town Planning, University of Wales College of Cardiff, Cardiff.

Webster, C. (1990a) *Approaches to Interfacing GIS and Expert System Technology*, Technical Reports in Geo-Information Systems, Computing and Cartography No. 22, Wales and South West Regional Research Laboratory, Department of Town Planning, University of Wales College of Cardiff, Cardiff.

Webster, C. (1990b) *Object-Oriented Programming, Database and GIS*, Technical Reports in Geo-Information Systems, Computing and Cartography No. 23, Wales and South West Regional Research Laboratory, Department of Town Planning, University of Wales College of Cardiff, Cardiff.

Webster, C. (1991) *Expressing Regional Planning Propositions as Complex Database Objects*, Paper Given at the Meeting of the British Section of the Regional Science Association, at Mansfield College, Oxford (September).

Webster, C., Oltof, W. and Berger, M. (1992) Pattern Recognition Techniques for Smart Urban GIS, *Proceedings of the EGIS '92 Conference*, Munich, pp. 930–9.

Webster, C. and Omare, C.N. (1990) *An Object-Oriented Data Model for Regional Planning*, Technical Reports in Geo-Information Systems, Computing and Cartography No. 27, Wales and South West Regional Research Laboratory, Department of Town Planning, University of Wales College of Cardiff, Cardiff.

Webster, C. and Omare, C.N. (1991) A Semantic Data Model for Regional Planning, *Proceedings of the Second International Conference on Computers in Urban Planning and Urban Management*, Oxford, pp. 213–31.

Webster, C., Oltof, W. and Berger, M. (1991) *Using Co-occurence Texture Statistics in the Interpretation of a High Resolution Satellite Image of Harare*, Technical Reports in Geo-Information Systems, Computing and Cartography No. 36, Wales and South West Regional Research Laboratory, Department of Town Planning, University of Wales College of Cardiff, Cardiff.

Webster, C., Ho, C.S. and Wislocki, T. (1989) *Text Animation or Knowledge Elicitation?: Two Approaches to Expert System Design*, Technical Reports in Geo-Information Systems, Computing and Cartography No. 18, Wales and South West Regional Research Laboratory, Department of Town Planning, University of Wales College of Cardiff, Cardiff.

Werschlein, T. and Weibel, R. (1994) Use of Neural Networks in Line Generalization, *Proceedings of the EGIS/MARI '94 Conference*, Paris (March 29–April 1), Vol. 1, pp. 76–85.

Westmacott, S. (2001) Developing Decision Support Systems for Integrated Coastal Management in the Tropics: Is the ICM Decision-making Environment Too

Complex for the Development of a Useable and DSS?, *Journal of Environmental Management*, Vol. 62, No. 1 (May), pp. 55–74.

Wharton, S.W. (1987) A Spectral Knowledge-based Approach for Urban Land-cover Discrimination, *IEEE Transactions on Geoscience and Remote Sensing*, Vol. 25, p. 272.

Wharton, S.W. (1989) Knowledge Based Spectral Classification of Remotely Sensed Image Data, in Asrar, G. (ed.) *Theory and Application of Optical Remote Sensing*, Wiley, New York, pp. 548–77.

Xiang, W.-N. (1997) Knowledge-based Decision Support by CRITIC, *Environment and Planning B*, Vol. 24 (January), pp. 69–79.

Yang, Z. and Sharpe, D.M. (1991) Design of Buffer Zones for Conservation Areas and a Prototype Spatial Decision Support System (SDSS), *Proceedings of the GIS/LIS '91 Conference*, Atlanta, Georgia (October 28–November 1), Vol. 1, pp. 60–70.

Yazdani, R. (1993) The Expert System for an Effective Decision Making in Resource Management, *Proceedings of the Canadian Conference on GIS '93*, Ottawa, Canada (March 23–5), pp. 179–88.

Ye, C., Feyen, J. and Lode, H. (1992) Development of a Decision Support System for Irrigation Scheduling Management, *Proceedings of the EGIS '92 Conference*, Munich (March 23–6), Vol. 2, pp. 1022–31.

Ying, L., Levine, S.P. and Tomellini, S.A. (1987) Self-training, Self-optimizing Expert System for Interpretation of the Infrared Spectra of Environmental Mixtures, *Analytical Chemistry*, Vol. 59, pp. 2197–202.

Zeng, T.Q. and Zhou, Q. (2001) Optimal Spatial Decision Making Using GIS: A Prototype of a Real Estate Geographical Information System (REGIS), *International Journal of Geographical Information Science*, Vol. 15, No. 4 (June), pp. 307–21.

Zhan, F.B. and Buttenfield, B.P. (1995) Object-Oriented Knowledge-Based Symbol Selection for Visualizing Statistical Information, *International Journal of Geographical Information Systems*, Vol. 9, No. 3, pp. 293–315.

Zhu, X. and Healey, R. (1992) Towards Intelligent Spatial Decision Support: Integrating Geographical Information Systems and Expert Systems, *Proceedings of the GIS/LIS '92 Conference*, San Diego, California, pp. 877–86.

Zhu, X., Healey, R.G. and Aspinall, R.J. (1998) A Knowledge-Based Systems Approach to Design of Spatial Decision Support Systems for Environmental Management, *Environmental Management*, Vol. 22, No. 1 (January/February), pp. 35–48.

Part II

Building expert systems (with and without GIS) for impact assessment

II.1 INTRODUCTION

The picture developed in the previous chapters suggests that IA is evolving in a way that might benefit from increased automation. At the same time, computer technology is becoming more adaptable and user-friendly for practical problem-solving.

Good practice and expertise in IA seem to be now well established in the UK, as indicated by the establishment of accepted standards of content and procedure in Environmental Statements (DoE, 1995, 1996), and also the appearance of a "second generation" of publications – the new IA regulations (DETR, 1999), new editions of classic texts like Glasson *et al.* (1999) and Morris and Therivel (2001) – all suggesting that IA seems to be reaching what we could call its *maturity*.

Expert systems combine rather elegantly the ability to crystallise accepted expertise and a degree of user-friendliness which make them good vehicles for *technology transfer* when applied to the solution of specific problems, such as those that appear in IA. On the other hand, expert systems cope best with relatively small problems, and the complexity of these systems can grow with the complexity of problems only up to a point. Beyond a certain degree of complexity, rather than having an expert system to deal with all the issues, experience suggests (Rodriguez-Bachiller, 1991; Hartnett *et al.*, 1994) that a natural "division of labour" between expert systems (or parts of an expert system) exists, and a "modular" approach to ES design is likely to work better. Some expert systems can be designed to deal with specific problems, while other ("control") systems can deal with the general management of the problem-solving process. Such control systems can be themselves expert systems, or they can be part of a more flexible decision support system (DSS), depending on the degree of flexibility needed and on whether "what-if" evaluations are required or not.

GIS are powerful databases which can be useful in dealing with some spatial aspects of IA, especially in the general areas of environmental monitoring and management, which provide the backcloth for the more technical core of IA. Experience also seems to indicate that for GIS to perform more technical tasks going beyond the role of "data providers", they require a considerable amount of expertise and/or programming. This suggests that GIS also can benefit from being linked to other systems (like expert systems) that "manage" their performance. GIS can be used by such systems as data providers, or their functionality can sometimes be used to help solve specific problems in IA.

If we add to this picture the traditional instrument used in the technical core of IA – simulation modelling – the full picture that emerges shows a top-level system (an expert system or a DSS) controlling lower-level problem-solving modules (expert systems are also good candidates for these low-level tasks). These in turn manipulate lower-level tools (like models or GIS routines) to perform specific tasks, relying on data sources provided by databases of various kinds (GIS being one of them).

II.2 STRUCTURE

In Part II we discuss these issues of expert system design applied to specific areas of IA, from project screening and scoping to the treatment of specific impact areas, to the review and assessment of Environmental Statements. We can follow the classic view of ES design in stages as summarised by Jackson (1990) from Buchanan *et al.* (1983). Jackson's summary stages refer specifically to "knowledge acquisition" but, in fact, also correspond to the initial stages needed for general ES design:

- *identification*: identifying problem characteristics
- *conceptualisation*: finding concepts to represent knowledge
- *formalisation*: designing the structure to organise knowledge
- *implementation*: formulating rules to embody knowledge
- *testing*: validating rules that organise knowledge.

Beyond these stages, there is "prototyping" (building the first full system) followed by testing, and then successive cycles of refinement. Our discussions in Part II will extend in most cases to the formalisation stage, and only occasionally will go further into implementation or prototyping. In most cases, we shall go as far as what can be best described as "designing a paper-ES", describing verbally and graphically the structure an expert system would have and indicating how it could be formalised.

To progress in this direction, a *knowledge acquisition* stage was organised in a well-established fashion, based on the two-pronged approach of consulting written documentation and consulting established experts personally. Some of the manuals and textbooks used have already been referred to, and will be mentioned. With respect to knowledge acquisition *from experts*, two main sources of expertise were used: (i) academic experts with practical experience, in particular, academics in the Impact Assessment Unit (IAU) in the School of Planning at Oxford Brookes University; (ii) practicing impact-assessment professionals, in particular, specialists employed by an internationally recognised firm of consultants, Environmental Research and Management (ERM), with one of their branches in Oxford and another in London. The choice of experts from these sources was made on grounds of superior expertise and the resulting breakdown of experts and topics was:

- *project screening*: Joe Weston, IAU
- *scoping*: Joe Weston, IAU
- *socio-economic impacts*: John Glasson, IAU
- *air pollution*: Roger Barrowcliffe, ERM (Oxford)
- *noise*: Stuart Dryden (Oxford)
- *terrestrial ecology*: Nicola Beaumont, ERM (Oxford)
- *fresh water ecology*: Sue Clarke, ERM (Oxford)
- *marine ecology*: Dave Ackroyd, ERM (Oxford)
- *soil/geology*: John Simonson, ERM Enviroclean (Oxford)
- *waste*: Gev Edulgee, ERM (Oxford)
- *traffic*: Chris Ferrary, ERM (London)
- *landscape*: Nick Giesler, ERM (London)
- *environmental statement review*: Joe Weston, IAU.

Also, consultation of a more general nature about IA was carried out with two of the managers of ERM: Karen Raymond and Gev Edulgee. Repeated interviews were carried out with these experts by Rodriguez-Bachiller, and the "protocols" of these interviews were later amalgamated with relevant technical documentation into the material that provides the basis for the discussion of different aspects of IA in the next few chapters. This first amalgamation was undertaken by the following graduates from the Masters Course in Environmental Assessment and Management at Oxford Brookes University:

- Mathew Anderson: *soil/geology*
- Andrew Bloore: *landscape, air pollution, marine ecology*
- Duma Langton: *socio-economic impacts*
- Owain Prosser: *terrestrial ecology, fresh water ecology*
- Julia Reynolds: *traffic*
- Joanna C. Thompson: *noise.*

The list of impact types included in Chapter 1 could be used as a guiding principle for the discussion in this Part, but it is preferable to structure the discussion in the next few chapters grouping these areas of IA into themes and/or approaches, relating to the potential ways in which ES, modelling and GIS technologies relate (or could relate) to these impact assessments.

The sequence of chapters follows an overall framework of IA stages, starting from screening and scoping, then moving on to impact assessment as such – at this stage the discussion "branches out" into various areas of impact – and finishing with the review of Environmental Statements. We start in Chapter 6 with the two related issues of project screening and scoping, which are highly regulated and relatively "easy" subjects for expert systems treatment. In Chapter 7 – the first of the impact assessment chapters – we go to one extreme by discussing areas of impact characterised by "hard modelling", using air pollution and noise as examples. In contrast, Chapter 8 examines areas where modelling has a lesser role to play: terrestrial ecology and landscape impacts. Subsequent chapters explore "mixed" areas of IA, where modelling is complemented (sometimes replaced) by more low-level techniques: Chapter 9 looks at socio-economic and traffic impacts, Chapter 10 discusses hydrogeology and water ecology. Finally, returning to the main IA process, Chapter 11 applies the same reasoning process to the question of Environmental Statement review. These discussions will help raise some general issues of ES design and GIS use which, together with Part I, provide the material for the concluding Chapter 12.

REFERENCES

Buchanan, B.G., Barstow, D., Bechtal, R., Bennett, J., Clancey, W., Kulikowski, C., Mitchell, T. and Waterman, D.A. (1983) Constructing an Expert System, in Hayes-Roth, F., Waterman, D.A. and Lenat, D.B. (eds) *op. cit.* (Ch. 5).

Department of the Environment (1995) *Guide on Preparing Environmental Statements for Planning Projects*, HMSO, London.

Department of the Environment (1996) *Changes in the Quality of Environmental Impact Statements for Planning Projects* (Report by the Impact Assessment Unit, School of Planning, Oxford Brookes University) HMSO, London.

DETR (1999) *The Town and Country Planning (Environmental Impact Assessment) (England and Wales) Regulations 1999*, Department of Environment, Transport and the Regions No. 293.

Glasson, J., Therivel, R. and Chadwick, A. (1999) *Introduction to Environmental Impact Assessment*, UCL Press, London (2nd edition, 1st edition in 1994).

Hartnett, J., Williams, R. and Crowther, P. (1994) Per Pixel Reasoning Using a GIS Closely Coupled to an Expert System to Produce Surface Classifications Based on Remotely Sensed Data and Expert Knowledge, *Proceedings of the EGIS/MARI '94 Conference*, Paris (March 29–April 1), Vol. 1, pp. 677–83.

Jackson, P. (1990) *Introduction to Expert Systems*, Addison Wesley (2nd edition).

Morris, P. and Therivel, R. (2001) (eds) *Methods of Environmental Impact Assessment*, UCL Press, London (2nd edition, 1st edition in 1995).

Rodriguez-Bachiller, A. (1991) Diagnostic Expert Systems in Planning: Some Patterns of System Design, in Klosterman, R.E. (ed.) *Proceedings of the Second International Conference on Computers in Urban Planning and Urban Management*, School of Planning, Oxford Polytechnic (July).

6 Project screening and scoping

6.1 INTRODUCTION

Project screening, to decide if a project needs to go through the EIA procedures (making an Environmental Statement and assessing it) in support of a planning application, is the "gateway" into EIA. It has two important characteristics: first many projects being screened are likely to be found *not* to require EIA. Therefore, the number of projects screened is likely to be much higher than the number of projects eventually subjected to EIA, and screening is likely to become a routine procedure to which more and more projects are subjected. Second, the pressures of project-screening cut across the public–private divide and affect agents on both sides of the development control system. It is engrained in the system that (public) controlling-agencies have the need for adequate project screening, but also private developers can benefit from similar capabilities to "try out" different project configurations and find out if they require extra EIA work, before entering the complicated and expensive development control process.

These two characteristics already suggest the potential benefits of some form of automation of the screening process – for example using ES technology – to alleviate the pressure on both public and private organisations. In addition, project screening also shares some of the typical pre-conditions of "sensible" ES application discussed in Chapter 2:

- the screening process is mostly a *regulated* one (DETR, 1999a,b);
- expertise consists mostly of the *knowledge of the published regulations and guidelines*, with relatively minor contributions from experience, in borderline cases or in "grey areas";
- this problem is virtually "*routine*" for experts, while it is too complicated for non-experts;
- it is a relatively *simple* problem, taking an expert a few hours at most to determine the grounds on which a project may – or may not – require IA.

For all these reasons, project screening is a good "testing ground" for ES technology, and it is no coincidence that (together with impact "scoping") it has attracted considerable attention from the ES research community, as discussed in Chapter 5.

6.2 THE LOGIC OF PROJECT SCREENING

Project screening in IA is very similar to the determination of "permitted development" in development control (whether a development requires planning permission), which has also attracted attention in the ES literature in the 1980s (Rodriguez-Bachiller, 1991).

To develop an ES to help with project screening, we must first look at the overall logic of the process. When IA was first adopted in the UK in 1988, screening was based on a two-tier system replicating the earlier European Directive of 1985, which classified EIA projects into those always requiring impact assessment (Schedule 1 projects) and those for which it is required only if they exceed certain thresholds (Schedule 2 projects) or are likely to produce significant impacts, significance being judged on three criteria:

- the scale of a project being of "more than local importance";
- the location being "particularly sensitive";
- the project being likely to produce particularly "adverse or complex" effects.

The new Regulations of 1999 (DETR, 1999a) and the associated Circular (DETR, 1999b) added further considerations within Schedule 2: *minimum* project characteristics (defined in the Regulations as "minimum exclusion criteria") below which an impact study *will not* be required, and a set of *maximum* indicative thresholds likely to trigger an impact study when considered in conjunction with the criteria (listed in Schedule 3) of impact significance, which are similar to those used before:

- characteristics of the development (size, use of natural resources, quantities of pollution and waste generated, risk involved, etc.);
- environmental sensitivity of the location (land use, absorption and regenerative capacity, etc.);
- characteristics of the potential impacts (magnitude, duration, reversibility, etc.).

There is potentially a "grey area" in Schedule 2 (Weston, 2000a), where the indicative thresholds are supposed to be applied not in an "exclusionary" way (as they were in the previous regulations) – to narrow down the band of uncertainty – but as additional criteria in conjunction with the other criteria listed above (project characteristics, location, magnitude of impacts) of potential impact significance. The Circular provides a diagram of the *sequence* of how all these criteria are to be applied to a project, from which we gain an idea of what the intended *order of priority* should be when considering which criteria of significance to apply (Figure 6.1).

- first, consider if the project is included in Schedule 1 (if yes, IA is required);
- if not, see if it is in any of the categories listed in Schedule 2 (if not, IA is not required);
- if it is, first check if the location is in an area designated as environmentally sensitive;
- if not, see if it exceeds the Circular's indicative thresholds (if not, IA is not required);
- if the project is in a sensitive area or it surpasses any of the indicative thresholds, the likely significance of the impacts must be assessed from:

  (i) the size/characteristics of the project;
  (ii) the sensitivity of the location;
  (iii) the characteristics (magnitude, risk, etc.) of the impacts.

In *practice* however (Weston, 2000b), the indicative thresholds in the Circular are used in a more exclusionary way both by developers and by

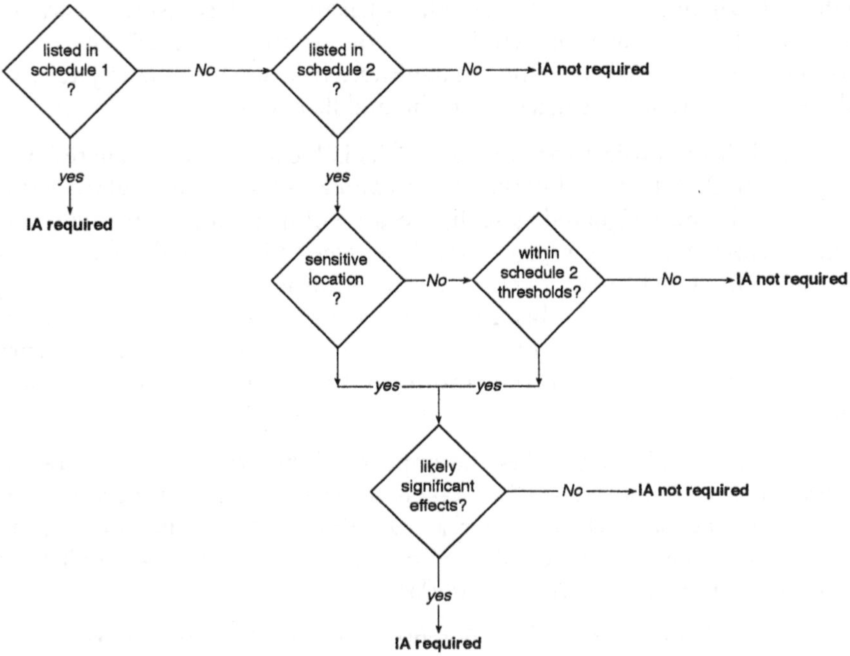

Figure 6.1 The scoping sequence.

control authorities, and the sequence of checks takes a slightly different form:

1 First, consider if the project is included in Schedule 1, i.e. if any aspect of its development is listed under Schedule 1 in the Regulations and, when thresholds of size or duration are specified, check if they are reached or exceeded. If the answer to this question is yes, IA *is* required, and the screening is finished.

2 If it is not a Schedule 1 project, start checking if it falls under Schedule 2: first, see if a part or all of the project is located in an area *designated* as environmentally sensitive, such as those listed in the Regulations (areas of special scientific interest, nature conservation areas, national parks, etc.). If it is, IA *is* required.

3 If the answer to the previous question is also no, see if any part of the project falls into any of the categories listed in Schedule 2 and, if so, check if it falls below the *minimum* thresholds indicated. If the project is not listed or its characteristics fall below these thresholds, IA *is not* required.

4 If the project cannot be "excluded" this way, check if any part of it exceeds the Circular's *maximum* indicative thresholds. If any of those thresholds are exceeded, IA is considered – in practice – to be *required*.

If the answer to the previous query is still no, we enter the grey area where the significance of potential effects has to be judged using the criteria in Schedule 3: the size/characteristics of the project, the sensitivity of the location and the characteristics (magnitude, risk, etc.) of the impacts. Of these, Weston (2000b) suggests that the middle one is considered first.

5 Check if the project (or any part of it) is located in a designated area (e.g. Green Belt, National Park, AONB) *perceived* to be particularly sensitive (even if not designated as such), because of its land use, its low regenerative capacity, or the type of area it is (wetlands, coastal area, forests, densely populated areas, etc.). If the location is sensitive, IA is usually considered necessary. *If* this check is negative and the question is still undecided, then the other two sets of Schedule 3 criteria (scale, importance of impacts) come into operation, not necessarily in any particular order.

6 Being of a scale that makes the impacts of the project likely to be "of more than local importance" is taken to refer to "effects beyond their immediate locality, which give rise to substantial national or regional controversy, which may conflict with national (or regional) policy on important matters..." (Weston, 2000a).

7 Under "characteristics of the potential impact" there are different types of criteria, some relating to risk, others to irreversibility of the impacts, and others to the general obnoxiousness/danger of the impacts. Weston (2000b)

suggests that a rule of thumb applied in practice is to consider under these categories any projects (or parts of) which would normally require authorisation from pollution control agencies (IPPC, Waste Management Licence, Hazardous Waste Licence, etc.).

This general sequence can be translated into a step-by-step diagram (Figure 6.2), using the same symbols as before (but swopping the directions of the "yes" and "no" options to fit the page). We can say that this diagram represents the overall logic of the way experts screen a project organised like this, from a combination of what is in the legislation and experience, which is used to fill the gaps in the regulated procedures and, sometimes, to simplify them. Such logic can be translated into an "inference tree" of the kind discussed in Chapter 2, ready for ES treatment (an "arc" between two branches implies an "and" conjunction between them, the absence of an arc implies an "or" relationship) as in Figure 6.3.

But what characterises an expert is the fact that, even if there is a sequential logic to this problem-solving process, the expert "knows" the whole approach from the start, giving him/her the possibility to "short-cut" steps and go directly to the crucial issues, or even to see the overall answer from the outset. This "gestaltic" perception of a problem and its solution map – and the possibilities it opens for so-called "strategic" decisions and changes in the direction of enquiry – is characteristic of top-level expertise, and has been a classic target for critics of artificial intelligence since the early days (Dreyfus, 1972); ES are no exception.

This is one of the simplifications that ES design imposes: while the expert "sees" the whole solution map from the outset (like an expert chess player can sometimes anticipate the end-game ahead), it has to be formalised for the non-expert as a sequential, step-by-step process which explores all the possibilities and does not leave the non-expert any room for error. An inference tree like the one above may provide a vehicle for such formalisation, but it still cannot deal with some of the problems of having to "sequentialise" something "synchronic". For example, the first check on a project will be to see if it is included in that part of Schedule 1 which does not have thresholds, including projects that, by definition, will require an impact study. This check is easy and the list of such projects is rather short, so we can expect a first enquiry about whether the project is of a type such as:

- crude-oil refineries;
- installations for gasification and liquefaction of coal or bituminous shale;
- nuclear power stations or reactors;
- installations for the production/processing/disposal of nuclear fuel;
- integrated chemical installations for the production of explosives or chemicals;
- hazardous waste disposal installations;

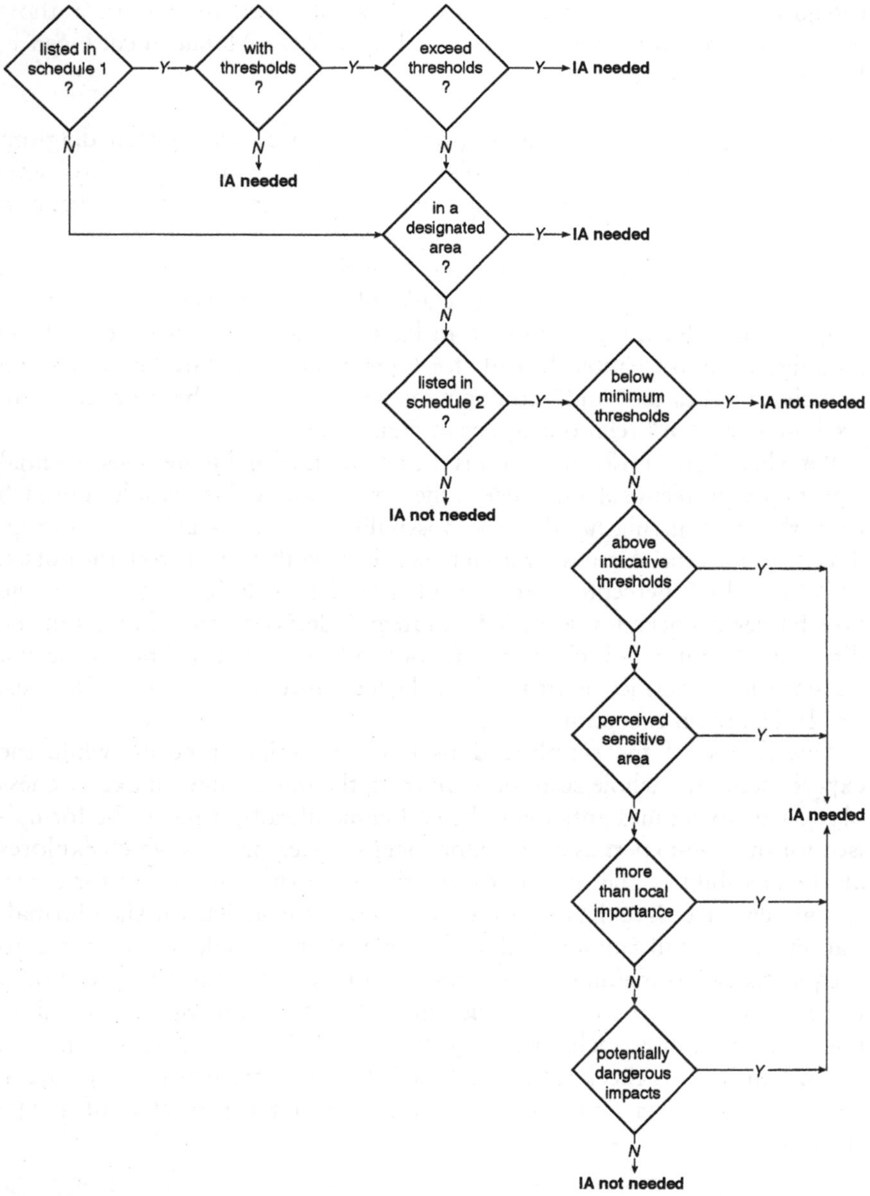

Figure 6.2 Step-by-step scoping.

- non-hazardous waste disposal installations by incineration;
- industrial plant for the production of pulp from timber or similar materials.

The next check will be if the location is in a designated area. If either of these two tests is positive, the problem is solved but, if not, the next checks

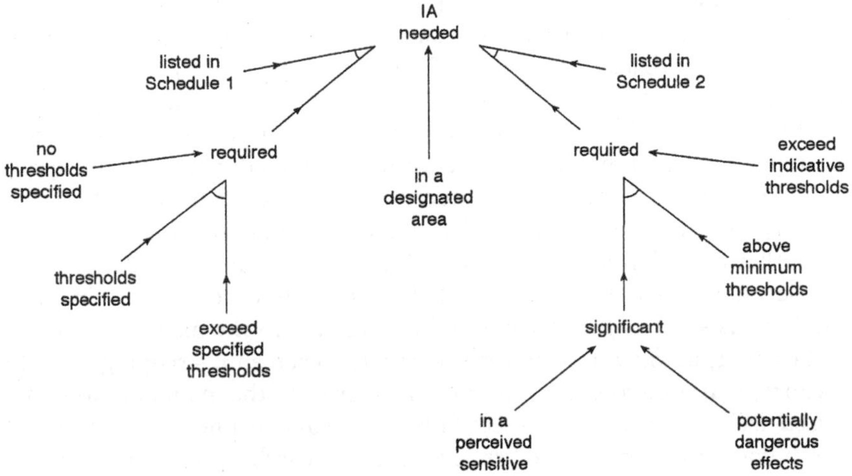

Figure 6.3 The scoping inference tree.

will be if the project includes activities listed in Schedule 2. In fact, Schedules 1 and 2 involve largely overlapping lists of activities[19] where, in many cases, it is only the thresholds that are different. For practical purposes, we can treat this as *one* series of activities (we have to check if the project contains any) and *three* sets of thresholds concerning them: (i) minimum Schedule 2 thresholds below which IA is not needed; (ii) indicative Schedule 2 thresholds above which IA is needed; and (iii) Schedule 1 thresholds above which IA is needed. For example, if our project contains or consists of a "thermal power station", its size must be compared with various thresholds: if it occupies an area of 0.5 hectares or less, IA is not needed (minimum Schedule 2 threshold); if it has a thermal output of more than 50 MW, IA is required (indicative Schedule 2 threshold); and also if its heat output is 300 MW or more, IA is required (Schedule 1 threshold).

This part of the problem-solving logic may be expressed as a logically related set of issues to consider – as in the inference tree above – but, in terms of the information needed to consider them, it can be reduced to two sets of questions to put to the user: (i) *does the project include "listed" activities*? and, if so, (ii) *at what level/intensity*? to see which thresholds (if any) they exceed. As the above example shows, not all the thresholds related to the same type of development are always of the same kind – some may refer to physical size while others refer to output or quantity – which may require additional questions relating to the various types of size indicators, but the logic is still quite simple. Then, it is only a question of

19 Similar but not identical, as some Schedule 2 activities are not mentioned in Schedule 1.

deducing from the answers to these questions what category the project belongs to: not requiring IA, requiring it as a Schedule 1 project, as a Schedule 2 project, etc. In order to formalise such a very simple sequence, a simpler alternative to a "backward-chaining" inference tree (see Chapter 2) shown above can be a "forward-chaining" approach: the relevant sequence of questions is defined – some questions being conditional on the answers to previous ones – and the overall conclusions are then derived and the appropriate messages to the user are produced. This was the approach chosen for the design of the system which is discussed in the next section.

Another typical aspect of ES design which often comes into conflict with the elegant simplicity of backward-chaining inference trees – that stop searching as soon as they have found the answer to the main question – is the fact that, in *diagnostic* expert systems (and screening–scoping is in this category) you need to know not only the answer to the main question (*does it need IA*?) but also, if a project fails, you want to know *all the reasons* why it does, and not just the first one in the "search" process that made it fail. For example, if a project involves a new road going through a residential area, a chemical plant with an incinerator, and the generation of dangerous waste, *all* these elements must be investigated in order to establish if any of them violate the legal thresholds (probably they will all fail), and knowing that one of these elements does infringe the legal limits is not sufficient: we must know what the situation is with each of these elements. Taking the analogy of medical diagnosis, when using diagnostic ES we want to know *all that is wrong* with the patient, and not just his/her most important ailment.

Returning to IA, the reason is that, in addition to the screening objective (knowing *if* it fails) there is the scoping objective of knowing which impacts will need investigation, and in order to know this we need to know all the aspects of the project which would make it fail. It is for this reason that diagnostic ES – even if taking advantage of the elegance and simplicity of inference-trees – often have to involve "lists" of aspects to consider (each case involving a different list) *sequentially*, and the logic-driven search of inference trees only apply within each element of such lists.

6.3 THE SCREEN EXPERT SYSTEM AT OXFORD BROOKES UNIVERSITY

As part of the research project from which this book results, an expert system (SCREEN) to deal with project screening in the UK was developed at Oxford Brookes University in the mid-1990s, and it is a useful example to illustrate the issues of expert-system design being discussed. The first version was "stand-alone" and a later version was connected to a GIS, although only the first version has been used regularly for demonstration and teaching. The SCREEN system is based on the "first generation" set of EIA regulations and guidelines (DoE, 1988, 1989), with a logic similar to the more recent ones but slightly simpler, in that there was only one set of Schedule 2 thresholds and they were exclusionary rather than indicative. However, the general logic was the same as that of the 1999 regulations, and the points of ES design discussed previously also apply. The user is asked about those factors which the regulations identify as crucial to determine if a project requires an Environmental Statement, in three typical stages:

(i) First, the project's *location*: some locations (like an Area of Outstanding Natural Beauty) carry the automatic obligation to have an impact study, and some locations (like Enterprise Zones designated before July 1988) used to carry automatic exemption from it.
(ii) Second, the project's *characteristics*: the different types of operations involved, compiled from the "lists" of Schedules 1 and 2.
(iii) Finally, the *magnitude* of the different project activities must be determined, for comparison with the relevant thresholds.

For instance, after sorting out if the location of the project is special or not, the user is asked what the proposed project involves:

Does the project involve any of the following types of operations?
– *agricultural operations*
– *extractive industry*
– *energy industry*
– *chemical processing*
– *buildings*
– *road construction*
– *etc.*

If, for example, our project involves pig farming, we select the "agricultural operation" option (we can select several) and the system next asks about the nature of the operation, to narrow it down:

What type of agricultural operation?
– *afforestation*
– *new planting*
– *dealing with farm animals*
– *etc.*

In our example, we would select the third option, and the system would then ask for more details:

What type of farm-animal operation?
– *poultry rearing*
– *pig rearing*
– *knackers' yard*
– *etc.*

When we select the second option (pig rearing), the system enquires about the size of the operation:

For processing how many animals each year?

Also (it would have asked this earlier) it will enquire about where the project is to be located with respect to any housing in the area, since this is central to the consideration of the potential "significance" of this type of project:

How far is the nearest housing?

Having gathered the information required, the system classifies the project into a category with respect to the requirement for an Environmental Statement and advises the user. This advice is compiled combining elements of "canned text", key phrases and words retrieved from an archive in a particular order as required, wording the general conclusions and also listing the *reasons* for the advice. For instance, if an Environmental Statement were required, the *reasons* for this could be varied:

- the *nature* of the operation, for instance if it were a "knackers' yard";
- its *size*, if the project has more than 50 000 places for processing animals;
- its *location*, if it were to be built within a short distance – for instance 500 m – from existing housing.

Additionally, the system occasionally gives advice on the need to consult certain organisations when information is missing or uncertain (as reflected in answers from the user of the "I do not know" type, available in some questions). The advice is displayed on the screen at the end of the consultation, and a copy is also put into a digital Report file which can be printed later by the user. An example of such advice could be:

Project XXXX in location YYYY

This development will REQUIRE an Environmental Statement, as included in Schedule 2 of the Environmental Assessment Guide (Department of the Environment, 1991) because it exceeds some of the criteria specified in that document.

REASONS:

– it has more than 50 000 places for processing animals

6.4 SCOPING

Identifying the types of impacts that should be investigated – and establishing which of them are likely to be *key* to the acceptability of the project – has been traditionally presented in EIA manuals as an essentially *technical* exercise based on tables and matrices (Wathern, 1988; Petts and Eduljee, 1994; Petts, 1999; Morris and Therivel, 1995, 2000) which link different types of projects to types of impacts likely to be produced by them. Also, scoping can be seen as following directly from screening, in that the reasons why a project should need IA could also be interpreted as indications of what types of impacts to expect. This has been the traditional approach.

In addition to this, the new Circular (DETR, 1999b) has added a new regulatory contribution, by including in the discussion of the "indicative thresholds" (associated with Schedule 2 projects) clear indications of which types of impacts ought to be studied, with phrases like (from the section on "wind farms"): *The likelihood of significant effects will generally depend upon the scale of the development, and its visual impact, as well as potential noise impacts.* Such key impacts will be those that will decide the overall significance of the effects of the project, and the Circular "lists" them for us.

6.5 THE SCOPE EXPERT SYSTEM AT OXFORD BROOKES UNIVERSITY

The SCREEN system described above is linked to another similar system (SCOPE) for project scoping in the UK, and the two systems can run consecutively: after a project has been "screened", the user may want to finish the consultation (for instance if an Environmental Statement is not needed) or he may choose to "scope" it. However, the reverse is not true, because SCOPE uses information obtained at the screening stage and, in order to run the scoping system, the screening system must be run first.

The Oxford Brookes scoping system also follows from the previous regulations of 1988 and 1989. After a project has been screened, if an Environmental Statement is required, the scoping "module" can be applied. As a starting point, it uses much of the information already gathered at the screening stage on which the need for IA was based:

(i) If the *location in a sensitive area* contributed to the screening decision, the nature of that sensitive area (a National Park, an area of archaeological importance, etc.) should point in the direction of different types of impacts requiring study:

- location in a natural landscape or conservation area makes it practically obligatory to investigate landscape and land-ecology impacts;
- location in an archaeological area or an urban conservation area or historic centre suggests the need to investigate heritage impacts;
- location in wetlands suggests the need to investigate land-ecology impacts, soil and hydrogeology issues;

- location on or near a water system suggests automatically the need to study water ecology impacts;
- etc.

(ii) Certain *types of projects* have associated with them certain types of impacts: a road construction project will need traffic impacts to be studied, an industrial project with an incinerator will require an air pollution study, etc. *Matrices* are used here (in the traditional approach often suggested in impact assessment manuals) like the example in Figure 6.4.

(iii) In a similar way, if it is the *magnitude or intensity* of certain project activities that determines the significance of their impacts, the same reasoning applies, and this will tell us also about some areas of impact that require study (a similar "matrix" is used).

In addition to the information derived from the screening stage, the SCOPE module also adds questions about other specific aspects of the

TYPE OF PROJECT	Socio-economic	Landscape	Heritage	Traffic	Noise	Waste	Air	Soil	Land Ecology	Geology	Hydrogeology	Freshwater Ecology	Estuarine Water Ecology	Marine Water Ecology	Etc.
a nuclear reactor	●	●		●	●	●	●		●	●			●	●	
a hydroelectric energy production installation	●				●			●	●	●	●	●			
a wind generator installation	●	●			●										
a test bench for engines, turbines or reactors				●	●	●									
glass manufacturing	●	●		●	●	●	●								
afforestation		●							●			●			
water management for agriculture		●						●	●		●	●			
reclamation of land from the sea for agriculture		●									●		●	●	●
new drainage/flood defence works		●						●	●	●	●	●			
a new poultry-rearing installation	●			●	●	●	●	●	●						
a new pig-rearing installation	●			●	●	●	●	●	●						
a knackers' yard				●	●	●	●	●	●						
off-shore marine salmon farming		●												●	●
an installation for the rearing of salmon		●										●	●		
a salmon hatchery		●		●	●	●	●					●	●		●
extraction of peat		●		●	●		●	●							
deep drilling		●			●	●	●			●	●				
extracting minerals by underground mining	●	●		●	●	●	●			●	●				
etc.															

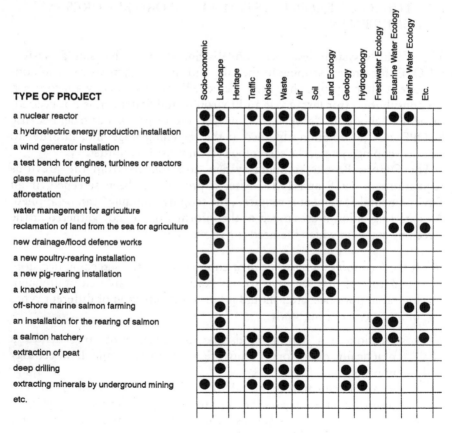

Figure 6.4 Type of impact by type of project.

project, in order to determine the complete range of impacts that should be studied:

(a) Even when the *location* of a project is not directly inside an environmentally sensitive area – and therefore it is not a decisive factor at the screening stage – it may be near, upwind or upstream from one such area, suggesting that it may be advisable to investigate certain pollution and/or noise impacts.
(b) Even away from special environmental areas, the location of projects can suggest the need to study certain impacts: for instance, if next to a water system, or if located on good agricultural land.
(c) Some *aspects of all projects* have associated with them potential impacts which may need study if they exceed certain thresholds – even if they were not crucial when deciding the need for an impact study – like the labour force of a project requiring a socio-economic impact study if the numbers are large, or certain physical features of the project (buildings/structures) requiring a visibility impact study if they exceed a certain height.
(d) Finally, projects involve different *stages*: most projects include a *construction* stage which requires especial investigation, because it can be very different from the *operational* stage of the project, and it can generate impacts specific to that stage which may be unrelated to the effects considered when screening the project (which tend to be associated only with the project's operational stage). The so-called "decommissioning" stage (involved in only certain projects) may also require a similar investigation.

The SCOPE system asks extra questions to cover these additional aspects, not derived from the project screening, as they apply to each of the *stages* in the life of the project: construction, operation, and decommissioning (if applicable). The system then applies to all this information a series of "matrices" which derive impact types from types of locations, projects and activities, as they apply to each of the stages of the project, and derives the range of *direct impacts* requiring study. The list of impacts that the system uses in these matrices is:

1 *Social*
- gains
  - housing
  - social facilities
- pressures
  - housing market
  - social facilities
  - cultural/psychological pressure

2 *Economic*
employment
gains
losses
retail, goods and services
gains
losses
3 *Landscape*
destruction of landscape resources
damaging the view
obstruction
interference
topographic change
4 *Cultural heritage*
archaeology
urban Conservation Areas and historic built environment
listed buildings
5 *Traffic*
traffic generation
persons
materials, fuels, waste, etc.
amenity loss
6 *Noise*
through air
reradiated noise
vibration
7 *Air*
chemical pollution
odours
dust and particulate matter
8 *Waste*
disposal
treatment
9 *Soil and land*
loss of agricultural land
soil contamination
10 *Landuse and planning*
plans and planning policies
11 *Material assets and resources*
minerals
buildings and property
12 *Blight*
risk from hazardous substances
interference with social/economic processes and markets
perception of interference or risk

13 *Geology/hydrogeology*
geology
geological suitability to withstand the action
hydrogeology
surface water run-offs
surface water contamination
drainage patterns
underground water
levels and flow
water contamination
14 *Terrestrial ecology*
plants
land animal species
birds
15 *Water ecology*
fresh water
river ecology
aquatic species
birds
lakes/dams/reservoirs ecology
aquatic species
birds
estuarine ecology
aquatic species
birds
marine ecology
aquatic species
birds
16 *Water as a resource*
rivers
water quantity
water quality/contamination
loss of leisure use
lakes/dams/reservoirs
water quantity
water quality/contamination
loss of leisure use
estuarine/marine water
water quality/contamination
loss of leisure use.

Also, the system uses these direct impacts to generate a range of *indirect impacts* also needing investigation (using another "matrix" linking direct and indirect impacts): these are *impacts derived from other impacts* – like noise effects derived from traffic – and can be sometimes as important as the direct ones, and cannot be ignored.

As a result, the scoping module produces – using "canned text" – another Report for the user (displayed on the screen and also copied onto a file for later printing), with a list of the different types of impact that should be studied. If the screening module was also used, the scoping report is appended to the screening report already produced. An example of the scoping part of such a report could read like this:

In the CONSTRUCTION/PREPARATION stage of the project, the following possible impacts are likely to require investigation:

1 **SOCIO-ECONOMIC:**
 – *fears of the local population*
2 **TRAFFIC:**
 – *traffic generated by the movement of materials, fuels, waste, etc.*
3 **NOISE:**
 – *noise transmitted through air*
 – *the effects of vibration*
4 **AIR QUALITY:**
 – *the effects of dust and particulate matter.*

The areas of impact likely to require investigation with respect to the OPERATIONAL stage of the project are:

1 **SOCIO-ECONOMIC:**
 – *fears of the local population*
2 **TRAFFIC:**
 – *traffic generated by the movement of materials, fuels, waste, etc.*
3 **NOISE:**
 – *noise transmitted through air*
 – *the effects of vibration*
4 **AIR QUALITY:**
 – *the effects of dust and particulate matter*
5 **LAND USE AND PLANNING:**
 – *potential discrepancies with the local plans and planning policies*
6 **GEOLOGY AND HYDROGEOLOGY:**
 – *the geological suitability to withstand the project*
 – *effects on the drainage patterns of the ground*
 – *effects on the level and flow of underground water.*

In addition to the direct impacts which need investigation with respect to the operation of this project, another area that needs investigation is that of INDIRECT impacts – produced by other impacts – like, for instance, traffic producing noise impacts that require investigation even if the original project is not noisy.

The INDIRECT IMPACTS requiring investigation with respect to the CONSTRUCTION and OPERATIONAL stage of the project are:

1 **BLIGHT:**
  – *possible blight of property markets.*

The DECOMMISSIONING STAGE of the project is likely to require investigation with respect to some types of impact, not too different from those investigated for the construction/preparation stage:

1 **SOCIO-ECONOMIC:**
  – *fears of the local population*
2 **TRAFFIC:**
  – *traffic generated by the movement of materials, fuels, waste, etc.*
3 **NOISE:**
  – *noise transmitted through air*
  – *the effects of vibration*
4 **AIR QUALITY:**
  – *the effects of dust and particulate matter.*

The SCREEN and SCOPE systems are used mainly for demonstration and teaching purposes, and also for practical project work by students applying IA to particular projects as part of their courses. The systems are successful both as tools to apply IA guidelines to specific projects, and as vehicles for learning the logic of screening and scoping (as well as that of expert systems), and a future project under consideration is the adaptation of these two systems to the new regulations of the late 1990s. Both systems use the same inference engine, which was programmed in "C" by Rodriguez-Bachiller using an approach where information is gathered from relevant sequences of questions and the conclusions are derived as a result.

6.6 ADDING GIS TO THE SCREEN-SCOPE SUITE AT OXFORD BROOKES

We have seen how GIS include in their functionality the capacity to: (i) count and measure the size of spatial features; (ii) measure distances; (iii) construct buffer zones; and (iv) identify and measure spatial overlaps. In the context of environmental ES, we can use these capabilities to provide automatically information which otherwise would have to be obtained by asking questions to the user questions and, in this way, simplify the consultation process and make it even easier for the user (one of the aims of ES).

In particular, this applies to questions related to the *location* of certain elements of a development, and questions related to the *extension* of certain features of the project, crucial for some stages of project screening and/or scoping:

Screening

- Some projects (like industrial estates or infrastructure projects) require an IA study if their area reaches a certain *size*, and a GIS can calculate this from a map of the project.
- Often, it has to be established if a project lies *within* environmentally sensitive areas, and using GIS to "overlay" a map of the project and of the relevant sensitive areas will establish this.
- It is normal to have to determine if certain projects are adjacent to or within a certain *distance* from a certain type of feature (like roads from residential areas), and "buffering" can be used to answer such a query.

Scoping

- If the location of a project impinges on good-quality agricultural land, agricultural and soil impacts will need to be studied, and this can be determined by "clipping" and measuring the areas involved using GIS.
- Similarly, the need for a study of the impacts on a sensitive area (like an archaeological site) could arise from determining that the project is within a certain distance of the area in question, easily determined with a GIS using its buffering capabilities.
- If a project which produces discharges pollutants is close to a water system, a study of water pollution will be necessary, and the "buffering" function in a GIS can easily determine this.
- If the project involves emissions into the atmosphere and is located upwind from a nature reserve (this is more difficult to do automatically with GIS, but is still feasible), an air-pollution study and an ecological study must be carried out.

These are only examples, but show that the potential for GIS use at the screening scoping stage is considerable. What is needed in order to apply these GIS functions is to operationalise some form of communication between the ES and the GIS. As we have discussed in Chapter 4, this can be achieved by "embedding" one into the other – by programming the ES within the GIS using the latter's own programming language (like Arc-Info's AML, or "C" in the case of GRASS) – or it can be done by programming the ES externally to the GIS and "linking" the two. The latter approach was used with the Oxford Brookes system, whose inference engine – for this version – was programmed in

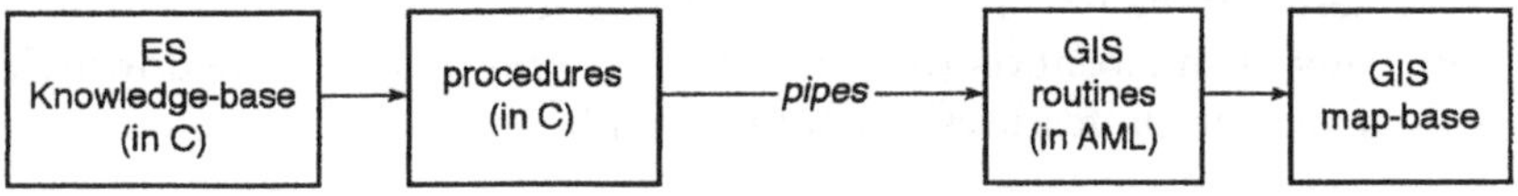

Figure 6.5 GIS-expert system connection.

"C",[20] and the link with the GIS (Arc-Info) is achieved in several intermediate steps:[21] (i) specially programmed "procedures" (routines in C) are called from the ES as required; (ii) these procedures establish communication with Arc-Info through "pipes"; (iii) through these "pipes", the procedures run Arc-Info routines (programmed in AML); (iv) the AML routines apply GIS functions to the map base; and (v) provide answers (through the "pipes") to the original procedures, which would relay them to the ES (Figure 6.5).

The additional work that this required was the programming of the "procedures" (in C) and of the GIS routines (in AML).

6.6.1 Procedures linking ES and GIS

The procedures used in the Oxford Brookes system varied in complexity, from performing simple "single-instruction" GIS operations, to applying more complicated procedures to parts of the map base:

(i) delivering single instructions to Arc:

- file management (listing or describing the structure of maps, deleting or copying maps);
- spatial analysis (appending two maps, making a buffer around all features in a map).

(ii) measuring/counting:

- measuring the extension of features in a map (one or several, finding the maximum size);
- counting the number of features of a certain kind;
- adding up the value of an item (like floorspace) in several features.

(iii) spatial analysis:

- selective buffering (making a buffer around specific features in a map);

20 By Anthony Prior-Wandesforde, from the Computer Services Department at Oxford Brookes University.

21 Programmed by Agustin Rodriguez-Bachiller.

- clipping (seeing if a buffer overlaps with any features in a map, seeing if the features in two maps overlap);
- downwind location (finding if the features in a map are downwind from certain features of another map).

To this list, should be added a range of procedures which are *combinations* of those listed above, especially combinations of spatial analysis with measuring/counting: measuring the overlap between the features in two maps, adding up the value of certain items in the features in a map within the area (or a distance) of another, counting the number of features in a map within the area (or a distance) of another. In fact, most procedures are combinations of several others, and to perform automatically a relatively simple spatial-analysis operation can involve rather lengthy chains of simple GIS procedures. For instance, in our previous example, using the GIS to answer the question about the pig-farming operation being within a certain distance of housing or of a sensitive area would involve: (i) finding a map of existing housing and a map of the project; (ii) making a buffer around the pig-farming operation; (iii) overlaying this buffer with the housing map; and (iv) checking if there are any dwellings caught within the buffer area. The logical sequence could be:

- identify the project map and, in it, identify the feature(s) which fall in the "offending" category;
- identify the land-use map for the area and, in it, identify the feature(s) which fall in the "sensitive" category;
- extract from the land-use map a sub-map containing only the sensitive features;
- make a buffer at the critical distance from the "offending" features in the project map;
- clip the buffer map with the land-use map;
- measure on the clip map, the extent (if any) of the overlap between the buffer and the sensitive areas;
- if the overlap is zero, the answer passed on to the ES is "no", if the overlap is positive, the answer is "yes".

This is just one way of achieving this relatively simple result – used here to illustrate the logic and the way a task can be broken down into GIS steps – but there are also other ways: for example, after "extracting" a map of sensitive areas from the land-use map, we could apply a *near* Arc-Info command (which computes the nearest distances between features on different maps) between the "offending" features in the project map and the sensitive features in the land-use map, and then check if any distances between the project features and the nearest sensitive areas were within the critical distance.

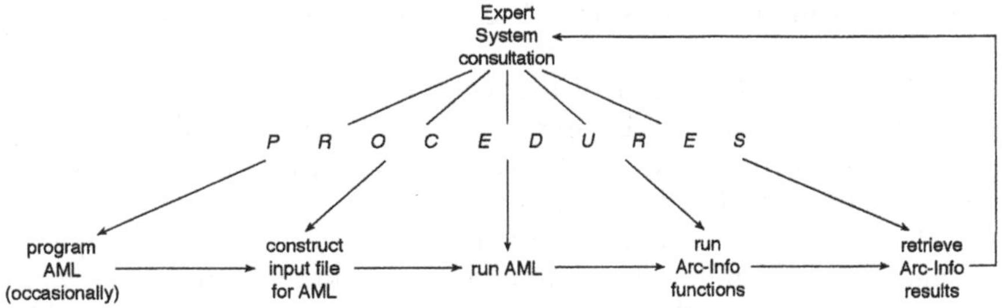

Figure 6.6 The range of GIS-control procedures.

The whole process of communication with the GIS is controlled by the procedures programmed in C, whose function is multiple (Figure 6.6): (i) using data obtained from the ES consultation to construct a *data input file* for the AML to use, containing map names and feature names, critical distances, etc.; (ii) occasionally, *programming* the AML routines, when its programming and not just the data depends on the particular case being investigated; (iii) *running* the AML routines, which in turn run the various Arc-Info functions needed; and (iv) retrieving the relevant *results* from the GIS run, to be fed back into the running of the ES.

6.6.2 Evaluating the GIS links

As we have seen, most of these are simple-enough procedures with which a considerable range of GIS operations can be performed automatically without the user having to retrieve the information themselves, which theoretically simplifies the consultation of the system by the user. But there are also added inconveniences in the process, both in terms of added complexity to the programming of the ES engine and its *knowledge base*, and in terms of the *running* of the system.

First of all, the ES "engine" must be complemented with the *additional programming* of the various procedures and AMLs, although this should be counted as a "sunk" design-cost incurred only once, which will benefit all the systems using that "engine".

Second, there is an added design-cost which will affect *each* different type of ES to be run with the same "engine": the *knowledge base* for the non-GIS system must be expanded considerably for the GIS-linked system. For example, comparing the knowledge bases used by the SCREEN–SCOPE suite in its two versions (with and without GIS), the "impact" of adding GIS becomes apparent: of the 8,260 lines which make up the knowledge base (GIS version), only 4,610 are for dealing with the ES consultation, and the other 3,650 lines (44 per cent) are there only because of the GIS. The breakdown of these GIS-related additions is also interesting:

- 2,370 lines (29 per cent of the whole knowledge base) are used for "learning" (by asking the user) the general structure and composition of the map base: maps available, their names, information they contain, names for their features.
- About 1,260 lines (15 per cent of the knowledge base) are for acquiring additional specific information about some local features (like water systems) and for the application of the various procedures which communicate with the GIS; of these 1,260 lines, 830 are for SCREEN and 430 for SCOPE.

What this indicates is that the greatest additional cost in terms of programming the knowledge base comes from the need to know in detail the structure and composition of the map base which the system is going to use. This will also be reflected in the third type of additional cost we are considering, what we could call "running" costs.

Running costs in ES can be measured in terms of time and effort for the non-expert user – since one of the aims of this technology is to ease the task for non-experts – and these can be used as comparative measures of efficiency between systems. In the case of the SCREEN–SCOPE suite in Oxford Brookes, adding GIS to the system also added running costs of various types:

First, the user must be *asked* all the information needed about the map base and whether maps exist containing the information required:

- the names of those maps;
- the items of information contained in those maps relevant to the consultation;
- the names of each of those items of information.

This may take up a considerable proportion of the user-consultation time: if almost 30 per cent of the knowledge base is taken up by these enquiries, we can expect a similar proportion of the user-consultation session to be taken up by this. Also, this approach assumes that the user *knows* all this information, which may not be true in all cases, and the user may have to seek advice from others who are closer to the map base or who were instrumental in its creation, thus detracting from one of the objectives of ES. These problems are going to arise every time the system is run, every time a new project is processed for the same area. This suggests that one way to alleviate these problems – *not* used in the Oxford Brookes system – is to separate the enquiry about the map base from the enquiry about the project, and "store" the former for later (repeated) use applied to other projects in the same area. On the other hand, the effectiveness of this solution depends on the expectation of having to deal with future projects relating to *the same* area.

Second, to activate some procedures, additional questions to the user are occasionally required (like for instance about the directions of the dominant winds) which lengthen the consultation.

Finally, the GIS-handling procedures themselves take time: communicating with the GIS takes time and performing some of the GIS functions required for the various enquiries also can be time consuming. To evaluate this aspect of the performance of the GIS link, various procedures were timed and, in particular, the various GIS operations (the individual AML steps triggered off by the procedures) were also timed.

In order to carry out these tests, a hypothetical project was defined in the Arc-Info GIS – consisting of various buildings and structures, with parking space, etc. – and a hypothetical setting was also defined – a rural area with a nature reserve, a river running through, etc. In addition, the project required an access road, and it also discharged to the river.

One of the problems of trying to time the performance of the various procedures was that using "absolute" time as a measure of efficiency would be dependent on: (i) the type of hardware used; (ii) the "load" on the system (the number of other users) at the time of the tests; and (iii) how complicated the particular maps (the project's and the area's) were. A more reliable measure of time had to be used, one that was not dependent on the speed of the system, and one also valid for projects of varying degrees of complexity. The approach used was to use as a *unit of reference* the length of time that a "standard" GIS operation took, in particular, *the time it took the GIS to copy the project map*, which also brings into consideration the relative complexity of the particular case. Average timings were used, and tabulating the results for some of the typical operations (Table 6.1) is quite revealing, especially the *ratio* figures, as the absolute-time figures are hardware-specific and also likely to become outdated with the progress of computer speed.

This is only a comparative evaluation, and it will probably be made irrelevant by future increases in computer speed, when even the slowest of these operations can be performed almost instantly. However, until this happens, this list suggests that, while there are fairly innocuous GIS operations which "cost" relatively less to an ES consultation (describing maps, adding/deleting items from maps), others do add considerable time, precisely the GIS operations used most frequently in IA: *buffering, clipping, extracting sub-maps*. In practice, this means that, at some stages in the consultation, the process almost comes to a halt for the user, as some procedures can take almost one minute. If, for instance, the system has to check if the project is within a certain distance of a series of sensitive areas, this involves repeated buffering and clipping, once for each of the possible areas being tested, and the total time can add up to minutes rather than seconds, totally interrupting the natural flow of the consultation and going "against the grain" of the idea of expert systems as support tools.

Table 6.1 Times added by GIS procedures

	Time (s)	*Ratio* (*to "copy"*)
Arc-Info commands		
Copy (unit of comparison)	9	1
Moving around		
Arc (starting Arc-Info)	5	0.6
Arcedit (moving into the map editor)	33	3.7
Tables (moving into the Tables section)	18	2.0
Quit (exiting Arc-Info)	8	0.9
GIS functions		
Describe (describing the components of a map)	7	0.8
Frequency (frequency distribution of map features)	24	2.7
Statistics (basic descriptive statistics of map features)	15	1.7
Dropitem (deleting an item from a map)	8	0.9
Buffer (making a buffer map)	52	5.8
Clip (clipping one map with another)	34	3.8
Reselect (extracting a sub-map from a map)	32	3.6
Cursor (accessing individual features)	9	1.0

6.7 CONCLUSIONS

We have seen how project screening and scoping – being quite regulated areas of IA – are good testing grounds for ES technology. Even the simple logic of these two stages in the IA process already points out some of the critical areas of ES design. For such reasons, screening and scoping provide a good gateway into the discussion of the practical application of ES to different aspects of IA. Also, we have seen that these two IA tasks are so closely linked that an ideal arrangement for their ES treatment is to design inter-linked systems that feed from each other, as the Oxford Brookes suite shows.

The practical examples discussed here also have provided the opportunity to test the idea of linking ES with GIS – one of the central themes of this book. This discussion has thrown up an interesting tentative conclusion, that the evaluation of the GIS link seems to produce a slightly negative net balance of advantages and disadvantages, because of the additional time it takes for the GIS link to work and due to the disproportionate complication of having to describe the structure and composition of the map base for the area each time the system is used (all the environmental maps, their features, etc.). This is a point similar to the one made in Chapters 3 and 4 when discussing GIS use in general: when arguing that the costs of setting up a map base for a "one-off" use were probably the most important deterrent to the use of the – otherwise extremely powerful and appealing – GIS technology. In a way, that same point re-emerges now when we discuss linking ES to GIS: unless there are simple ways of transmitting to the ES the general cartographic information needed to apply the system to a particular project, such application is going to be overloaded with the need to use a lot of time and effort simply to "find out" what maps are available and what their contents are.

On the other hand, this slightly negative conclusion brings up an issue which will appear again in our discussion, about the lack of good, standardised, nationwide environmental data in digital form. To the extent that digital environmental data – initially provided in a rather disjointed way by different branches of the various agencies concerned with the environment (see O'Carroll, 1994 for an early review) – becomes available in a standardised format for all agencies and covering the whole national territory (in the image of Ordinance Survey), these problems, including the problem raised in this chapter, do not arise or are greatly reduced. ES would still have to find out in their runs about the *availability* of maps (easy enough to cover with a few questions in the general consultation) but, once determined that a certain map is available, nothing further would need to be investigated about its structure, which would be known in advance and could be "interrogated" directly by the procedures in the ES at practically no extra time-cost to the user.

REFERENCES

Department of the Environment (1988) *Environmental Assessment*, Department of the Environment Circular 15/88 (Welsh Office Circular 23/88), 12 July.

Department of the Environment (1989) *Environmental Assessment: A Guide to the Procedures*, HMSO, London.

Department of Environment, Transport and the Regions (1999a) *The Town and Country Planning (Environmental Impact Assessment) (England and Wales) Regulations 1999*, DETR No. 293.

Department of Environment, Transport and the Regions (1999b) *Environmental Impact Assessment*, DETR Circular 02/99.

Dreyfus, H.L. (1972) *What Computers Can't Do: The Limits of Artificial Intelligence*, Harper & Row, New York.

Morris, P. and Therivel, R. (eds) (1995) *Methods of Environmental Impact Assessment*, UCL Press, London (1st edition).

Morris, P. and Therivel, R. (eds) (2001) *Methods of Environmental Impact Assessment*, UCL Press, London (2nd edition).

O'Carroll, P. (1994) (with Rodriguez-Bachiller, A. and Glasson, J.) *Directory of Digital Data Sources in the UK*, Working Paper No. 149, Impact Assessment Unit, School of Planning, Oxford Brookes University.

Petts, J. (ed.) (1999) *Handbook of Environmental Impact Assessment*, Blackwell Science Ltd, Oxford.

Petts, J. and Eduljee, G. (1994) in *Environmental Impact Assessment for Waste Treatment and Disposal Facilities*, John Wiley & Sons, Chichester (Ch. 8).

Rodriguez-Bachiller, A. (1991) Expert Systems in Planning: An Overview, *Planning Practice and Research*, Vol. 6, No. 3 (Winter), pp. 20–5.
Wathern, P. (ed.) (1988) *Environmental Impact Assessment: Theory and Practice*, Routledge, London.
Weston, J. (2000a) *Screening, scoping and ES Review Under the 1999 EIA Regulations*, Working Paper No. 184, School of Planning, Oxford Brookes University.
Weston, J. (2000b) Personal communication.

7 Hard-modelled impacts

Air and noise

7.1 INTRODUCTION

After discussing in the previous chapter issues of ES design applied to some of the initial stages of IA – screening and scoping – we are now going to move into its "core": the prediction and assessment of impacts. The prediction of specific impacts always follows variations of a logic which can be sketched out as in Figure 7.1.

Different areas of impact lend themselves differently to each of these steps and give rise to different approaches used by "best practice". We are going to start this chapter by looking at some areas of impact prediction characterised by the central role that *mathematical simulation models* play in them. As we shall see, this should not be taken to imply that the assessment is "automatic" and that judgement is not involved, far from it: issues of judgement arise all the way through – concerned with the context in which the models are applied, their suitability, the data required, the interpretation of their results – but the centre stage of the assessment is occupied by the models themselves, even if the degree of understanding of their operation can vary. When these models are run by the experts themselves – who know their inner workings and understand the subtleties of every parameter – they can be said to be running in "glass-box" mode. On the other hand, in a context of "technology-transfer" from experts to non-experts – which expert systems imply, in line with the philosophy of this book – models can be run in "black-box" mode, where users know their requirements and can interpret their results, but would *not* be able to replicate the calculations themselves. It is this *transition* from one mode of operation to another – the explanation and simplification needed for glass-mode procedures to be applied in black-box mode with maximum efficiency – that we are mainly interested in.

Of all the areas of impact listed in the last chapter, two stand out as clear candidates for inclusion in this discussion – air pollution and noise. Their assessment is clearly dominated by mathematical modelling, albeit with all the reservations and qualifications that will unfold in the discussion.

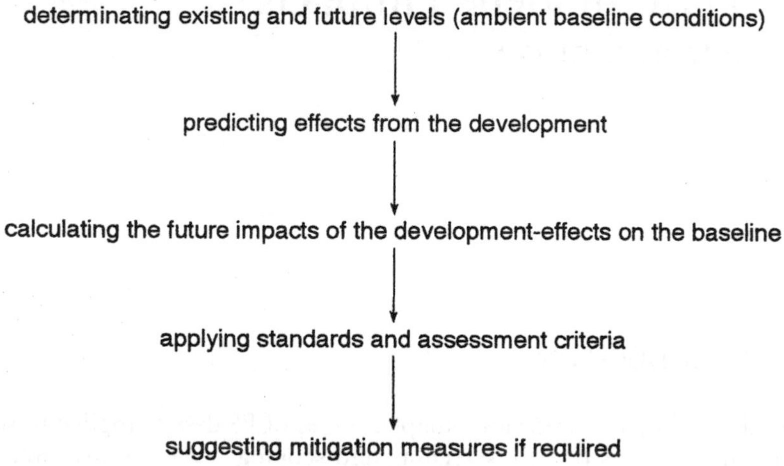
determinating existing and future levels (ambient baseline conditions)
predicting effects from the development
calculating the future impacts of the development-effects on the baseline
applying standards and assessment criteria
suggesting mitigation measures if required

Figure 7.1 The general logic of impact prediction.

7.2 AIR POLLUTION

In common with other impacts, the prediction of the air pollution impacts from a development can be applied at different *stages* in the life of the project (e.g. construction, operation, decommissioning), and at different stages in the IA:

- consideration of alternatives about project design or its location
- assessment (and forecasting) of the baseline situation
- prediction and assessment of impacts
- consideration of mitigation measures.

The central body of ideas and techniques is the same for all stages – centred around simulation models – but the level of detail and technical sophistication of the approach vary considerably.[22]

7.2.1 Project design and location

At the stage when the precise characteristics of the project (equipment to be used, types of incinerators, size and position, etc.) as well as its location are

22 The knowledge acquisition for this part was greatly helped by conversations with Roger Barrowcliffe, of Environmental Resources Management Ltd (Oxford branch), and Andrew Bloore helped with the compilation and structuring of the material. However, only the author should be held responsible for any inaccuracies or misrepresentations of views.

being decided, it would be possible to run full impact prediction models to "try out" different approaches and/or locations – testing *alternatives* – producing full impact assessments for each. However desirable this approach would be (Barrowcliffe, 1994), it is very rare as it would be extremely expensive for developers. Instead, what is used most at this stage is the *anticipation* of what a simulation would produce – based mostly on the expert's experience and judgement – as to what the model is likely to produce in varying circumstances, applying the expert's "instant" understanding he/she is capable of, as mentioned in the previous chapter. The range of such circumstances is potentially large; however, in practice, the most common air pollution issues are linked to the effects of buildings and to the effects of the location. To the expert's judgmental treatment of these issues are also added questions of acceptability and guidance, to be answered by other bodies of opinion.

With respect to the effect of *buildings*, the main problem is that the standard simulation models used for air dispersion do not incorporate well the "downwash" effects that nearby buildings have on the emissions from the stack (although second-generation versions are trying to remedy this, as in the case of the well-known Industrial Source Complex suite of models). Her Majesty's Inspectorate of Pollution (HMIP) produced a Technical Note in 1991 (based on Hall *et al.*, 1991) discussing this issue for the UK, and a rule-of-thumb that is often used (Barrowcliffe, 1994) simply links the relative heights of the stack and the surrounding buildings, stating that the height of the stack must be at least 2.5 times that of nearby buildings.

The crucial location-related variable concerning the anticipation of air-pollution impacts at this stage is the height and evenness of the *terrain* around the project, as air-dispersion simulation models find irregular terrain (which make local air flows variable) difficult to handle. Such situations can be "approximated" using versions of the standard model – like the Rough Terrain Diffusion Model (RTDF) (Petts and Eduljee, 1994, Ch. 11) – with its equations modified for higher surrounding terrain. However, the effect of irregularity in that terrain is still a problem, until more sophisticated simulation models are produced and tested, and looking at previous experiences in the area is often still the best source of wisdom. This also applies to another location-related issue: the possible compounding of impacts between the project in question and other sources of pollution in the area, through chemical reaction or otherwise. This connects with the general area of IA known as "cumulative impact assessment", an example of which can be found in Kent Air Quality Partnership (1995) applied to air pollution in Kent. This is possibly the only aspect at this stage where GIS could play a role, albeit limited, identifying and measuring proximity to other sources of pollution.

Finally, in addition to these technical "approximations" – short of running the model for all the alternative situations being considered – consultation

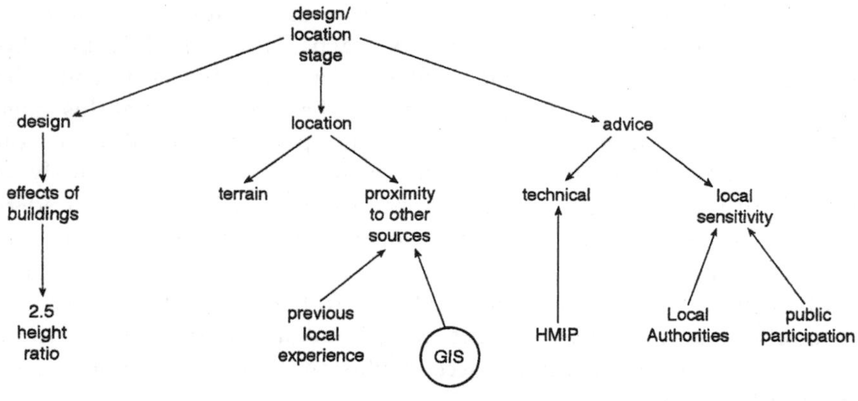

Figure 7.2 Information about project characteristics and location.

with informed bodies of opinion must be used. On the one hand, there may be technical issues of project design on which responsible agencies (like HMIP/Environment Agency) can give opinion and guidance. On the other hand, and more important at this stage, the relative *sensitivity* of the various locations must be assessed in terms of public opinion, and local authorities and public opinion are often the best source for this information (Figure 7.2).

7.2.2 Baseline assessment

Assessing the baseline situation with respect to a particular impact usually involves, on the one hand, assessing the present situation and, on the other, forecasting the situation *without* the project being considered. Baseline assessment is a necessary stage in IA. However, with respect to air pollution, it does not seem to exercise the mind of experts beyond making sure to cover it in their reports. This maybe due to the fact that this stage does not really involve the use of the technical tools (models) and know-how which characterises their expertise.

The first task, assessing the present situation, does not involve any impact simulation, but simply the recording of the situation with respect to the most important pollutants (for a complete list, see Elsom, 2001). These can be grouped as follows:

- chemicals (sulphur dioxide, nitrogen oxides, carbon monoxide, toxic metals, etc.)
- particulates (dust, smoke, etc.)
- odours.

This recording could be done directly by sampling a series of locations and collecting the measurements following the techniques well documented in manuals. In developed countries this is rarely done, as it is possible to get the information from local authorities and environmental agencies who run well-established *monitoring programmes* for the relevant pollutants (particularly chemicals and particulates). In the UK, various short-term and long-term monitoring programmes for different types of areas (see Elsom, 2001, for more detail) are also made available via the National Air Quality Information Archive on the Internet. This is not the place to discuss in detail such agencies or programmes, but only to mention these sources for the interested reader. The point of interest to us is that this aspect of baseline assessment does *not* involve any impact simulation nor any running of the model. It is enough to know which agencies to contact and which chemicals to enquire about:

- Local authorities are the first-choice sources (Barrowcliffe, 1994); it is common for them to have well-established air-quality networks covering traditional pollutants (such as smoke or nitrogen and sulphur dioxides) but also covering sometimes other pollutants. It is always good practice to contact them for data that may represent better the environment local to the project site rather than national surveys and networks.
- The National Air Quality Information Archive Internet site provides information about concentrations of selected pollutants for each kilometre-square in the country (Elsom, 2001).
- The Automatic Urban Network (AUN) provides extensive monitoring in urban areas for particulates and oxides.
- For other chemicals, agencies can be found running more specific monitoring programmes, like the one for Toxic Organic Micropollutants (TOMPS) in urban areas.
- More *adhoc* monitoring programmes can also be found in previous Environmental Statements for the same area.
- If the area is not covered by any on-going or past monitoring, on-site pollution monitoring may be required at a sample of points, as the lack of credible baseline data may compromise the integrity of the air-quality assessment (Harrop, 1999).
- Odour measurement is a difficult area, it can be undertaken scientifically by applying gas chromatography to air samples, but the method most commonly used in the UK is by olfactory means using a panel of "samplers".

For the second task, forecasting the future air pollution without the development, future changes can refer to two sets of circumstances: (i) the whole area changing (growing in population, businesses, traffic, etc.); (ii) specific new sources of pollution being added to the area (new projects in the pipeline, an industrial estate being planned, etc.).

The pollution implications of expected changes – if any – in the whole area, can be forecast with the so-called "proportionality modelling" (Samuelsen, 1980) which assumes changes in future pollution levels to be proportional

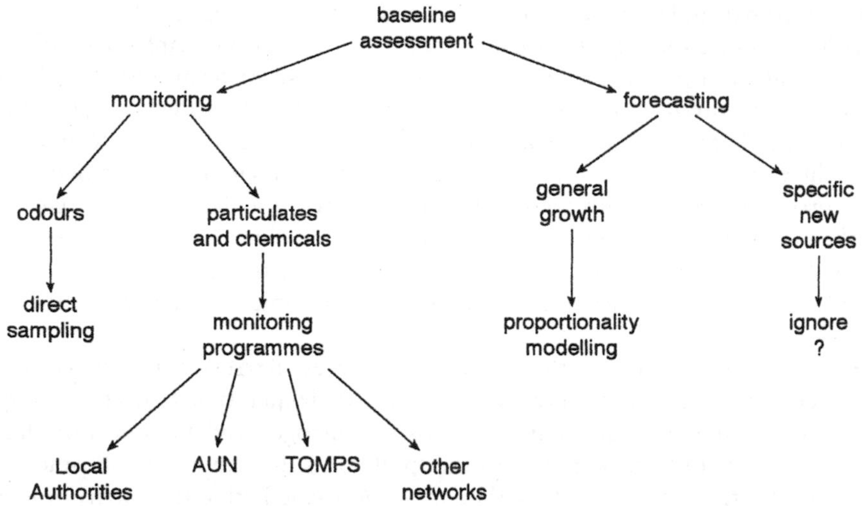

Figure 7.3 The logic of baseline assessment.

to changes in the activities that cause them, and future pollution levels can be estimated by increasing current levels by the same rates of change expected to affect housing, traffic, etc. As indicated by Elsom (2001), DETR (2000) provides guidance to local authorities on projecting pollution levels into the future. With respect to forecasting pollution from *specific* new sources expected in the area, these sources are not included in the general growth counted in a proportionality modelling exercise – as their effects are likely to be localised and not general – and, in practice, this forecasting is not done, the reason being the very low real usefulness of such forecasts, were they to be produced. The accuracy of air-dispersion models (the most commonly used type of model) is quite low and, as we shall see in the next section, the results can be inaccurate by a factor of *two* (equivalent to saying that they can be out by 100 per cent) at short range, and even more at long distance. This has repercussions when it comes to forecasting air pollution from the project, but it has even more crucial repercussions when forecasting the baseline. The baseline forecast is supposed to provide the basis for comparison of the predicted impacts from the project, but if that basis can be out by up to 100 per cent, any comparison with the predicted impacts becomes meaningless (Figure 7.3).

7.2.3 Impact prediction and assessment

As textbooks and manuals show, the approach that has dominated this field from the 1980s (Samuelsen, 1980; Westman, 1985; Petts and Eduljee, 1994; Harrop, 1999; Elsom, 2001) is based on the so-called "Gaussian dispersion model" which simulates the shape of the plume (assumed to settle into a steady-state shape) as it bends into its horizontal trajectory and then disperses and oscillates towards the ground downwind from the source. At any point, the cross-section of the plume is assumed elliptical, with elliptical "rings" showing varying concentrations of pollutants – stronger towards the centre and weaker towards the edges. The distribution of the levels of concentration *between* rings is assumed to be "normal", in the statistical sense of the word ("Gaussian"), bell-shaped, both horizontally and vertically, and becoming "flatter" in both directions with distance from the source, making the sections of the plume larger (Figure 7.4). The rates at which these cross-sectional distributions of pollution concentrations become "flatter" with distance in the horizontal and vertical directions,[23] making the section of the plume bigger, are crucial to the behaviour of the plume and to the variation of its impacts with distance. The vertical spread in particular is crucial in the estimation of the concentrations of pollution that will "hit" the ground (the ultimate objective of the simulation) at different distances. These rates of spread, in turn, vary with the atmospheric

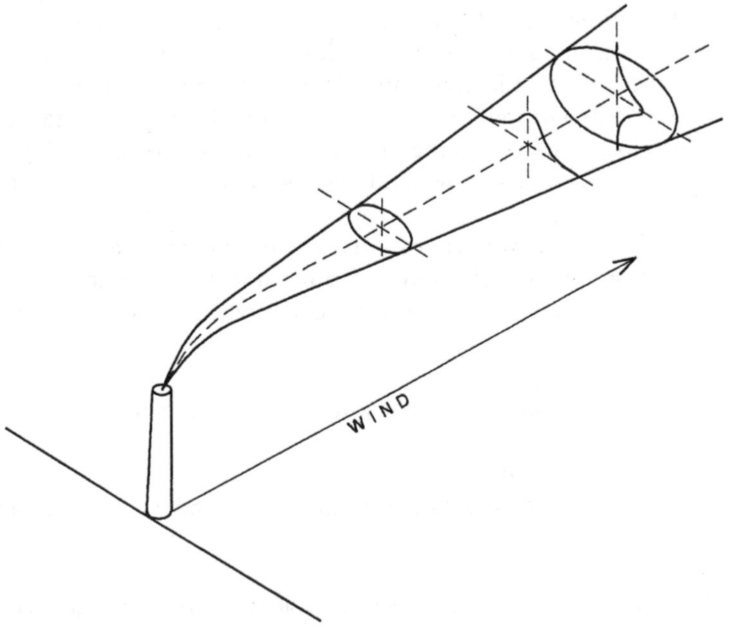

Figure 7.4 The Gaussian pollution-dispersion model.

23 These rates are usually measured by the Standard Deviations σ of the horizontal and vertical Gaussian distributions of pollution concentration.

conditions[24] – determined by wind speeds, temperatures at different distances from the ground, etc. – which become the crucial variables determining the behaviour of the model.

The mathematical details of this model are well documented (Barrowcliffe, 1993; Samuelsen, 1980; Westman, 1985) and what interests us more is not how the model works, but *how it is used*. Were this model to be used in "glass-box" mode, its equations would be applied to all combinations of wind speeds and directions relevant to the area, in the various atmospheric conditions that affect the area, applying different "rates of spread" at different distances, etc. In practice, however, the model is most commonly used in "semi black-box" mode – which corresponds better to the philosophy underlying our discussion – so that the equations have been programmed into a computer model (see Section 7.2.3.2 below) and all these variables (wind, atmospheric conditions, spread) are usually already combined in the meteorological data fed into that computer model. In the UK, the standard data-set provided by the Meteorological Office has already been *pre-processed* to suit this kind of use; it consists of a multi-variable frequency distribution, over a 10-year period, of wind directions,[25] wind speeds and atmospheric conditions that apply to the area being investigated.[26] If there is a weather station very close by, the data for the frequency distributions will come from that station. If there are no weather stations nearby, the pre-processing of the data will include (at extra cost): (i) selecting from the nearest surrounding stations those whose conditions (topographic, etc.) are more like those of the area of interest; and (ii) calculating weighted averages of the data from different stations, using as weights the inverse distances from each station. In any case, it is the provider of the meteorological data who takes care of the complications, and the model-user runs the model with that data.

This model runs on two sets of *data*: meteorological data as discussed, plus information about each source of pollution. In the simplest case, it is a *point source* involving a stack (the most common case), and the information required refers to:

- geometry of the source (stack height, internal diameter, area)
- temperature of emissions
- concentration of pollutants
- emission rate (velocity, volume before and after the addition of warming air).

24 So-called "Atmospheric Stability Conditions", classified originally by Pasquill and Gifford into six types (A, extremely unstable; B, moderately unstable; C, slightly unstable; D, neutral; E, slightly stable; F, moderately stable) and often simplified – for example by the Meteorological Office in the UK – into only three categories: unstable, neutral and stable.

25 16 sectors.

26 Quantifying the proportion of the total recorded period in which each combination of wind direction, wind speed and stability condition, was present.

When, instead of information about the emissions, there is only information about the processes producing the pollutants and their engineering (type of process, type of incinerator, power, etc.), we must go to documentary sources to translate such information into the data needed for the model. Sometimes we can get the "destruction efficiency" of a process (an incineration, for instance) which, by subtraction, will give us the emission rates of the residuals.

This type of information must be normally provided for a variety of pollution sources, some point sources with stacks, others of a totally different nature or shape (area sources, traffic line sources, dust) all to be simulated in their effects. Harrop (1999) lists the typical emissions from a variety of projects, from power stations to mining and quarrying. For impact assessment, an overall *emissions inventory*[27] should catalogue each source and provide for it the relevant emission data to be combined with the atmospheric data for the simulation. The final set of data which is needed in some special cases to run these models – as we shall see in the next section – is about the terrain (altitudes, slopes, etc.) and the built environment (buildings nearby, heights, etc.) if applicable. It is only in the provision of such data automatically that GIS can have a role to play at this stage (Figure 7.5).

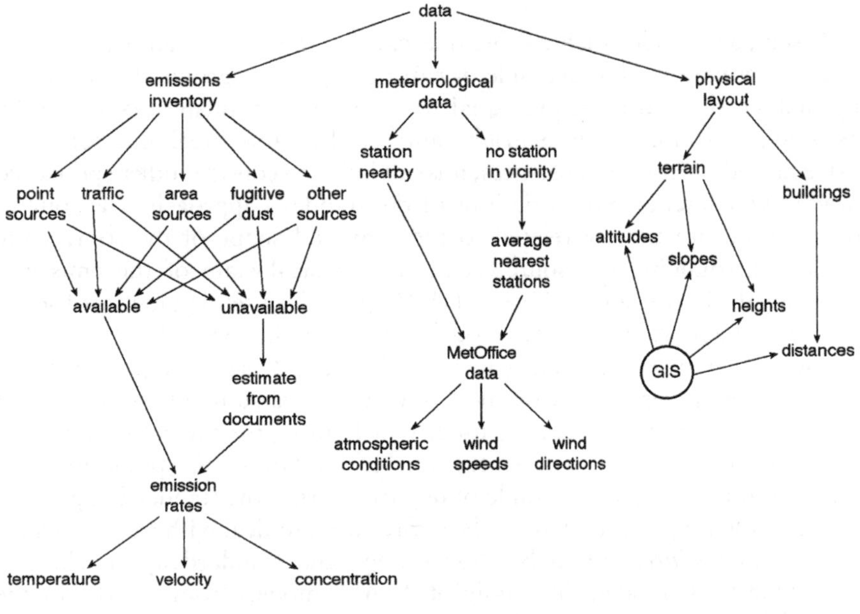

Figure 7.5 Data requirements for the pollution-dispersion model.

27 Harrop (1999) argues that the investigation of emissions should be directed at any pollutants with health risks, and not just those which are regulated.

7.2.3.1 *Variations in the modelling approach*

The model described above represents the cornerstone of air-pollution impact assessment – as it applies to gaseous emissions from a point source into the atmosphere – and it is by far the most frequently used, with versions of it available in different countries, like the ADMS collection in the UK (Elsom, 2001). Harrop (1999) also contains a useful list of computer-based air-dispersion models. Most of these models try to replicate and improve on the performance of the classic example from the US Environmental Protection Agency, the "Industrial Source Complex" model, which incorporates all the features discussed above, and which has also been improved over the years to provide additional flexibility in addition to the standard approach (ERM, 1990) with:

- versions of the model for long-term and short-term averages (1–24 h);
- consideration of an urban or rural environment;
- evaluation of the effects of building waste;
- evaluation of the dispersion and settling of particulates;
- evaluation of stack downwash;
- consideration of multiple point sources;
- consideration of line, area and volume sources;
- adjustment for elevated terrain.

A standard model such as this one can be adjusted to simulate a wide range of situations. For example, it can be applied to ground-level sources by making the source height equal to zero, or to a small area source by assuming a source of zero height and of the same area. But for more extreme and precise circumstances, it is advisable to consider other specialised models which tend to be variations of the standard approach. The sources of variation are usually related to the type and shape of the source, the terrain surrounding the source, and the physical state of the emission. The Royal Meteorological Society (1995) provides useful guidelines for the choice of the most appropriate model (quoted in Harrop, 1999).

Sources can be *multi-point*, which can be treated as several point sources and dealt with separately, or models (such as versions of the Industrial Source Complex model) can be used, which allow for several sources and consider the separation between them in its simulations. Air pollution from *traffic* is another typical example of departure from the standard approach, and a whole range of models has been produced to deal with this particular type of *line source*, often by "extending" the standard approach, like the Dutch CAR model, the family of "CAL" models from the US, or the AEOLIUS collection developed in the UK (Elsom, 2001). For example, the PREDCO model (Harrop, 1999) produced in the 1980s by the Transport Research Laboratory in the UK divided up the line sources (each road) into segments, and represented each segment by an equivalent point source, whose effects were simulated in the standard way using data about traffic flows and speeds to calculate emission rates.

To incorporate the effects of *higher terrain*, the standard model can be modified by subtracting the height of the terrain from the stack height – when the height of the terrain is no greater than the stack – (version 2 of the Industrial Source Complex can do that) or the whole trajectory of the plume may be assumed to change direction and "glide" above the hills when the height of the terrain is greater than that of the stack. A typical example of such a model is the Rough Terrain Diffusion Model (Petts and Eduljee, 1994, Ch. 11), including topography as high or higher than the release height, and also varying slopes of the hills or ridges. However, such a model requires additional information about terrain height between the emission source and every receptor of interest. If this data is not given, the model runs as a flat-terrain model.

Sometimes the variation from the standard approach is due to the physical state of the release (the standard model is ideal for gaseous emissions). One typical case is when the emissions are *dense gases* (gases heavier than air) which fall and spread on the ground rather than rise and disperse with the air. Specialised models have been built for this case, such as the DEGADIS model quoted in Petts and Edulgee (1994, Ch. 11), after Havens and Spicer (1985). Another typical case is that of "particulates" (*dust* specifically), which are not buoyant in the air like heavier gases, but travel in it carried by any wind blowing at speeds above 3 m/s (approximately 10 km/h). Larger particles will travel shorter distances (up to 100 m) and lighter particles will travel longer, depending on wind speeds. The model that experts apply is much simpler than the dispersion model, expressed as a mathematical relationship between distance travelled, wind speeds and particle size (ERL, 1992; ERM, 1993). This approach starts with the location of any potentially *sensitive receptors*, and then the use of wind data (similar to the data for the standard model) to work out what proportion of time winds will be able to carry dust certain distances away in the direction of those receptors, so that the impact of the heavier dust particles – if any – can be established. Smaller particles will be transported further away only by stronger winds, and the meteorological data will indicate what proportion of the time they are likely to be present in the directions towards the receptors. This is a typical approach used to assess the impacts of the *construction stage* of most projects, when dust pollution is one of the most important effects, and commercial models like the Fugitive Dust Model (developed by USEPA) are routinely used in the UK.

The impact of *odours* is also problematic to predict and requires a departure from the standard modelling approach. Very short-term concentrations are sufficient for an unpleasant impact but once the emission escapes from the source, it is diluted in the atmosphere at a rate which increases rapidly with distance (ERM, 1993). For these reasons, low wind-speeds (typical of the two opposite extreme atmospheric-stability conditions A and F, see Note 3) will be the ones conducive to higher odour concentrations. This means that, in practice, a similar approach is used for odours and for dust: (i) sensitive receptors are identified; (ii) the frequency of *extreme* stability conditions with winds in the direction of the receptors are identified in the meteorological data; and (iii) travel distances for sufficient concentrations of the odour-producing substances are determined and checked against the distance of the sensitive receptors.

At the extreme end of buoyant emissions, *flares* pose special problems because of their extreme buoyancy, and usually require special treatment. Finally, another challenge for experts is the prediction of *fugitive emissions* (other than dust) such as leaks from the equipment, valves, and release of pollutants at ground level as a result of handling. These also require special treatment and are extremely difficult to reduce to a simple set of rules which can be dealt with in "black-box" mode.

All these models use the same type of data (frequency of wind speeds and directions in different atmospheric conditions), but the choice of model is not trivial, and is an important part of the expertise, which has gradually replaced the ability to work out the equations by hand (which would express the expertise in a "glass-box" world) (Figure 7.6). Again, the possible role of GIS in these considerations is quite small, probably limited to identifying the kind of terrain where the experiment is being carried out.

Elsom (2001) argues that these models should not be used as "black boxes" and we can see from our discussion that the user needs to exercise judgement and understanding – even when using off-the-shelf software – in order to:

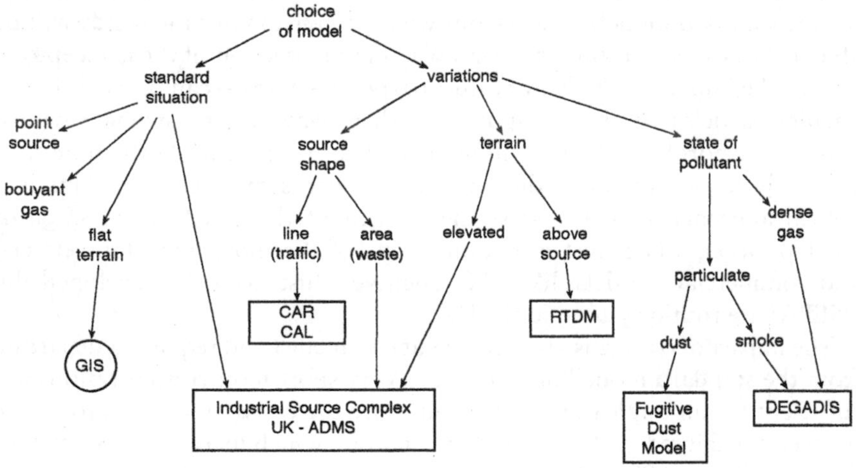

Figure 7.6 Choice of air-dispersion model.

- recognise the different situations when different models best apply;
- know when to use different "modes" and parameters available in the models;
- understand the outcomes of the models;
- understand the limitations and inaccuracies of the models;
- recognise the "boundaries" of the situations when models perform less well, and other approaches might be more effective.

7.2.3.2 Model output and accuracy

Irrespective of the model used, two impact scenarios are typically used for the predictions: (i) the most "representative" case, the most frequently encountered situation; and (ii) the worst case, the worst "peaks" of impact. In practice, these scenarios are represented "by proxy": the most representative situation is measured by a long-term average (usually an *annual average*) of ground-level concentration of pollutants, and the worst peak by a short-term average (usually a *hourly average*, which can be extended up to 24-hour averages). These averaging times are directly connected to the standards of air quality normally used, often derived from either EC directives or from the World Health Organisation (WHO). EC standards tend to be expressed as yearly averages, while WHO standards (revised in 1997) also use shorter averages (hourly or shorter, daily, weekly), and the UK National Air Quality Strategy has adopted both approaches since 1997. Elsom (2001) contains good summaries of all three sets of standards for the WHO, EC and UK, and Harrop (1999) also contains a useful international comparison of standards.

These averages are calculated automatically by the model, different values are estimated for different directions and distances (in an area within about 10 km around the project), and the results are normally presented in a variety of forms: (i) as maps showing the spatial distribution of values (especially for annual averages), often in the form of *contour maps* of total predicted pollution, after adding to the model predictions the baseline values at different locations; (ii) as profiles of distributions of values with distance for different atmospheric conditions (especially for short-term averages); and (iii) as sets of maximum values (some extracted from the previous profiles and maps) to be compared to the relevant standards. Because, in these models, ground-level concentration is directly proportional to the emission rate at the source (assuming the other parameters are the same), the results can be easily *scaled up or down*. The model produces a pattern of ground-level concentrations of any gaseous pollutant emitted at a certain speed and temperature. To adapt the results from one pollutant to another – or from one level of operation of the equipment to another – we only have to multiply the results by a factor reflecting the relationship between the new conditions and the original ones (for instance, if one chemical is emitted at half the rate of another, its simulated levels of concentration will also be halved).

After simulating the dispersion of pollutants – Harrop (1999) lists the air pollutants that have health effects on humans – and, the assessment of impacts should only require, theoretically, a comparison of the expected ground concentrations with the various standards (usually expressed in μg[28] or in mg/m^3) available for a whole range of pollutants:

- Sulphur dioxide and suspended particulates (which can act in synergy);
- Nitrogen oxides (except N_2O, which is usually harmless);
- Carbon monoxide and dioxide (mainly from fossil-fuel consumption);
- toxic/heavy metals (lead, nickel, cadmium, etc.) when relevant;
- Chlorofluorochlorides (CFCs) related to ozone depletion;
- Photochemical oxidants (like low-level ozone);
- Dioxin;
- asbestos;
- dust;
- smoke;
- odours.

These standards are regularly extended and refined, and come typically from three types of sources (Bourdillon, 1995, Ch. 2): the World Health Organisation, the European Community, or UK legislation (often derived from the other two). These different sets of standards are not always expressed in the same way. For some pollutants (like CO, from the WHO) there is no standard for a yearly average; for some (like SO_2 or NOx) both the EU and the WHO provide standards for annual averages but only the WHO has one for hourly averages, and the standard most appropriate for each case should be identified, preferably following the aforementioned list of organisations *in reverse order*: look first for a British Standard, if unavailable, look for an EU norm, and then look at the WHO. A good source for an up-to-date version of the standards as used in practice is always current Environmental Statements, although they tend to be limited to the pollutants relevant to the particular case, and a good compilation of those most commonly used in the UK can be found in Elsom (2001).

When air-quality standards relevant to a case are not available, Occupational Exposure Limits (OELs, published annually by Health and Safety) can be used. These limits are normally defined for workers who are in an environment for a number of hours (8 hours 5 days per week) and, to translate them for use in IA they are normally lowered considerably by multiplying them by a safety factor of 1/4 to account for increased exposure time (maybe 1/10 for sensitive individuals), and this can reach extremes of 1/100 for certain chemicals as an added safety precaution. With carcinogenic chemicals, "cancer potency factors" have been calculated (for instance by

28 μg = millionth of a gramme; mg = thousandth of a gramme.

USEPA) and can be used to calculate carcinogenic risk, although they tend to use a worst-case scenario for the variables in the formula (location, duration of exposure, emission rates, absorption rates by individuals, etc.). These factors are normally corrected downwards according to more realistic circumstances in which the project will operate, adjusting downwards the expected levels of ground-level concentration, and introducing in the calculation a variable reflecting how many days in the year (out of 365) the plant is likely to be in operation.

Even when the predictions are below the normal standards, the concept of "secondary standards" can be used to consider effects on human welfare (as opposed to human health covered by the "primary" standards). Also, the evaluation of effects can extend beyond humans, and consider effects on ecosystems, including both effects on flora/fauna; and long-term depositions (of heavy metals, for instance) which could enter the food chain. These areas of evaluation, however, are normally considered beyond, or on the limits of, the normal expertise of air-pollution experts, and are usually referred to experts in other fields (ecologists, etc.).

But the basic problem of comparing any of these standards with the output from these models is the latter's generally acknowledged *low level of accuracy*. The model's accuracy will always be compromised by its inherent uncertainties, arising from a certain degree of idealisation introduced in the model and from inherent atmospheric variability and/or errors in the data. For example, it is assumed that wind direction and speed will be constant during the averaging period, and that there will be *some* wind: zero or very low wind speed makes the model's equations virtually meaningless. For these and other reasons, the accuracy of these diffusion models has been found to be quite low, as Jones (1988) showed:

- annual average concentrations and maximum hourly concentrations (independent of location) are likely to be out by 100 per cent (reality can be between half and double the prediction) at short distances – within 10 km;
- at longer distances, predictions can be out by up to 300 per cent (from one third to three times);
- if specific locations are considered (specific receptors for instance) the error factors can be much higher.

More recent research in the UK (Wood, 1997, 1999) has shown a more promising picture after auditing the air-pollution predictions for two projects: in one case, the difference between the worst predicted annual average of NOx and the worst measurement encountered (irrespective of locations) was an overprediction by about 20 per cent and, when specific locations were considered, they also were systematically overpredicted by about 20 per cent. In the other case (Wood, 1997) the R-square between predictions and actual measurements was 0.82, with small differences between predictions and measurements at all the locations. The study of this aspect of impact prediction is receiving increasing attention but, until more extensive and systematic tests are carried out, this whole approach to impact prediction will remain vulnerable to strong criticism such as that by Wallis (1998).

Hypothetically, these models could be calibrated and their errors estimated each time before applying them to a particular project, using them to simulate the sources in the area and then comparing the model's simulations with the actual baseline. The errors identified could then be used as "corrective factors" for the results of subsequent simulations by the model in that same area. Unfortunately this is impractical, as it would be impossible to identify all the sources, and even more so to collect all the information needed to simulate them. What this means in practice is that a normal statistical treatment of results – using the confidence levels attached to them to calculate the probability of overlap with the "danger zones" defined by the standards at different locations – presents problems. Good practice (Barrowcliffe, 1994) adjusts to these problems by applying some rules-of-thumb (Figure 7.7):

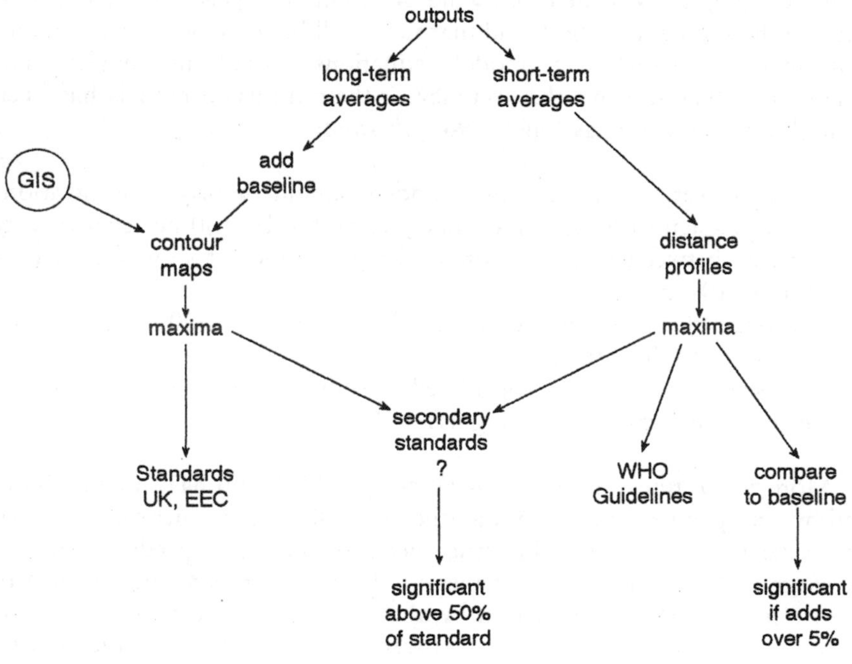

Figure 7.7 Air-dispersion model outputs and their significance.

- In the first place, and very significantly for the subject of this book, specific locations are ignored and what is taken from the simulation models' runs is only the *maximum levels* of ground concentrations, *irrespective of location* (even if the results are usually presented in map form).
- For maximum short-term averages (hourly most often, sometimes 24-hour averages), they will be considered to enter the danger zone when their level exceeds *50 per cent of the standard* as, with an error factor of 2, it could be over the limit.
- For long-term (annual) averages, in addition to the standards, *the baseline* level of concentration of the pollutant in question is used, and a project's impact is considered excessive if it adds to that level more than 5 per cent, irrespective of the standard (this rule works usually well below the levels dictated by the standards).

The model results are traditionally presented in the form of maximum values, distance profiles and maps, despite the previous comments about the unreliability of location-specific values. The maps are produced to give an indication of the general *direction*, rather than precise spatial reference, in which the worst effects will be felt, to help identify the type of area, rather than specific locations, likely to be affected (rural, urban, the coast, etc.). In a similar way to data inputs, data output in map form can also use GIS. The values for ground concentrations can be fed into a GIS and its functionality can be used to: (i) draw contour maps; (ii) superimpose them on background maps; and (iii) produce printouts. However, the relatively minor importance of the location-specific information puts the contribution of GIS also in perspective.

7.2.4 Mitigation measures

In theory, the best mitigation measures could be identified by rerunning the simulation of impacts with the particular mitigation and comparing the results with the unmitigated predictions. In practice, however, only some types of mitigation measures may require rerunning the models, as many relate to parameters in the model which we know will affect its performance *proportionally* (like emission rates), hence we can anticipate what the changes will be without running the models:

Reducing dust from traffic (both in the construction and operation stages) through:

- limiting vehicle speeds on unhardened surfaces;
- sheeting vehicles carrying soil;
- washing vehicles' wheels before leaving the site;
- spraying roads and worked surfaces.

Reducing emission rates by:

- reducing the concentration of pollutants (filters, or a variety of control systems);
- using dust-suppression equipment;
- mixing and batching concrete wet rather than dry;
- placing screens around working areas;
- covering or enclosing dumpers and conveyor belts;
- minimising drop heights for material transfer activities (unloading, etc.);
- sheeting stockpiles;
- installing filters in storage silos;
- keeping tanks and reaction vessels under negative pressure;
- installing scrubbers and odour-control units on tank vents.

To anticipate the effect of these measures we only need to quantify by how much the emission rates will be reduced, and we know that the model simulations will be reduced proportionally.

Another set of measures affects the *shape* of the emissions (especially from stacks) by, for example:

- raising the stack height;
- increasing the velocity of emission;
- raising the temperature of the emission;
- aligning stacks to increase chance of plumes merging and increasing buoyancy.

To anticipate the effect of these measures, we would either need to know the inner workings of the relevant model so that we could reconstruct the effect that the changes would have on its equations, or we would have to rerun the model with the changed parameters. Finally, another set of mitigation measures is directed to altering the *plume-diffusion* itself:

Controlling and redirecting the diffusion through:

- routing vehicles away from sensitive receptors;
- roadway trenching, embankments;
- using walls and trees;
- widening narrow gaps between buildings;
- changing the height and layout of buildings;
- roofing of open spaces.

Changing temperatures and micro-climates through:

- choice of building and road-surface materials;
- consideration of building layout in relation to areas of sunshade;

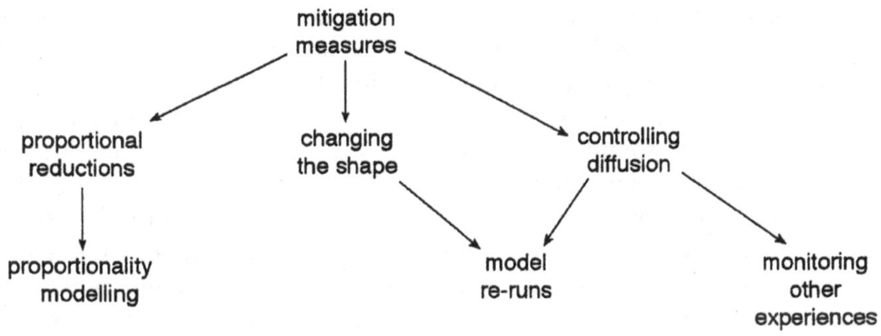

Figure 7.8 Air-pollution mitigation measures.

- tree planting and landscaping;
- preventing frost pockets with openings in embankments;
- controlling areas of standing water nearby.

The effects of these measures are even more difficult to quantify, given that models are not sensitive enough to simulate many of these changes. In some cases, such as introducing changes in the size and layout of buildings, a rerun of the models might yield results but, for changes which do not change the model's inputs or parameters, precise assessments may require using monitoring data from past experiences where similar measures have been applied (Figure 7.8). As we see, it is the capacity to generate simulations which gives strength to the whole process of air-quality assessment. While the simulations are *actual* in the core of the assessment (impact prediction), they tend to be just *hypothetical* in the design and mitigation stages, when experience and good knowledge of the model makes it possible to anticipate the expected results from the simulations without having to carry them out. In any case, what dictates what to do at any stage is the choice of model, being able to run it properly with the correct data, and being able to interpret its results in terms of impact assessment in accordance with the right standards. As we shall now see, in the field of noise impact prediction things are not too different.

7.3 NOISE

Noise impact assessment is also centred around a highly technical predictive approach, but the modelling of noise propagation is not based on a probabilistic simulation model of the type used for air pollution, but on a scientific model based on an understanding of the physics behind the phenomenon of sound. This results from a long-standing scientific tradition – the accuracy of which is well established and does not become an issue when using these models for prediction. A good account of the scientific treatment of sound modelling can be found in the classic reference by Mestre and Wooten (1980), and Petts and Eduljee (1994, Ch. 14) and Therivel and Breslin (2001) also provide useful summaries of its application to impact assessment. These and other sources illustrate how the mathematical complexity of the treatment of sound derives from the requirements to measure it using a meaningful scale. Sound can be measured in terms of its "power", "intensity" or (the most common) sound *pressure*, using very similar formulae for all three. These formulae measure sound level as a *ratio* between the actual sound and a minimum audible level. Because the resulting numbers are very high – in the formula for sound pressure the ratio is also raised to a power – the *logarithms* of these ratios are used instead. The logarithmic form of these formulae means that the resulting units ("decibels", Db) cannot be added directly. For instance, if there are two identical sources, their sound levels are added together by adding 3 Db to the sound from the single source. If we have ten such sources, the number of decibels to be added is 10, and the other intermediate values follow the curve in the Figure 7.9.

If the sources being added are not identical, the decibels to be added (to the noisier source) depend on how different the two sources are, ranging from 3 Db if both sources are equal, to 1 Db if the second source is 6 Db below the first, to virtually zero if the second source is about 20 Db

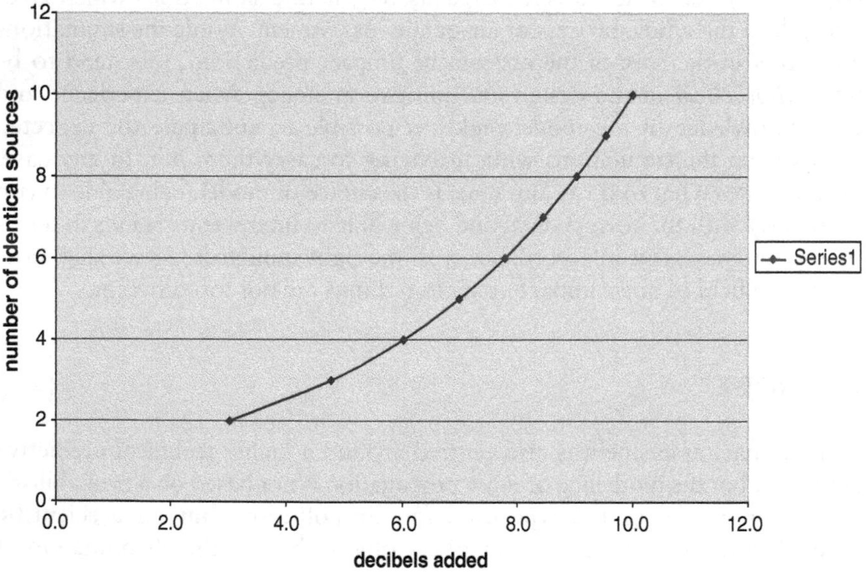

Figure 7.9 Accumulation of noise with multiple sources.

below.[29] This complication also arises when adding sound levels *over time*, for instance to calculate average levels over a certain period which, as we shall see, is central to the assessment of noise impacts.

Another mathematical complication is related to the frequency at which a sound is emitted. The perception of sound varies with its frequency and, for most part of the hearing spectrum (up to 4000 Hz), a sound at a certain number of decibels and a given frequency will be perceived as being as loud (in "fons")[30] as another sound at a *lower* frequency and a *higher* number of decibels. This means that often, in order to reduce different sounds to comparable scales of "perceived" loudness (and in order to compare them to the relevant standards), all the sounds emitted at a variety of frequencies must be converted into their equivalent at a standard frequency (usually 1000 Hz). The conversion is normally done by adding (or subtracting) to the sound level at each frequency a number of decibels, normally calculated from the so-called "A" curve, the iso-loudness curve corresponding to 40 fons. Sometimes, logarithmic aggregations are combined with this conversion process, for instance when we need to calculate the sound level from a complex source emitting at several frequencies: in order to calculate the "perceived" overall level, we must first convert the sound levels at each frequency to their 1000 Hz-equivalent, and then all the equivalent levels can be added logarithmically.

These apparent complications in the calculations are really used to adapt a complex theoretical framework necessary to understand sound propagation and perception so that it can be used in practical situations and with a realistic amount of information. As in air pollution, the modelling of noise impacts is a compromise between scientific soundness and practicability, and an important part of the expertise in this field is to be aware of how such compromise and simplification may affect the results or their interpretation.

As in air pollution, noise-impact assessment can be applied at various stages in the life of the project and/or of the impact study although in its own peculiar way. Also, various types of impacts (noise, vibration, and "re-radiated" noise transmitted through solid materials) are included under the general heading of "noise", and they present quite different challenges and require very different approaches (Figure 7.10). The following sections adopt a similar framework to that used for air pollution, adapted to these variations.[31]

29 For graphs showing this relationship as a continuous curve, see any technical references like Mestre and Wooten (1980) or Therivel and Breslin (2001).

30 The fon-level of a sound is equal to its decibels at 1000 Hz.

31 The knowledge acquisition for this part was greatly helped by conversations with Stuart Dryden, of Environmental Resources Management Ltd (Oxford branch), and Joanna C. Thompson helped with the compilation and structuring of the material. However, only the author should be held responsible for any inaccuracies or misrepresentations of his views.

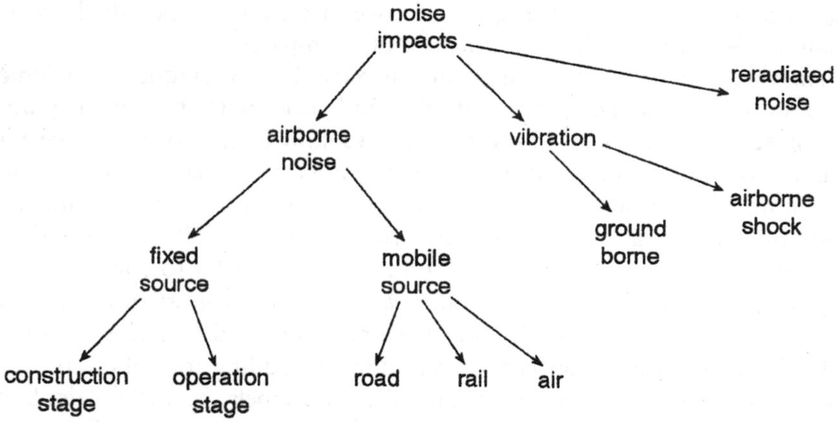

Figure 7.10 Types of noise impacts.

7.3.1 Project design

As with air pollution, noise experts can advise project designers on the basis of their "anticipated" impact calculations resulting from one design or another, not necessarily based on actual calculations or "runs" of a model, but on their expert knowledge of the mathematics involved and on their experience. Because of the nature of noise – an oscillatory phenomenon travelling in straight line in all directions – the main considerations in terms of noise impacts tend to be associated with the relationship and proximity between noise sources and receptors deemed to be potentially sensitive (e.g. housing, schools, hospitals, libraries). In particular, advice at the design stage involves:

- First, the broad identification of potentially sensitive *receptors* nearby (GIS can help with this)[32] anticipating a more systematic search to be carried out for the baseline and impact assessments.
- Second, the advice usually refers to the possible *repositioning* of noise sources and/or with the interposition of *barriers* between them and the potential receptors (very much like "anticipated" mitigation measures).

Repositioning of noise sources can be the basis for advice on possible alternative *locations* for the project further away from sensitive receptors, or it can be the basis for changes of *position* within the project: to a different

32 For example, GIS functionality can be used to identify the nearest building of certain type of use.

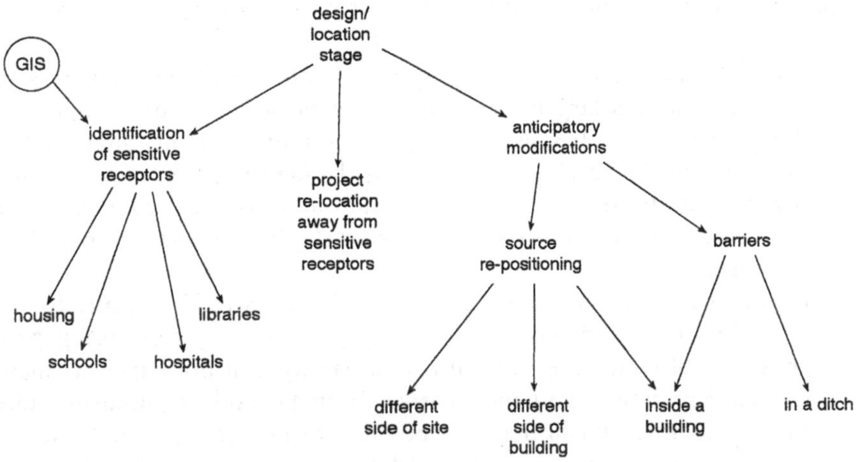

Figure 7.11 Project information for noise-impact assessment.

side of the site, from outside to inside a building, to another side of a building. Encasing a source inside a building can be similar to erecting a barrier between the source and the receptors, but other types of barriers, such as solid, vegetation or sinking the source into a ditch, can also be used (Figure 7.11).

Some of these anticipatory modifications are no different from possible mitigation measures, only here they are considered before the details of the project have been finalised. In practice, one of the problems of "anticipating" noise impacts at the project-design stage is that noise experts are often consulted at the wrong time (Dryden, 1994). It can be too early in the design process, before the developers know the details and location of all the ancillary equipment that is going to be used, before they know the location and nature of all the noise sources. Or it can be at the other extreme, *after* everything has been decided and the design finished, when changing anything can have many knock-on effects requiring further changes. The ideal situation would be if developers consulted the noise experts *while the design is taking shape.*

7.3.2 Noise baseline assessment

As in air pollution, the determination of noise levels in the area where the new project is likely to have an impact is central to the impact assessment, and the baseline situation must be measured in a way that can be compared with the relevant standards for the particular project. Again, this relates to *what* is being measured – noise in this case (no baseline study is made for vibration or re-radiated noise) – and also to its *type of manifestation*, particularly its concentration over time: averages, peak values, etc. But the approach differs in other respects, derived from two intrinsic differences:

- Air-dispersion modelling is highly inaccurate – making references to specific locations largely irrelevant – but noise modelling is not, and this gives the noise baseline assessment a totally different "shape": The first stage in the baseline assessment is the *identification of potentially sensitive receptors*, their nature, location and distance from the noise sources, and it is at those receptors that the baseline situation will be measured.
- Pollutant concentrations are difficult and expensive to measure for an individual case and we tend to rely upon existing monitoring programmes, of which there are sufficient variety. But noise measurement is relatively simple and inexpensive, with portable equipment quite easy to operate, which makes it possible to identify specific locations considered relevant to our study (the sensitive receptors) and, in the second stage of the baseline assessment, to carry out the monitoring directly.

The search for receptors starts around the project in ever-widening circles, and keeps increasing to about 500 m (only exceptionally beyond 1 km) until receptors are found in a sufficiently wide range of directions, but this may be related to the type of project. For instance, in projects involving new roads, a crucial distance for receptors is 300 m – properties within this distance can be entitled to compensation[33] – or, for projects expected to run also during the night (like a railway line, or a power station), the search distance can be much greater. The search for receptors can go further than any noise is likely to reach, but as a general rule, locating the receptors, rather than specifying the distance, is the issue (Dryden, 1994). What is crucial is to identify the *front line of receptors*. How deep beyond this front line the baseline assessment needs to go is largely determined by the expectation of success or failure in the mind of the expert, largely based on experience, of the noise-impact measurements. If the project "looks like" not violating the standards at the front-line of receptors, the search does not need to go further, as noise decreases with distance. Also, if infringements of the noise standards are clearly expected, again the measurements do not have to extend further. Only when the noise impact is expected to be marginal – violating some standards but not by much – that the identification of receptors needs to extend over a fringe of a certain *depth*, determined by the additional distance expected to make the impacts fall below acceptable levels. That distance will depend upon the source, the topography,

33 For example, in the UK, properties within 300 m of new roads are eligible for noise insulation paid for by the developer responsible, and GIS "buffering"can be used to identify them.

and a number of other factors such as the *anticipated opportunities for mitigation*. How easy and practical a reduction of noise impacts would be by mitigation are at the forefront of the expert's considerations even at this early stage.

Usually the search – or at least the "planning" of the search – is based on a map of the area, and GIS could be used to do it automatically. A field visit is useful to check local features, variations in the topography, or the potential receptors themselves, and such a visit may change our priorities over which receptors to study. For example, what is thought to be a building, a potential candidate for "receptor", from the map may be just a half-demolished shed.

Sometimes the sensitivity of the receptors is compounded by the state of local opinion, in which case the Local Authority is an important source of information as to local sensitivity and even the location of any foci of concern. It might therefore be an advantage to carry out measurements close to the properties where the occupiers are known to be concerned, in order to clarify the situation. On the other hand, if the situation has reached the point of potential conflict, baseline recordings may be limited to public rights of way, or to land owed by the same developer behind the project. Sometimes developers start tentative enquiries about possible impacts at an early stage in the design of the project, before the details are known. This can have the effect of sensitising public opinion against the project and, in such situations, recording baseline noise at some locations over extended periods of time may encounter local opposition, and a modification of the schedule of recordings may be needed.

In terms of the types of measurements to use for baseline assessment, particularly their time dimension, the same general criteria apply as in air pollution, including an idea of the general *average* levels of what is being measured (noise), plus some idea of the *peaks* to be expected over time. In sound-level measurement, there are some well-established types of indicators ranging from those which measure "averages" to those which measure "peaks":

- The "equivalent continuous noise level" (L_{eq}) is the level of steady noise which would have produced over a period of time the same energy as the various noise levels present over that same period (combined logarithmically using the A-weighting curve).
- The "background noise" is usually measured by the noise level exceeded over most (90 per cent) of a period of time (L_{90}), undertaken by recording the duration of each noise level and defining the cut-off level of the worst 90 per cent by simple statistical analysis.
- Other similar indices like L_{50} (using only the worst 50 per cent of the period, somewhere in between measuring averages or peaks) are used much less frequently.

- To measure "peaks" it is common to use an approach similar to the last two, but narrowing the fraction of the period considered to the worst 10 per cent (L_{10}).

It is widely accepted that the L_{eq} is the most useful of these indices, and most measurements (at source or at the receptors) tend to be expressed in this form. The L_{10} "peak" is also used frequently, mostly for noise measurements from mobile sources (such as road traffic or railways), as we shall see below.

These peaks or averages are measured over a finite *time period* (1 h, 12 h, 24 h, etc.) which must be selected to fit the period when the noise impacts are more likely to occur. This involves two types of considerations: (i) the *length* of the period selected: if a noise is to be produced during normal business hours and we measure its baseline as an average over 24 h, this will exaggerate the difference between the impact and the background noise; (ii) the *time of day* when the baseline measurements are taken, so that the recording periods correspond to those periods where the impacts are likely to be critical. Such periods are normally defined in the noise standards, and they tend to oscillate around the notions of *daytime* (7 a.m.–7 p.m.), *evening* (7 p.m.–10 p.m.) and *night* (10 p.m.–7 a.m.) – these are the periods used for the construction stage of a project – which are adapted slightly to different types of sources. For instance, noise from moving vehicles at the operational phase is usually measured over a different set of periods: road traffic from 6 a.m. to midnight, aircraft noise from 7 a.m. to 11 p.m., railway traffic uses 24 h for existing railways and the same periods as road traffic (6 a.m. to midnight) for new railways.

Baseline assessments result from the combination of these options, certain types of peaks or averages, over certain periods, at certain times of day, with a focus on the "worst case scenario". For example, if a project is going to produce noise 24 h a day, we do *not* compare its noise to an average background noise over 24 h, but to an average over the night period, when the impact is more likely to be significant. An interesting variation is the measurement of baseline noise when we are expecting noise from road or rail traffic: the standard approach is to measure the so-called "18 h L_{10}", which is the average (statistical, not logarithmic) of all the "1 hour L_{10}" values over the 18 worst hours of the day, from 6 a.m. to midnight. In this case, this approach is dictated by the regulations, but it serves to illustrate the need to tailor the measurement periods to the characteristics of the project and the noise impacts expected. As a project in all its stages involves a combination of many noise sources with many different baseline-measurement requirements, what is done in practice for most projects (Dryden, 1994) is to cover all the possibilities with a combination of long-term and short-term measurements:

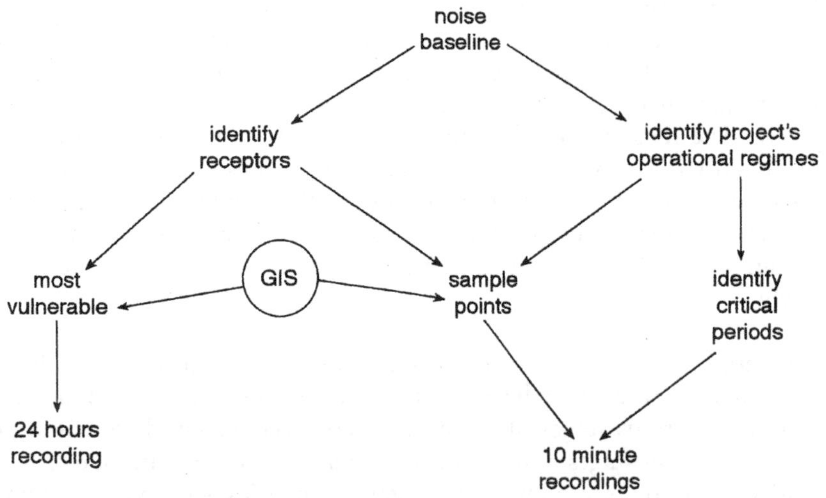

Figure 7.12 The noise baseline study.

- identify the receptor that is most likely to be severely affected;
- do at least a 24-h measurement at that receptor;
- supplement that with short term (10 min) sample measurements at other locations carried out in the daytime, the evening and the night time.

The 24-h measuring equipment is set up and, while it is running, a portable set is used for the 10-min measurements at the other sample sites. The measurements are basically the same, noise levels and durations, irrespective of what indices are going to be used later. When special measurements (like L_{10}) are required, the computer is used later to calculate the right indices from the measurements taken in the field (Figure 7.12).

7.3.3 Noise-impact prediction

The simulation of noise propagation is based on point-to-point calculations. Hence the simulation of noise impacts does not consist of simulating the noise propagation in all directions and measuring how it affects the surrounding area, as we do with air-diffusion models, but on identifying first the points of interest, and then applying to each of them relatively simple equations that reflect their circumstances. This is one of the reasons why the simulation does not run in distinct stages as in air dispersion – collecting the data, running the model, interpreting the results – but is much more *interactive*, with many "mini simulations" being done separately and then put together if necessary. Comprehensive computer packages to simulate noise propagation do exist (for roads, for railways, for surface mineral-workings and landfills) but their use is often constrained by the difficulty of getting all the information required to run them. Quite often, it is more convenient to adapt "general-purpose" computer tools like Spreadsheets (Dryden, 1994), putting in them the right formulae and seeing how different parameters, distances, etc., affect the results, quickly and interactively. The logic followed in these simulations is usually the same: (i) first, identify the noise sources and compute the noise levels at source; (ii) second, calculate whatever attenuations of those noise levels affect the different potential receptors (already identified in previous stages). The calculation of *noise levels at source* can be quite complicated, and it can involve:

- Identifying "bundles" of noise sources likely to be in operation during the same periods (like diggers and bulldozers, or turbines and incinerators) at various stages of the project: construction (itself consisting of very different stages in terms of noise generation), operation, etc.
- Identifying the noise levels and their frequencies – estimating them if not available – of each of the sources; these data are often available in catalogues and reports (even in some of the official Standards) and are provided already in L_{eq} form, which facilitates use.
- Adding together (logarithmically) the noise levels of all the sources in each bundle, sometimes having first to convert noise levels at different frequencies to the standard 1000 Hz equivalent using A-weighting (a Spreadsheet can be used quite effectively for this, with the advantage that all the individual values are accessible and can be altered easily).

With respect to the *propagation* of noise, noise energy at source can be used to calculate noise pressure (the usual way of measuring noise levels) by dividing the energy by the area over which it is being transmitted over a distance. This area will vary depending on the position of the source: if it is standing or moving in 3D space, noise can be assumed to propagate from it as a spherical wave (the area of which is $4\pi r^2$), and if it is on the ground it will have a semi-spherical shape (of half the area). The precise formula for the division of noise energy by area varies for different locations of the source, but this general approach gives us the basic component of *noise attenuation with distance*: as energy is applied over a wider area (the area of a larger sphere), its pressure decreases proportionally to the square of the distance by "wave-divergence" and the perceived noise level decreases. Accordingly, at double the distance, the noise pressure from the same source will be one fourth and – applying the same logic as when considering several identical sources in Section 7.3 – dividing by four the energy of a source is equivalent to deducting 6 Db, so we can derive the rule-of-thumb that, for every doubling of the distance, noise levels decrease by 6 Db. We can see how the accurate measurement of (straight-line) distance is critical, and GIS functionality can be used to measure it automatically.

In addition to this "normal" attenuation, we must consider various types of what is commonly known as *excess attenuation*, caused by:

- *Air absorption* – where a small part of the sound is extracted by the air and transformed into heat – over a distance, depending on *relative humidity* (per cent) and *temperature*; this type of attenuation is rarely used in impact assessment, because temperature and humidity cannot be predicted accurately.
- *Atmospheric precipitations* (rain, snow, fog), normally also left out of IA because of its comparatively small magnitude.
- *Atmospheric turbulence* – wind and temperature gradients – can reduce noise levels by tens of decibels in certain atmospheric conditions but, because such conditions cannot be guaranteed, it is considered more prudent – always looking for the "worst-case scenario" – to ignore this possible source of attenuation.
- *Barriers* in between the source and the receptor, so that sound does not reach the receptor in a straight line, but after *refraction*; this means on the one hand that the distance travelled will be greater (a source of some excess attenuation) and, on the other, that the laws of refraction apply and sounds of different frequencies will be refracted differently – the higher the frequency, the higher the attenuation will be – and the total attenuation from this kind of obstacle can vary from 7 to 24 Db.
- *Ground vegetation* (grass or shrubs) is sometimes introduced in the calculations (especially in the US) using a formula that considers the length of this type of surface *in between* the noise source and the receptor, and its attenuation is also directly proportional to sound frequency (for the standard 1000 Hz, the attenuation is approximately of 1 Db for every 40 m of this surface).
- *Trees* are often discussed under the same heading as the previous category, but their attenuation effect is modelled by a different formula (as used in the US), also a function of the depth of the forest and the frequency of the sound (although the former is much more influential than the latter), and at standard frequencies (1000 Hz) it is approximately equivalent to 1 Db for every 10 m of forest depth (for an average US forest).

We can immediately see that GIS functionality can be used to identify and measure the land uses involved in the last three cases (barriers, ground vegetation, forests) (Figure 7.13). Noise propagation predictions are not presented in a probabilistic way, as in air pollution, but an "informal" interval of confidence of ± 2 Db is normally used to qualify the results of the calculations (Dryden, 1994).

In terms of the assessment of the noise impacts, predicted noise levels at the receptors must be compared with the *appropriate standards* (for a summary, see Bourdillon, 1995, Ch. 3; Therivel and Breslin, 2001). Standards are

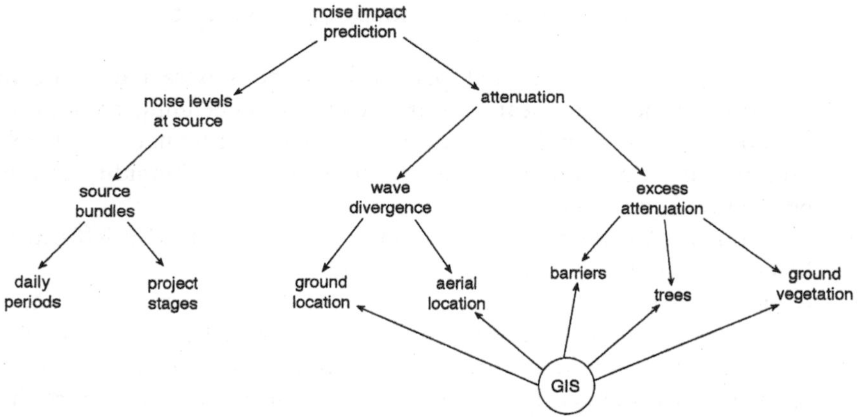

Figure 7.13 Noise-impact prediction.

sometimes defined in terms of how much new levels are allowed above existing noise levels, but usually they are expressed simply as *maximum acceptable levels* when received at the receptors. PPG 24 (DoE, 1994) defines a general standard of desirable noise levels inside dwellings (45 Db during the day and 35 Db at night) and also establishes the general concept of "noise exposure categories" (Therivel and Breslin, 2001, summarise the ranges of noise levels and recording times corresponding to each category) for sensitive receptors like dwellings or schools:

- A: noise is not a factor;
- B: noise control measures required;
- C: strong presumption against development, if alternative location is unavailable, insulation measures are required;
- D: planning permission normally refused.

These standards are normally defined either by the level of *nuisance* derived from different noise levels (interference with speech, interference with sleep, etc.), or by the type of *reaction* to be expected from the population exposed to the noise, for instance the likelihood and intensity of complaints. BS4142 (BSI, 1990) considers that anything up to 5 Db above the existing background noise is likely to be of only "marginal significance" and can be alleviated with simple measures, while if the additional noise is 10 Db above the existing levels "complaints are likely". These general standards are directly applicable to reasonably continuous noise. If a noise level is expected to be reached only occasionally (or only during emergency operations), the criteria of "interference with speech" should be applied.

Sometimes the acceptable levels are defined outside *buildings* (as "facade" levels) inside which the sensitive receptors are, and sometimes they are defined indoors, and a conversion from-outside-to-inside (or the reverse) is performed, on the assumption that walls will absorb some noise: from 10 Db with semi-open windows to 20 Db with closed single glazing. On the other hand, if the sensitive receptor is in front of a building, reflection from it can *add* to the noise level (about 1.5 Db) and the standard must be revised accordingly.

Another possible variation relates to the *type* of noise, as standards are defined for steady and relatively homogenous noise, but if it contains a discrete continuous note (whine, hiss, screech or hum) or distinct impulses (bangs, clicks, clatters, thumps), or it is very irregular (Bourdillon, 1995, Ch. 3) this is equivalent to adding as much as 5 Db to the noise-level, and the standards must be lowered.

The types of standards mentioned refer normally to the operational stage of a project and to the case of immobile noise sources. One case of interest can be when the nature of the project (landfill, or surface mineral workings) requires that, during operation, the noise sources (heavy drills or excavators, for instance) move around over considerable extensions and even in three dimensions, as the depth of the work varies constantly. One possibility is to "freeze" the movements of the noise source at key points – where their impact on the outside is likely to be greatest in various directions – and measure the noise impacts from each of them. Another possibility is to treat the project as a building site, and treat the operational stage in the same way as the construction stage using standards like BS5228 (next section), or treat it as for surface mineral works and apply Mineral Workings Guidance 11 (DoE, 1993) (Figure 7.14). To be on the safer side (Dryden, 1994) standards tend to be applied leaving a margin of about 5 Db, not so much because of statistical uncertainty (already covered by the ±2 Db confidence interval mentioned before), but because of future uncertainty, to leave room for future developments as part of the operation of the same project.

7.3.3.1 Construction noise

In terms of identifying "bundles" of noise sources likely to be operating during similar periods, the construction stage itself is normally broken down into up to four phases (Dryden, 1994; ERL, 1991):

- preliminary works, demolitions and site clearance, using breakers and earth-moving plant, lorries, mobile cranes, etc.;
- piling and foundations, using piling plant and excavators, loaders, concrete lorries, heavy cranes, etc.;
- building and erection of structures, using compressors, generators, concrete lorries, pumps, lifting equipment, etc.;

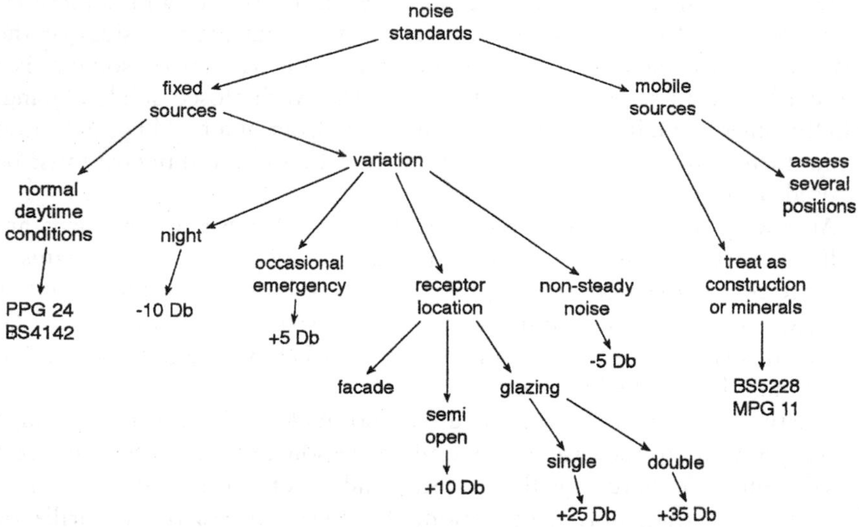

Figure 7.14 The significance of noise impacts.

- additional "fittings" (access roads, landscaping, etc.) for which excavators and rollers are used together with other general tools: compressors, hand tools, generators, lorries, etc.

A detailed schedule is also needed specifying the *time of day* when these phases will operate; in particular, if there will be construction work in the evening (after 7 p.m.) or at night. After identifying the plant likely to be used at each phase and time of day, two additional items of information are needed for each unit of equipment: (i) the *proportion of the time* that it is likely to be in use (usually in per cent) during that phase; (ii) information about the noise power of each unit, this can be obtained from catalogues or from previous experience, but it is normal to use the figures suggested in BS5228 (BSI, 1984).

From these two sets of data we can estimate the equivalent L_{eq} for continuous use of each item of equipment and, adding (logarithmically) the values for all the items of equipment likely to be in use simultaneously, we can calculate the equivalent overall noise level for that phase of the construction stage. Looking at all the phases together, we can identify the one which is the *worst offender* (always looking for the worst-case scenario) – either because it is the loudest or because it operates at sensitive hours – and it is on this basis that the impact assessment for the construction stage will be carried out (Figure 7.15).

Attenuation by distance (GIS can help) is applied in the direction of each of the sensitive receptors identified as well as any other relevant excess

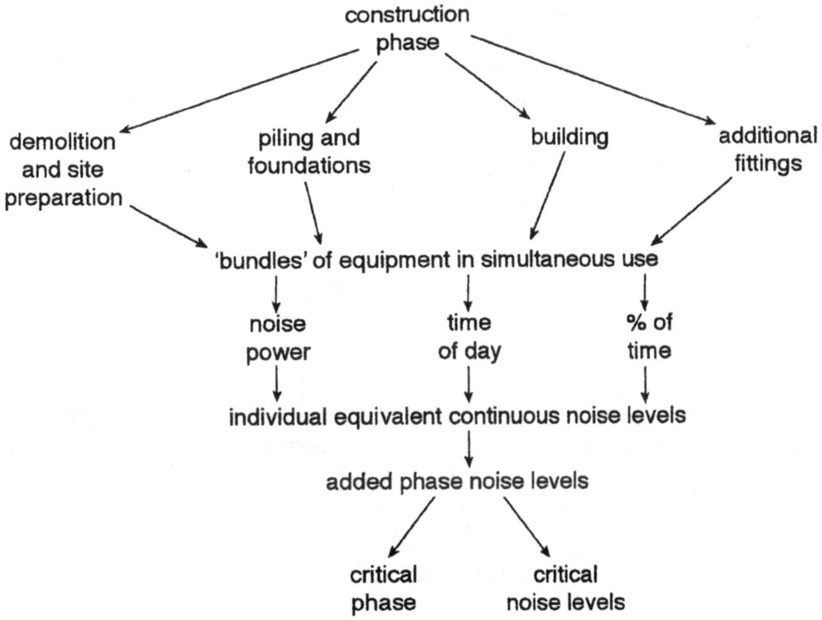

Figure 7.15 Construction noise.

attenuations (barriers, soft ground, etc.) and the relevant standards are applied (Bourdillon, 1995, Ch. 3; Therivel and Breslin, 2001), usually taken from BS5228 (BSI, 1984), the DoE Advisory Leaflet 72 (DoE, 1976), PPG 24 (DoE, 1994) with its standards for different "noise exposure categories", and the World Health Organisation guidelines (WHO, 1980). These sources provide the normal standards applicable to construction noise, as well as variations for special circumstances: for instance, if evening or night work is involved, both BS5228 and the DoE Advisory Leaflet 72 recommend lowering the day-time standard by 10 Db, and the World Health Organisation guidelines recommend a level below 35 Db inside buildings at night; if schools are affected, the Department of Education and Science guidance ("Acoustics in Educational Buildings", Building bulletin No. 15) can be applied, which recommends a maximum background noise of 35 Db inside a classroom (Figure 7.16).

7.3.3.2 Traffic noise

The case of mobile sources requires special treatment in several respects. *Road traffic* has the peculiar characteristics that while the noise sources (the vehicles) are mobile, traffic often runs in a semi-continuous flow and considerable noise comes from the friction with the road surface (a fixed

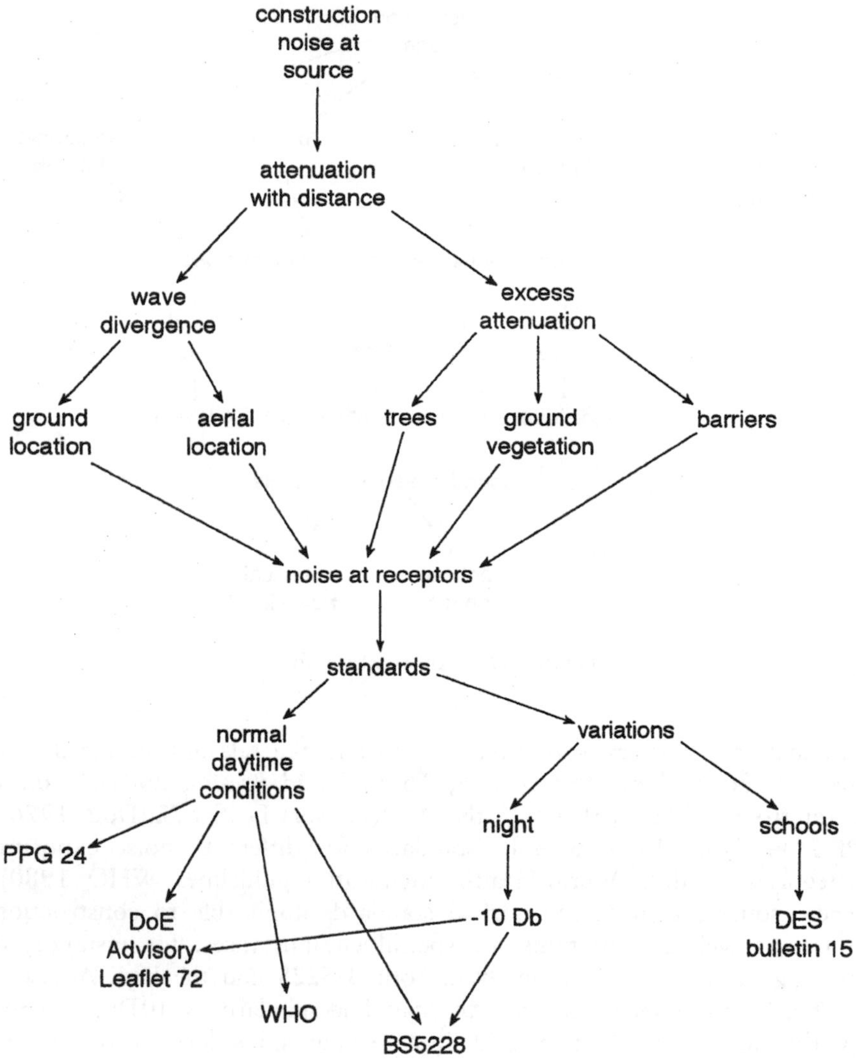

Figure 7.16 Construction noise impacts.

source). Advice on the general approach can be found in the section on noise in the "Design Manual for Roads and Bridges" (DMRB) (DoT, 1993), and we can start referring to the measurement of *ambient noise* (the baseline), which is influenced by the characteristics of the area where the potential receptors are. If ambient noise in that area is already dominated by traffic noise, the baseline should be determined by the 18-hour L_{10} as suggested by the guidance in "Calculation of Road Traffic Noise" (CRTN) from the Department of Transport (DoT, 1988). If the ambient noise is low or comprised of a combination of several undefined sources – as in a rural setting – there is no generally accepted approach, and the L_{90} (background noise) or the L_{eq} could be used, although the DoT's DMRB (DoT, 1993) recommends using the 18-hour L_{10} over *several days*. If the ambient noise is dominated by *other* traffic sources such as aircraft or trains, DoT (1993) recommends using the 18-hour L_{10} or the L_{90} over 18 h as well.

With respect to predicting road noise, DoT (1988) gives a simplified "model" to predict the likely noise levels at a distance (ranging from 4 to 300 m) from a road, using a step by step method which takes into account:

- traffic flow;
- speed;
- composition of traffic;
- road configuration;
- intervening ground between the road and the receptors;
- any screening from the road;
- the "angle of view" of the traffic from the receptor;
- possible reflection from buildings' facades nearby.

The calculations in the model extend to a distance of 300 m, on the assumption that beyond that distance, traffic noise is unlikely to have an effect except in rural areas, and also that its prediction becomes unreliable. If the development is in a rural setting, traffic noise may impact at more than 300 m from the road, and DoT (1993) recommends using the Transport and Road Research Laboratory Supplementary Report 425 "Rural Traffic Noise Prediction – An Approximation" (see Therivel and Breslin, 2001).

The assessment of the noise impacts from *new* roads (or roads being widened) is normally based on the Noise Insulation Regulations (HMG, 1975), which gives the standards of permitted noise levels, and also establishes the obligation to compensate property owners within 300 m to pay for the noise insulation of their properties (normally double glazing). When dealing with the normal increased traffic on existing unaltered roads, there is no explicit guidance in the UK to deal with its noise impact, and the norm is to fall back on the general criteria in PPG 24 (DoE, 1994), which uses the standard four "noise exposure categories" applied to traffic noise as it affects sensitive receptors (dwellings, schools). Other general criteria used are that an increase of 1 Db due to traffic noise will be "perceptible", and an increase of 3 Db is likely to cause annoyance (Figure 7.17).

For *rail* noise, the baseline is normally measured using the 24-h L_{eq} for existing trains, and for new trains it is more common to use the 18-h L_{10} as for road traffic. After that, the prediction of the noise produced by the trains is based on a method from the Department of Transport which follows a similar logic (like a "model") to that used for roads, based on a wide range of variables relating to the trains and their speed, the tracks, etc.:

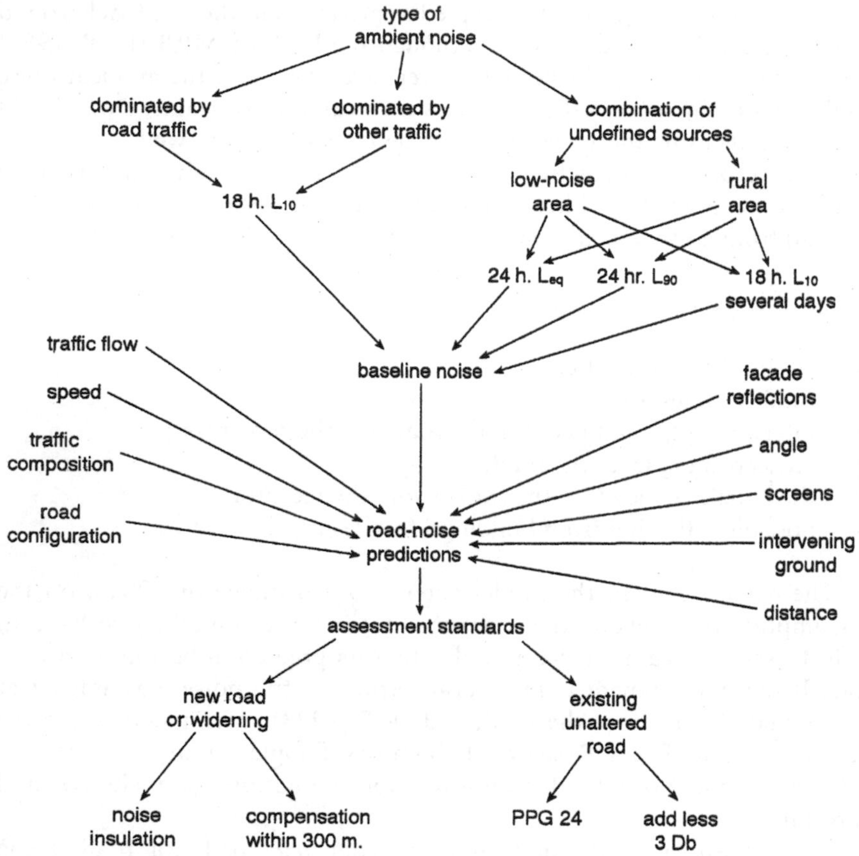

Figure 7.17 Traffic noise impacts.

- speed;
- type of locomotive (electric, diesel);
- type of brakes;
- type of cargo (freight, passengers);
- number of carriages;
- type of track (welded, joined);
- radius of curve in the track (if any);
- sleepers (concrete, others);
- presence of bridges (steel, concrete);
- track-side barriers;
- channelling of the track (cuttings, embankments, tunnels).

For the assessment of significance of the predicted noise, the "noise exposure categories" of PPG 24 (DoE, 1994) can be used. Before that source was available, ERM (1990) used the general criterion of 70 Db (24-h L_{eq}) as the limit of tolerability to railway noise, suggested by Walker (1988).

Air traffic noise is not continuous – as with railways – as the noise sources are also "event oriented" (Mestre and Wooten, 1980), and to predict noise levels we need to consider:

- the noise level of an event – usually a flyover, but also taxiing and engine testing – and its duration;
- the number of events;
- the time of their occurrence;
- the "mode" of each event, the directions – the "flight track", the projection on the ground of the flight path – of take-off and landing, specific to each type of aircraft;
- the trajectory or "flight path" the aircraft follows – in three dimensions – used to calculate the different "slant range distances" and angles at which the noise is received from different points on the ground.

Noise comes mainly from the aircraft's jet engines and is produced in various directions: (i) combustion noise is projected *sideways*; (ii) fan noise is projected *forwards*; and (iii) jet noise and various types of exhaust noises (from the fan and from the jet core) are projected *backwards*. The noise levels and directions are specific to each type of engine, and the number and position of engines is specific to each type of aircraft. In practice, complex computer models are used to calculate the "Sound Exposure Level" from different events (aircraft) at different slant distances, and an equivalent noise-level for day and night can be estimated (similar to the L_{eq}). For the assessment of significance, standards such as the "noise exposure categories" in PPG 24 (DoE, 1994) can be used.

7.3.3.3 Vibration

Vibration is a disturbance – usually low frequency – producing physical movement in buildings and their occupants, which can result in damage to buildings and/or annoyance to the occupants (DoT, 1993; Petts and Eduljee, 1994, Ch. 14). It usually comes together *with a noise* (produced by the same source) and can be transmitted through the ground or through the air, and the physical movement of the buildings or structures (or the ground under them) is measured in "peak particle velocity", in millimetres per second. It is normally associated with (i) the construction stage of most projects, especially if it involves sub-surface operations like tunnelling or piling; and (ii) the operational stage of many projects, for example those with a traffic component (roads, railways, air traffic).

The prediction of vibration levels has not been "modelled" to the same degree of accuracy as noise, and it is common practice to estimate vibration levels using measurements from similar equipment or traffic conditions in operation elsewhere (ERM, 1990). However, vibration impacts are often left out of Environmental Statements because they have been found to be negligible at relatively short distances from the source. Standards BS5228 (BSI, 1984), BS6472 (BSI, 1987) and BS7385 (BSI, 1993) identify the minimum perceptible vibration level with a peak velocity of 0.1 mm/s. Such level is only reached within a distance of 100 m from the source when the cause is heavy ground-hitting plant (like a percussive pile driver), and the distance is reduced to 20 m when the noise source is mobile construction equipment (see also ERM, 1990; or Bourdillon, 1995, Ch. 3). This means that, if the sensitive receptors relevant to our study are beyond these distances, which is easy to find out automatically using GIS, vibration is likely to be imperceptible and is not worth studying.

The minimum level of perceptible vibration mentioned above is also used as a limit of acceptability when sensitive equipment is concerned, but for people or buildings we have to go well above the thresholds of perception:

- The annoyance threshold for *people* inside buildings is considered to be from 0.2 to 0.4 mm/s during daytime, and 0.14 mm/s at night.
- For *buildings*, 5 mm/s is considered to be the maximum value compatible with the protection of the structure of a standard building, and 3 mm/s is used when listed buildings or potentially vulnerable buildings are involved.

In terms of distance, annoyance levels of vibration can be reached at just over half the distances of minimum perceptibility: for instance, vibration from a percussive pile driver will never reach the annoyance levels mentioned beyond a distance of 60 m. Also, it has been found (DoT, 1993) that annoyance from vibration (from traffic) is closely associated with the 18-hour L_{10} measurement for the traffic that generates the vibration; hence the latter can be used as a "proxy" for vibration impact, and assessed accordingly (Figure 7.18).

7.3.3.4 Re-radiated noise

Lastly, mention should be made of *re-radiated noise*, which can be associated with the same type of sources that produce vibration, but which is different in nature: it is a noise and not a movement as such, and it is measured in the same way as noise. It tends to be associated with subterraneous noise sources (often underground tunnels for trains or road traffic) and its transmission is influenced by:

- the rock-type and the geology of the different layers;
- the types of buildings and their foundations;
- the type of tunnel (size, depth, building material).

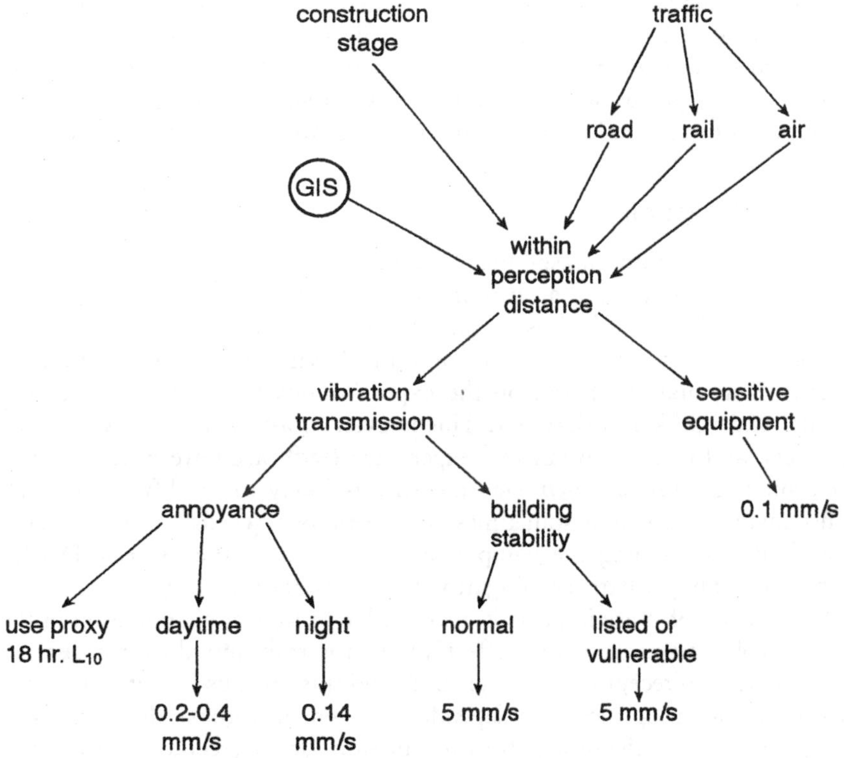

Figure 7.18 Vibration noise impacts.

Theoretically, the transmission of re-radiated noise can be modelled in the same way as with noise, estimating the noise energy absorption for each frequency of each geological layer the sound must cross to reach the receptor. However, in practice (Dryden, 1994) this can be difficult, and more *empirical* approaches are used, referring to other experiences in similar projects (as with vibration), although the same definitions and criteria are not always used, for instance between studies in the US and in the UK, and this can make the comparison irrelevant. Sometimes we can use vibration studies in the same medium (through the same geology) to estimate propagation distances, as the two phenomena behave in a very similar way. On the other hand, sometimes the normal airborne noise produced by the same source is at levels that make the re-radiated noise irrelevant: for instance, London Underground found that complaints about re-radiated noise only started at levels where the accompanying airborne noise was already in breach of standards, making the study of re-radiated noise unnecessary.

In general and in contrast with noise-simulation, where rigorous and well-established calculations apply, with re-radiated noise (and to some extent, with vibration) we are at the fringe of established expertise, much of which is still fluid and subject to scientific research. Practice tends to be quite improvisational and "opportunistic" following the availability of scientific sources and empirical information for circumstances comparable to those being assessed.

7.3.4 Mitigation

As in air pollution, noise-impact mitigation measures derive from the knowledge/experience that certain modifications will make the impact predictions change or be less significant, not necessarily requiring the simulations to be rerun with the mitigation, but having the character of "hypothetical" simulations based on the expert's knowledge of the simulation model and its likely behaviour. This is even more so in the case of noise impacts, as there is a wealth of experience (with accurate measurements) determining *by how much* each measure is likely to modify noise levels. This makes it possible sometimes to "work backwards" from the noise standards to deciding the equipment and layout of the project (Dryden, 1994), including whatever mitigation measures are required.

The nature of noise impacts is also similar to that of air pollution in that both involve a source of the "effect", which is transmitted over a distance, and impacts on receptors on the ground, and this gives us a "natural" breakdown of mitigation measures depending on which of those three stages the mitigation affects: the source, the transmission, or the receptor. Noise can be mitigated *at source* using three types of measures (Bourdillon, 1995, Ch. 3):

1 Engineering:
   - selection of an inherently quiet plant;
   - proper use of plant to minimise noise emissions;
   - use of insulation and silencers;
   - proper maintenance.
2 Site layout:
   - location of noisy plant as far as possible from sensitive receptors;
   - taking advantage of natural sound barriers;
   - building layout.
3 Administration:
   - avoiding construction work at night and restricting operating times;
   - making the contractor (during construction) adhere to the Code of Practice for Construction Working and Piling in BS5228 (BSI, 1984);
   - restricting activities;
   - specifying noise limits.

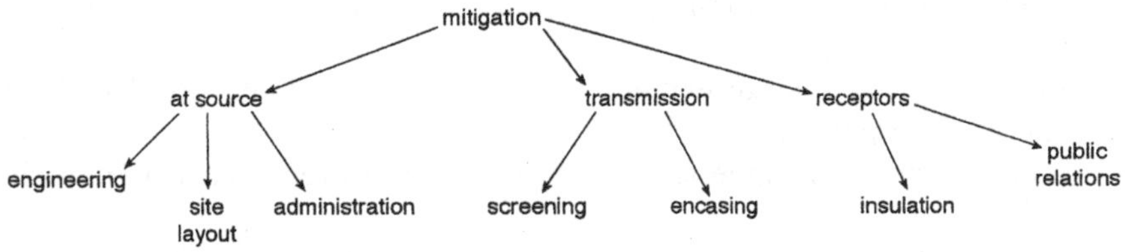

Figure 7.19 Noise-impact mitigation.

The most common way of mitigating the *transmission* of airborne noise is by some form of screening:

- encasing the noise sources inside buildings;
- erecting acoustic fences or screens;
- using buildings as screens;
- using spoil and stockpiles as screens (particularly during construction);
- using "bunding" to absorb/reflect rail noise;
- landscaping and tree-planting (often used in motorways).

At the *receptors'* end, the most common approach is to provide them with insulation (double glazing for instance), but "public relations" measures can also be effective: identification of a site liaison officer to deal with any complaints, informing local residents and Environmental Health if any particularly noisy operations are planned (Figure 7.19). For ground-borne vibration and re-radiated noise, mitigation is more difficult and has been less well studied, except for the fact that these effects are usually accompanied by airborne noise, and any at-source mitigation that can be used to reduce noise is likely to also reduce vibration and re-radiated noise.

7.4 CONCLUSIONS: EXPERT SYSTEMS FOR AIR-POLLUTION AND NOISE IMPACT ASSESSMENT

Expert systems to help with impact prediction are most likely to have a structure which follows the general sequence outlined in the introduction to this chapter (baseline, prediction, assessment, mitigation) which the assessment of most impacts follow. Within that general structure, we have seen how the emphasis on simulation modelling in areas of impact like air pollution or noise tends to "shape" the process in ways that suit the requirements of the operation of the models used, normally involving variations of the rather obvious sequence of data collection, model operation and production/interpretation of results. However, the nature and the choice of the models involved, and the availability of the relevant data, make these general modelling stages adopt different shapes and take different degrees of prominence for different impact types. For instance, in *air-pollution* modelling (Figure 7.20):

- models are to a large extent "pre-packaged" and the problem is to choose the right one;
- much of the (atmospheric) data come already pre-processed from the source;
- model runs are repeated (if at all) for *different sources* within the same project;
- the model results are only "directional" indicators with no locational accuracy.

On the other hand, in *noise* modelling (Figure 7.21):

- potentially sensitive receptors must be identified first of all (maybe using GIS);
- the raw data must be pre-processed (maybe using Spreadsheets) for all the noise sources;

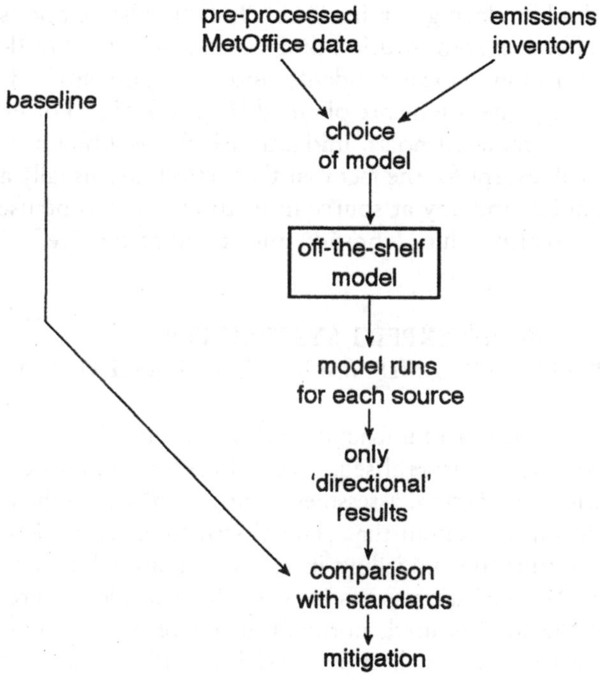

Figure 7.20 Using off-the shelf models for air-pollution impact assessment.

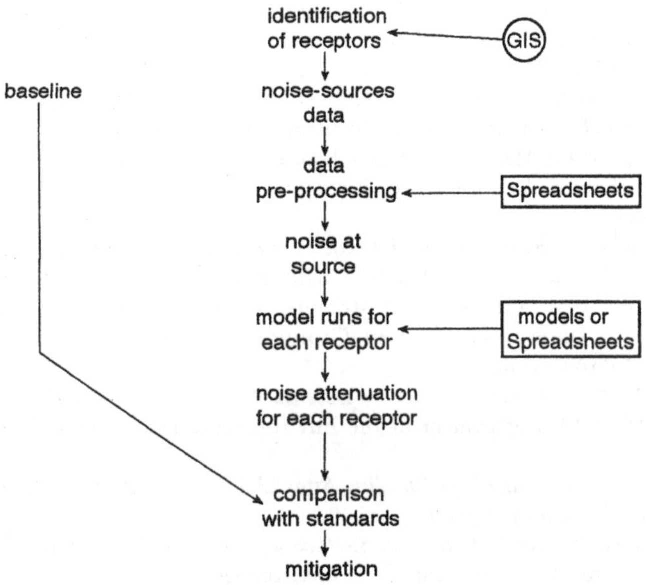

Figure 7.21 Models and subroutines in noise-impact assessment.

- standard – accepted – models of noise attenuation are run for the *different receptors*;
- the results are location specific.

The details within each of the "boxes" in the diagrams have already been discussed and are not repeated here. The spatial accuracy of the models used determines the relevance of using spatial tools like GIS which, albeit helpful, are not crucial to the outcome of the exercise, especially considering their cost. The degree of sophistication and "pre-packaging" of the models used and the degree of pre-processing of the data determine how *interactive* the simulations will be. Overall, the influence of modelling on the whole structure is determined by the reliability of the models and how well established they are in professional practice. This will be reinforced in the next chapter with discussion of areas of impact assessment where simulation modelling is far from being accepted, or even developed.

REFERENCES

Barrowcliffe, R. (1993) *The Practical Use of Dispersion Models to Predict Air Quality Impacts*, paper presented at the conference "Environmental Emissions – Monitoring Impacts and Remediation", Forte Crest Bloomsbury Hotel (June 10–11), London.

Barrowcliffe, R. (1994) *Personal Communication*, Environmental Resources Management Ltd, Oxford.

Bourdillon, N. (1995) *Limits & Standards in Environmental Impact Assessment*, Working Paper No. 164, School of Planning, Oxford Brookes University.

BSI (1984) BS5228: *Noise Control on Construction and Open Sites*, British Standards Institute, Milton Keynes.

BSI (1987) BS6472: *Evaluation of Human Exposure to Vibration in Buildings (1 Hz to 80 Hz)*, British Standards Institute, Milton Keynes.

BSI (1990) BS6841: *Measurement and Evaluation of Human Exposure to Whole Body Mechanical Vibration and Repeated Shock*, British Standards Institute, Milton Keynes.

BSI (1990) BS4142: *Rating Industrial Noise Affecting Mixed Residential and Industrial Areas*, British Standards Institute, Milton Keynes.

BSI (1993) BS7385: *Evaluation and Measurement for Vibration in Buildings. Part 2: Guide to Damage Levels from Groundborne Vibration*, British Standards Institute, Milton Keynes.

DETR (2000) *Review and Assessment: Pollutant-specific Guidance*, LAQM.TG3(00). Department of the Environment, Transport and the Regions, London.

DoE (1976) *Noise Control on Building Sites*, Advisory Leaflet 72, Department of the Environment (out of print).

DoE (1993) *The Control of Noise at Surface Mineral Workings*, Minerals Planning Guidance note 11, Department of the Environment.

DoE (1994) *Planning and Noise*, PPG 24, Department of the Environment.

DoT (1988) *Calculation of Road Traffic Noise*, Department of Transport, Welsh Office.

DoT (1993) *Design Manual for Roads and Bridges Volume 11: Environmental Assessment*, Section 3, Part 7 "Traffic Noise and Vibration", Department of Transport.

Dryden, S. (1994) *Personal Communication*, Environmental Resources Management Ltd, Oxford.

Elsom, D.M. (2001) Air and Climate, in Morris, P. and Therivel, R. (eds) *Methods of Environmental Impact Assessment*, Spon Press, London, 2nd edition (Ch. 8).

ERL (1991) *Environmental Statement for the Knostrop Sewage Treatment Plant* (Knostrop, West Yorkshire), Environmental Resources Ltd.

ERL (1992) *Environmental Statement for the Power Station at King's Lynn (no. 2)* (King's Lynn and West Norfolk BC), Environmental Resources Ltd.

ERM (1990) *Environmental Statement for the Jubilee Line Extension* (London), Environmental Resources Management Ltd.

ERM (1993) *Municipal Waste to Energy on the Process Plant Park, Billingham* (Cleveland), Environmental Resources Management Ltd.

Hall, D.J., Kukadia, V. and Emmott, M.A. (1991) *Determination of Discharge Stack Heights for Pollution Emissions*, Report No. CR 3445 (PA), Warren Spring Laboratory, Stevenage.

Harrop, D.O. (1999) Air Quality Assessment, in Petts, J. (ed.) *Handbook of Environmental Impact Assessment*, Blackwell Science Ltd, Oxford (Vol. 1, Ch. 12).

Havens, J.A. and Spicer, T.O. (1985) *Development of an Atmospheric Dispersion Model for Heavier-than-Air Gas Mixtures*, Report No. CG-D-23–85, US Coast Guard, Washington, DC.

H.M.G. (1975) *Noise Insulation Regulations*, SI 1975/1763.

Jones, J.A. (1988) What is Required of Dispersion Models and do They Meet the Requirements?, in *NATO-CCMS 17th International Technical Meeting on Air Pollution* Modelling, Cambridge (19–22 September).

Kent Air Quality Partnership (1995) *The Kent Air Quality Management System: Final Report* (September), Kent County Council.

Mestre, V.E. and Wooten, D.C. (1980) Noise Impact Analysis, in Rau, J.G. and Wooten, D.C. (eds) *Environmental Impact Analysis Handbook*, McGraw-Hill (Ch. 4).

Petts, J. and Eduljee, G. (1994) *Environmental Impact Assessment for Waste Treatment and Disposal Facilities*, John Wiley & Sons, Chichester.

Royal Meteorological Society (1995) *Atmospheric Dispersion Modelling: Guidelines on the Justification of Choice and Use of Models, and the Communication and Reporting of Results*, Policy Statement, May, 1995, Royal Meteorological Society, London.

Samuelsen, G.S. (1980) Air Quality Impact Assessment, in Rau, J.G. and Wooten, D.C (eds) *Environmental Impact Analysis Handbook*, McGraw-Hill (Ch. 3).

Therivel, R. and Breslin, M. (2001) Noise, in Morris, P. and Therivel, R. (eds) *Methods of Environmental Impact Assessment*, Spon Press, London, 2nd edition (Ch. 4).

Walker, J.G. (1988) A Criterion for Acceptability of Railway Noise, *Proceedings of the Institute of Acoustics*, Vol. 10, Part 8.

Wallis, K.J. (1998) Air Pollution: Use of Models in Air Pollution Assessment, *Impact Assessment and Project Appraisal*, Vol. 16, No. 2 (June), pp. 139–46.

Westman, W.E. (1985) Air and Water, in (same author): *Ecology, Impact Assessment and Environmental Planning*, John Wiley & Sons (Ch. 7).

Wood, G.J. (1997) *Auditing and Modelling Environmental Impact Assessment Errors Using Geographical Information Systems*, unpublished PhD thesis, School of Planning, Oxford Brookes University, Oxford.

Wood, G.J. (1999) Post-development Auditing of EIA Predictive Techniques: A Spatial Analytical Approach, *Journal of Environmental Planning and Management*, Vol. 42, No. 5, pp. 671–89.

WHO (1980) *Environmental Criteria 12: Noise*, World Health Organisation, Geneva.

8 Soft-modelled impacts

Terrestrial ecology and landscape

8.1 INTRODUCTION

In this chapter we are going to discuss areas of impact assessment representing in some respects the end of the spectrum opposite to those discussed in the last chapter, being areas where the presence of simulation modelling is virtually non-existent. To focus our discussion we have chosen the areas of terrestrial ecology (a heavily researched scientific area) and landscape, a more subjective area of impact assessment. The reasons for the absence of simulation modelling are very different for each of the two areas, but they have in common the fact that the logic of the thinking process is more dominated by the substantive content of their disciplines than by the logic of applying particular rules and simulation models of one level of sophistication or another, as was the case with the areas of impact discussed in the previous chapter. In addition, even if both areas conform in general terms to the stages and general sequencing sketched out before (baseline, prediction, assessment, mitigation), they each adopt very different approaches.

8.2 TERRESTRIAL ECOLOGY

The breadth and complexity of this field has been well introduced and discussed over the years in the corresponding chapters in well-known manuals: Hanes (1980) and Westman (1985) are good examples of "first generation" discussions of ecology in the context of impact assessment manuals. In more recent times, Petts and Eduljee (1994a), Morris (1995), Wathern (1999) and specially Morris and Emberton (2001) provide more up-to-date discussions which show the considerable complexity of this area of assessment, to the extent that it can be argued that the investigation of this area is so "open-ended" that it goes against the grain of an expert systems approach

(Beaumont, 1994),[34] which by definition requires a certain degree of closure. We shall not try to reproduce those expert discussions here, but seek to reduce that open-endedness as much as possible to a logic which could potentially be automated – with particular attention to the role GIS could play – in much the same way as in the previous chapter. For the practice of impact assessment, this area of study is broken down into two: *flora* and *fauna* and, in turn, their study can be applied to a terrestrial environment or to a water (fresh, marine, estuarine) environment. In this chapter we shall concentrate on the *terrestrial* case.

As in the case of noise or air pollution, ecological questions can be addressed at various stages in the compilation of an Environmental Statement: when considering alternatives for the project, when studying the baseline situation, when predicting future effects of the operation of the project, and when considering mitigation measures. However in the case of ecology it is the baseline study that dominates in the consideration of ecological issues and impacts. Although "the best mitigation is by site selection" (Beaumont, 1994), ecology is very rarely investigated when deciding the location of a project. Such investigation is usually not budgeted for by developers, and the most common situation is that impact assessors are called upon when the developer has already acquired the site. Although the detailed assessment of ecological issues is quite complex, the overall logic of ecological impact assessment is quite simple, resulting from the contraposition of the project and all its features with the ecological environment and all its qualities (Figure 8.1).

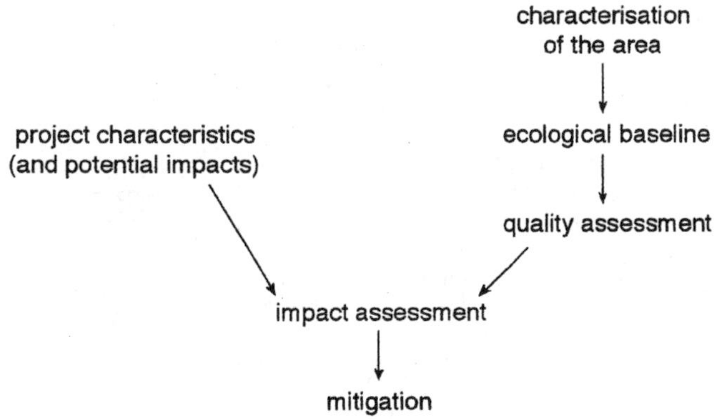

Figure 8.1 The logic of terrestrial ecology impact assessment.

34 The knowledge acquisition for this part was greatly helped by conversations with Nicola Beaumont, of Environmental Resources Management Ltd (Oxford branch), and Owain Prosser helped with the compilation and structuring of the material. However, only the author should be held responsible for any inaccuracies or misrepresentations of views.

8.2.1 Project characteristics and potential impacts

The potential ecological impacts from projects relate directly to the land affected by the project, considered under two main headings (Figure 8.2): the *land-take* by the project itself; and area affected by the *functional impacts* from the project. The project *land-take* relates to the area occupied by the various features of the project at various stages in its life. In terms of the information to be collected about the project, it can be seen as the three-dimensional combination (it can be thought of as a three-dimensional "table") of the project features, the project stages they apply to, and how temporary they are. First, the identification of what the project will involve can follow a typical checklist of all those *elements of the project* (the main plant and the infrastructure) that will disturb the natural environment in that area:

- demolitions to be carried out and safety areas around them;
- access roads (and areas around them needed for their construction);
- earth-moving/in-filling;
- storage areas (and possible safety areas around them);
- maintenance and repairs;
- buildings;
- structures (and safety areas around them);
- parking areas;
- areas paved for circulation or other reasons;

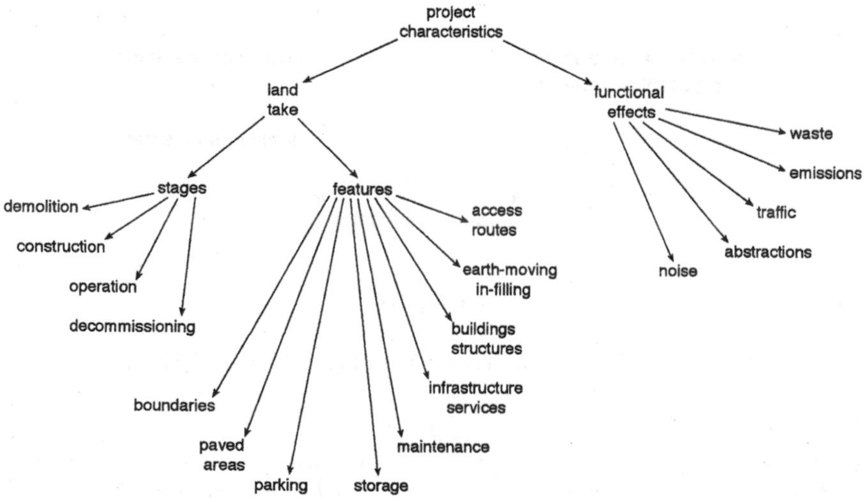

Figure 8.2 Project characteristics for terrestrial ecology impact assessment.

- infrastructure and services;
- boundaries (fencing, bunding).

Second, it is essential to determine at what *stages* of the life of the project the different features on the list will have an impact:

- pre-construction stage if there is one (demolition of existing structures, clearing the site);
- construction;
- operation;
- decommissioning (if relevant), although, for terrestrial ecology, it is common for this stage to have "positive" rather than negative impacts, as it can involve restoring the environment to its situation prior to the project.

Third, whether these features will be temporary or permanent. All these considerations are intended to define and measure the *presence and areal extension* of the various project features in its various stages. Occasionally, areas inside the perimeter of the project site can remain undisturbed and can be excluded from the consideration for impact assessment, as long as they remain connected to their natural surroundings and are not left as "islands" encircled by man-made areas, or are separated by fences from other natural areas.

The *functional impacts* from the project can also affect the ecology of areas which must be added to those affected by direct land-take. These can be derived from a standard short list of typical functional impacts:

- traffic;
- emissions into air, water or soil (from the project and from traffic);
- noise emissions (from the project and from traffic);
- water abstractions;
- waste generation.

Land-take is the most direct of the impacts on terrestrial ecology, but the land affected by some of the project's functional effects should also be part of the investigation, as it can involve impacts like depositions from emissions, or indirect impacts like those from traffic and access routes, all of these producing potentially dangerous effects (Petts and Eduljee, 1994a). The assessment of these functional and indirect impacts is part of the respective impact assessments studies (air pollution, noise, etc.), but the identification of the land affected should be an important consideration from the ecological assessment point of view. This should involve, for instance, extending the ecological baseline-study to areas where air pollution is going to impact on the ground, which in turn would mean either *postponing* the start of the ecological area characterisation until such impacts had been calculated, or *anticipating* (even if roughly) the extent of the area likely to be affected: for example, a few kilometres downwind in the dominant wind direction in the area. This can be more difficult for some impacts than for others, and is bound to be based on previous experience with similar types of projects, knowing for instance that industrial noise is likely to "carry" over several hundred metres, while motorway noise carries over several miles. As a logical corollary to this emphasis on the identification and measurement of the areal extension of different parts (and effects) of the project and how they overlap with the natural environment, we can already anticipate that the *mapping* of the project in its environment is going to be central to the whole process, giving a potentially pre-eminent role to GIS and similar technologies.

8.2.2 Area characterisation and ecological baseline

Ecological impact is one of the "classics" in impact assessment; therefore it is one of the first to be considered for inclusion in impact studies and is rarely added later as an afterthought. However, some level of *scoping* is necessary to determine the types of surveys needed, using a form of "scoping desk-study" (Beaumont, 1994) to plan the study and put the necessary team together, based on documentary evidence and second-hand information. This type of information very rarely refers to the specific site, but focuses more on providing a general picture of the area and its potentially sensitive spots in order to carry out the necessary surveys and collect first-hand information (Figure 8.3). Standard sources for this initial area characterisation are:

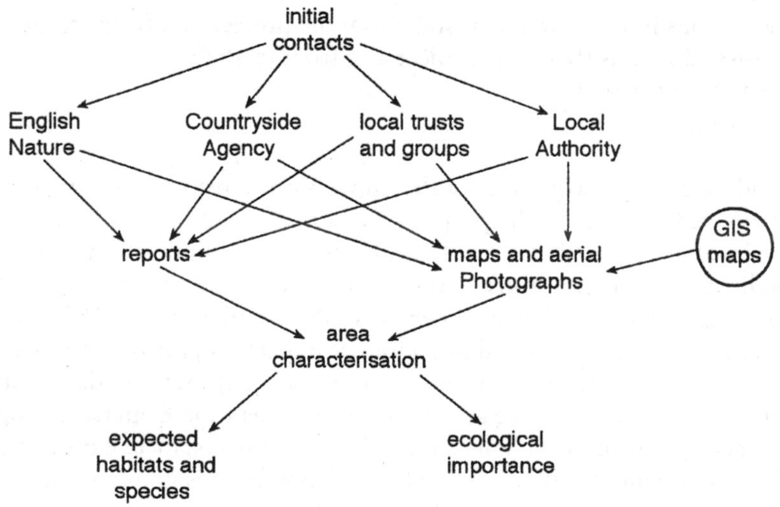

Figure 8.3 Area characterisation and ecological baseline.

- Preliminary contacts with relevant organisations to gather information about the site, particularly (in the UK) the local representatives of English Nature and the Countryside Agency if there are statutory sites in the area; documentary sources from these and other organisations – like English Nature reports with their lists of species present, or the Nature Conservation Review for nationally important sites – are essential, as most environmentally sensitive sites have been well documented since Victorian days (Appendix E in Morris and Therivel, 2001, contains a useful list of publications).
- The Institute of Terrestrial Ecology, which has produced a GIS map of the average environmental composition (extent and composition of habitats) for every kilometre square in the country.
- The local authorities, which should be contacted because of their knowledge of the area and of any local groups which may exist with ecological interests – and information – about the area.
- Local naturalist trusts and local interested people who are an invaluable source, as they often develop passionate ideas about sites, "take ownership of their sites" (Beaumont, 1994) and can know them better than the experts.

A problem with consulting organisations is that it is expensive in terms of obtaining data and in terms of time, and sometimes some organisations are left out for lack of resources. Contacting local people can also alert sections of the public to the developer's intentions – if not with full project details, at least with an idea of the type of development – and the developer may not be in favour of it. Ideally, the public should be contacted at the earliest opportunity, and the most useful and economical way to do this is through a public presentation of the development organised as an informal event: an exhibition with illustrations and a team of experts to answer questions (Beaumont, 1994). This can run in parallel with asking the local groups for their opinions. However, consultation with the public will happen only when client-confidentiality allows it, and it is common for developers to be cautious about making their intentions known. It is frequently the case that this type of public meeting only takes place after the planning application has been submitted, and often it is the Planning Authority that requests the developer to hold such meetings.

In addition to the existing environmental information about the area or the site, any published good-quality *descriptive* information, like small-scale maps or aerial photographs, is also an essential starting point. Ecological impact assessment is characterised by the fact that the baseline study (the description/assessment of the ecology of the area) serves at the same time the purpose of a "scoping" study (Wathern, 1999, makes a similar point), as it is the study of the existing situation that will dictate what types of ecological impacts to anticipate and study in depth. For this reason, an ecological baseline study has always two strands: (i) the "descriptive" strand, trying to find out *what* (in ecological terms) there is in the area; and (ii) the "evaluative" strand, trying to establish the *worth* (in ecological terms) of what there is. This is present from the start, and as the first stage in this baseline study, the idea behind the area characterisation desk study is to start anticipating what "valued ecosystem components" (Morris and Emberton, 2001) are likely to be present in the area:

- semi-natural habitats, like ancient woodlands;
- species or communities of nature-conservation importance at local, regional, national or international level;
- species particularly sensitive to disturbance.

This information will help to plan the surveys needed and can sometimes save resources: for example, it may pre-empt the need for multiple all year around surveys to detect seasonal variations if the information is already available. The "plan of battle" that follows usually consists of several phases progressing from the general and more "descriptive" to the particular and more "evaluative". For well-recorded sites, the area characterisation desk study can be sufficient to go directly into "Phase 2" surveys by specialists; for poorly recorded sites it is necessary to organise a "Phase 1" field survey first.

8.2.2.1 Phase 1

This first survey is necessary when documentary sources are not sufficiently rich but, even when they are, this type of survey is often carried out to confirm/expand the documentary information. First of all (Figure 8.4), the *area of survey* around the project site must be established (see Morris and Emberton, 2001) and, in practice, this radius will depend on the anticipated sensitivity of the area around the project site, as determined from the area characterisation (Beaumont, 1994):

- If areas or species of statutory interest have been detected within a radius of 5 km, the survey will be extended to include them.
- If no such areas/species have been anticipated in the preparatory stage, a radius of 1 km around the project in all its parts will be used.

The aforementioned question of functional and indirect impacts must be remembered here as it may have direct bearing on the definition of the study area: for instance, we saw in the last chapter that noise simulation usually extends to a few hundred metres (hence probably included in the "1 km rule" above, except for special projects like motorways or airports), but air pollution modelling may cover distances of up to 10 km downwind, and these considerations must be included in the definition of the study area.

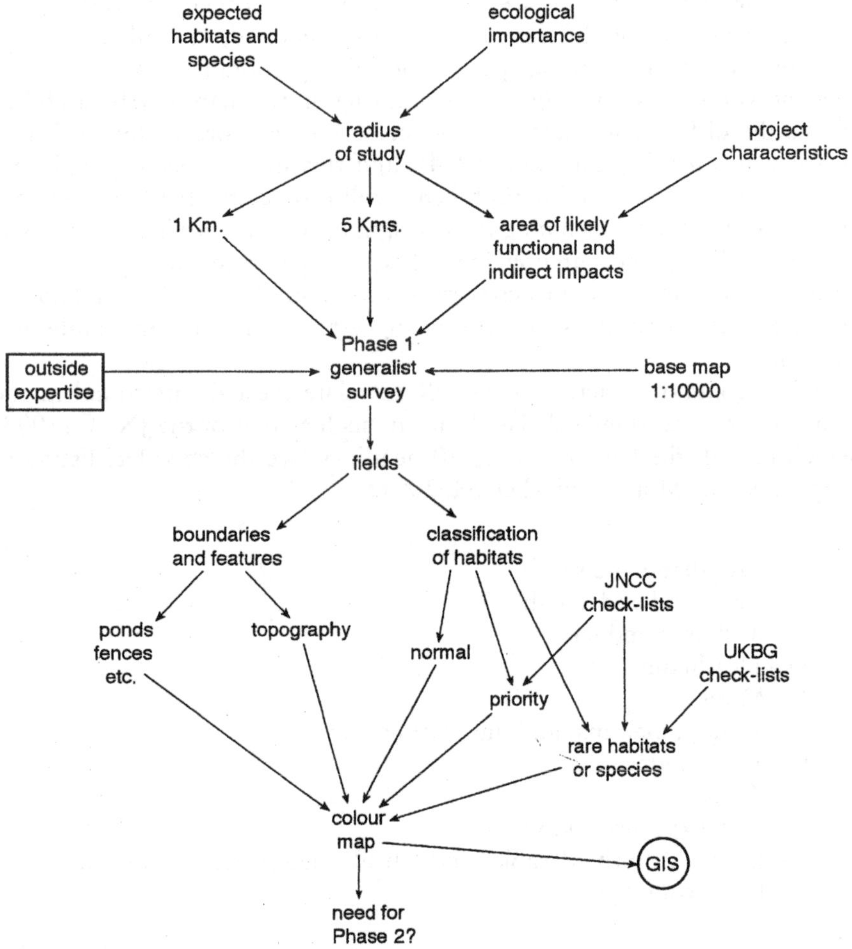

Figure 8.4 Phase 1 survey for terrestrial ecology.

To complement the documentation already collected for the area (including good maps or photographs), a "generalist" – usually an experienced botanist or herbatologist – is normally the first to walk into the area to survey it and identify the different habitats in it (type, location, extent) following the Nature Conservancy Council methodology (JNCC, 1993) summarised in Morris and Emberton (2001). Although a systematic approach based on transects of the study area can be used, in practice it can be best described as a "walkabout" (Beaumont, 1994) in which the expert goes anywhere he/she feels can provide the information necessary. On a 1:10 000 map, all the fields and their boundaries are identified: arable fields, semi-improved fields in use, non-improved fields (the most interesting usually). Any features present in the fields are also drawn (ponds, hedgerows, topography, management), all the different habitats are identified, and all the species detected are listed – making special note of any "notable" species like an ancient woodland or a badger set. Colouring on the map may be useful to define the different habitats. In practice, colour is more useful to denote *degrees of interest* (Beaumont, 1994), and this map – together with any species-lists identified and notes taken – will provide the basis for the next phases of the study – including the decision on whether a Phase 2 survey is needed. Although the map used in the field is paper-based, it may be a good idea after this first survey to convert it into digital form, allowing GIS use for subsequent impact assessment as suggested by Morris and Emberton (2001).

It is a common practice in the UK to relate the habitats found in the Phase 1 survey to standard classifications such as that of the JNCC (1993) which lists all the habitats under 10 headings (see the complete listing in Appendix F of Morris and Therivel, 2001):

A – Woodland and scrub
B – Grassland and marsh
C – Tall herb and fern
D – Heathland
E – Mires
F – Swamp, marginal and inundation
G – Open water
H – Coastlands
I – Rock exposure and waste
J – Other (disturbed/arable land, shrub, hedge, fence, built-up areas, bare ground, etc.).

This survey and checklist can serve well the descriptive purposes of this phase and can also be useful for the evaluation strand, identifying:

- semi-natural habitats, like ancient woodlands;
- habitats resulting from long-term management, like some grasslands, marsh, or heathland;
- sensitive habitats, like bogs;
- animal species which are rare or protected.

It may also be a good idea to classify the habitats found according to more evaluative checklists (also summarised in Appendix F in Morris and Therivel, 2001): the UKBG reports (UKBG, 1998, 1999) identified as part of their general list of habitats certain "priority habitats" which require particular attention and action. Also, the European Community lists in its so-called "Habitats and Species Directive" (EEC, 1992), habitats of special importance, and some of those habitats can be found in the UK. After the different types of habitats and features have been identified and mapped, a more in-depth "Phase 2" study can be carried out by more specialised experts.

8.2.2.2 *Phase 2*

After the "generalist" has identified the types of habitats present, more specialised experts usually visit the area to compile *species-lists* and identify ecological communities (Figure 8.5). Sometimes, lists of species typical of

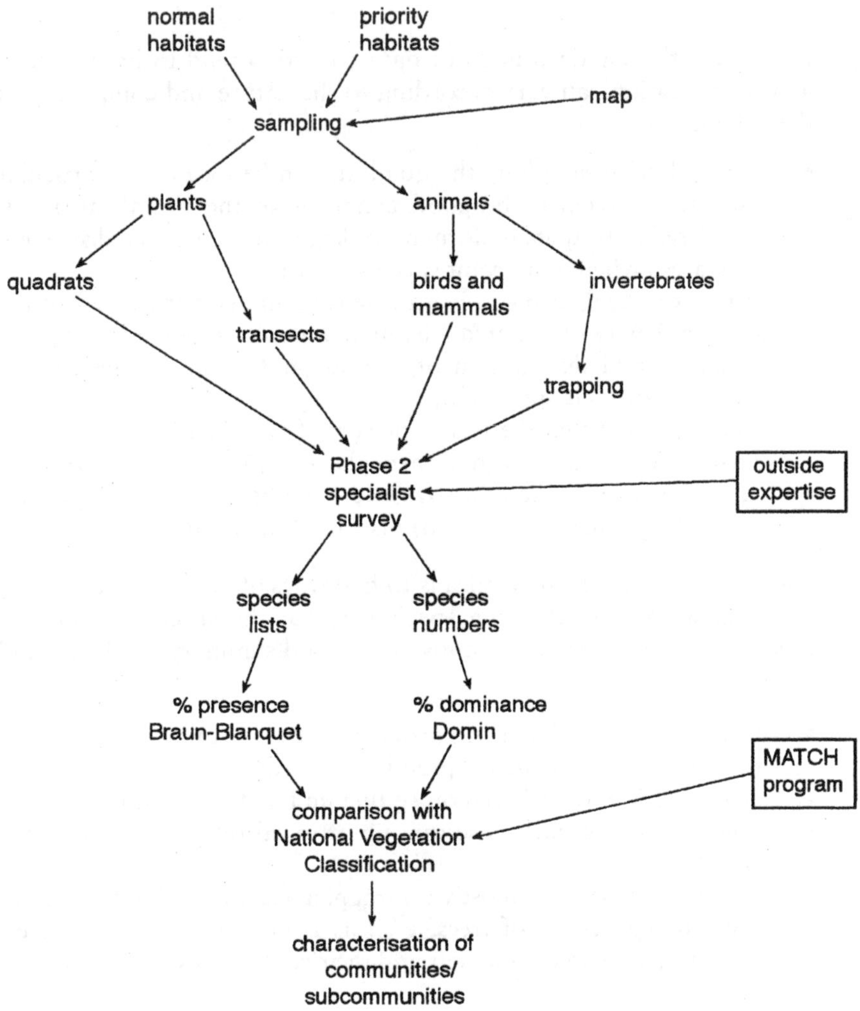

Figure 8.5 Phase 2 survey for terrestrial ecology.

various types of habitats may have been collected already from documentary sources, but it is this field survey that will confirm their validity and, sometimes, provide surprises with unrecorded additions (Beaumont, 1994). Although the simple "walkabout" approach is also used as the basis for data collection in this phase, it is often complemented with some form of *sampling* of habitats and species and some quantitative analysis of their composition. Morris and Thurling (2001) provide a systematic review of sampling methods commonly used, normally a combination of areal sampling using square sample areas ("quadrats") and linear sampling based on "transects" (lines) which the expert can walk recording – sometimes measuring – what he/she finds. In practice, the expert's perception of what requires sampling plays an important role:

1 For *plants*, the standard units of data collection tend to be *quadrats*, and their location can vary according to the nature and complexity of the habitats:

   - using selective sampling, the quadrats can be located in particular parts which seem to be good examples of the overall area – if considered to be quite uniform according to the expert's judgement – or in areas where important species occur;
   - or they can be distributed in an extensive and systematic way if the size/complexity of the area – as in an ancient woodland – suggests a more varied distribution of species (and if the budget for the study allows such approach);
   - also, if it is felt that there is some sort of "gradient" in the characteristics of the habitats in a particular direction as the distance to particular features increases, *transects* can be used (in the direction where the gradient is expected) instead of quadrats.

2 For *animal species*, the basic idea is to find ways of detecting (sometimes by trapping) and/or counting specimens, but the specific techniques used will vary for different groups (see a good summary in Morris and Thurling, 2001):

   - it may involve walking in particular patterns (following transects for instance) as is standard practice for birds;
   - sometimes it may be based on setting up fixed observation posts;
   - a more rare approach is setting up artificial refuges to attract the species;
   - sometimes it may rely mostly on trapping techniques (bottles, pitfall traps, sweep netting of trees, etc.) as is normal for invertebrates, where traps are sometimes left and their contents inspected later.

Although some experts can characterise communities directly, it is common at this stage to use a *quantitative* approach. This is done (Appendix F in Morris and Therivel, 2001) by, first, identifying the different species and establishing the intensity of their presence, usually based on two sets of criteria: (i) determining the "constancy class" of each species, the percentage of samples where it is present (the so-called Braun–Blanquet classification); and (ii) the "abundance" of each species in the samples, measured as percentage cover (the so-called Domin values). Once the presence of the different species has been quantified, the second step is to "match" their frequency distributions to tables describing the compositions of different standard communities and sub-communities (the National Vegetation Classification) as contained in the reports edited by Rodwell since 1991 (Rodwell, 1991–2000). This matching process can also be done by a computer program such as MATCH (Malloc, 2000). In practice, although rigorous sampling and a random statistical approach is often used, it is sometimes more important to characterise the area well and for the specialist to develop a "gut feeling" about its quality and the suitability of the habitats to support particular populations (Beaumont, 1994). It probably depends on the expertise available: if the field surveyor is an expert specialist, the intuitive approach will probably suffice; if he/she is a generalist or is not expert enough, the step-by-step quantitative approach is more advisable.

One of the problems with this type of survey is that it must be done by experts who can recognise the species by sight, and some argue that this in itself is a problem (Wathern, 1999) as most generalist ecologists can only recognise the most common species, hence the need to send specialists to the field at this stage. Another crucial issue raised by this type of data collection is its *timing*, as the time of the study may not coincide with the best seasonal time for certain species to be detected. This applies to habitats as well as to animal species surveyed in Phases 1 and 2, and Petts and Eduljee (1994a) and Morris and Thurling (2001) provide useful summary charts indicating the best sampling periods for various types of vegetation and fauna. Wathern (1999) also underlines this problem and its implications for the quality of many Environmental Statements, which fail to indicate explicitly whether the ecological sampling was done at suitable times or not. This is sometimes one of the reasons for some kind of "follow-up sampling", which leads into a possible "Phase 3".

8.2.2.3 Phase 3

Phases 1 and 2 are always included in an ecological impact assessment. Occasionally, a more detailed "Phase 3" is used, when certain rare species requiring special attention are found and detailed specialist surveys are required (like barn owl surveys, or bat surveys), or when some form of monitoring or seasonal-variation study is necessary. The main objective of this phase is to determine the numbers and importance of any rare species present, but it is necessary to put these "rarities" in perspective, and they should not be "looked for" specifically in Phases 1 and 2 (Beaumont, 1994); they may be found in those general surveys, and then a Phase 3 is organised to assess their status (Figure 8.6). These surveys provide the backcloth for the assessment of the ecological situation in the area, which normally constitutes the baseline study. However, in the context of impact assessment a complete baseline study should look not only at the present situation, but also should try to anticipate what the *future* is likely to be without the project. As Morris and Emberton (2001) point out, precise ecological forecasts are difficult and unreliable and, at best, only general predictions about patterns of "succession" between types of habitats can be hinted at. The sheer complexity of this field and its scientific foundation is again the problem when trying to reduce it to simple rules. Also, most ecological populations are very dependent on the weather (Beaumont, 1994). There can be 50–90 per cent fluctuations in bird numbers because of a good or bad winter, and this cannot be predicted.

Crucial to the study of the baseline in its various phases is the amount of resources (time and money) allocated. If ecology is one of the main issues involved (if the site involves a conservation area for instance), the magnitude of the budgets ought to reflect that. The complexity of the ecology and the variation of habitat types is going to be an important factor, but it is the *size* of the area that is usually most important (Beaumont, 1994). In a standard case when the site is not very big (under 10 hectares), ecology is not expected to be one of the main issues (like an area of improved grassland of no particular interest, for example), seasonal variations are not crucial and a Phase 3 is unlikely to be required, then a "3+3" time-schedule is typical: three days for the desk study (area characterisation) and another three days for the fieldwork. Then, for every day of fieldwork, an average of three days is needed to write up the corresponding report. If the site is large (above 10 hectares), the three days become five. On the other hand, the duration of the work may not be directly proportional to the size of the

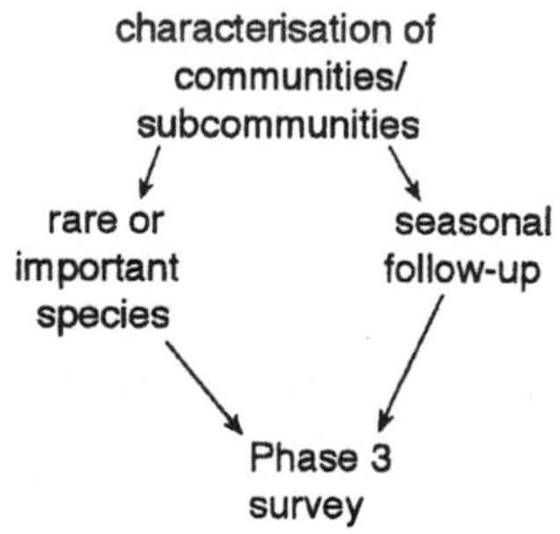

Figure 8.6 Phase 3 survey for terrestrial ecology.

area but to the complication of the work. A coastal site may take ten days because of its complicated ecology, or a survey of breeding birds may require eight to ten visits at dawn and dusk over a month. Also, if the site is on an administrative border (between counties, for example), the number of organisations to contact is twice the number, and the work must stretch over a much longer period, and the five days work may grow to over three weeks. These budgets are usually determined at the start and, if something requiring further investigation is found in Phases 1 or 2, it is usually necessary to go back to the client to recommend further work and re-negotiate the budget.

8.2.3 Quality assessment

The evaluation of ecological quality starts from the very beginning and carries on through all the survey stages, as the different surveys "add" value (see also Appendix D in Morris and Therivel, 2001, for a list as it applies to the UK):

1 The preliminary area characterisation already identifies (in documents or through local contacts):
   - sites with protected status;
   - sites hosting protected species.

2 The habitat survey in Phase 1 adds the identification in the field of:
   - priority habitats;
   - EU habitats of special importance;
   - rare habitats;
   - rare, distinctive or threatened species (animal or vegetal).

3 Phase 2, apart from the occasional "surprise" of finding species not detected in Phase 1, mainly deepens the detail of the evaluation of Phase 1:
   - quantifying the status of the different species;
   - identifying communities and sub-communities.

4 Phase 3 is really an extension of Phase 2, contributing to the assessment similar information but for quite rare animal species.

Finally (it could be seen as "number 5" on the list above), after the survey stage, an *overall evaluation* is usually conducted, using general criteria that apply to the whole area. One set of such criteria was first suggested by Ratcliffe (1977) and has been used universally in the UK ever since as the so-called Nature Conservation Review criteria (see a summary in Appendix D in Morris and Therivel, 2001). These are grouped into "primary" and "secondary".

Primary criteria are:

- *Size* of the ecological area (not of the project site), as a larger area is likely to be able to support a wider variety of species, and is also likely to be less vulnerable to change.
- *Diversity* in terms of variety (number) of different habitats present in the area and in terms of the different species present, although its measurement can become confused, as varieties are often related to the size of the area – the larger the area the more likely it is to contain a wider range of species – and the same number of different species will represent greater diversity in a small area than in a large one.
- *Rarity* of its habitats and/or species, by comparison with standard classifications such as those used in Phase 2.
- *Naturalness* refers to "the degree to which a habitat or community approximates to a natural state" as opposed to semi-natural or improved states.
- *Typicalness* refers to the degree to which a habitat or community "is a good example of those that are – or have been – characteristic of an area".
- *Sensitivity/fragility*, the susceptibility to environmental changes/impacts; again, this indicator is size-sensitive, as larger systems are likely to be more robust.
- *Non-recreatability*, related to some of the other indicators like naturalness and rarity, as the more natural a habitat is (for example, ancient woodland) the more difficult it is to recreate.

Secondary criteria are:

- *Recorded history* of an area which can increase its value for posterity and make it more worth preserving, as a model for the future, or for future research.
- *Position in a wider geographical or ecological unit*; the area may be part of a chain of similar or complementary areas, so that disrupting one will affect the whole set, and the area's degree of connectivity to the wider system will be crucial.
- *Potential* of the area, which may be enhanced by pro-active management actions or simply by future natural processes – hence likely to benefit from protection.
- *Intrinsic appeal* to the public, which can take various forms, usually including:

  (a) visual/landscape attraction;
  (b) social and amenity use;
  (c) educational value.

Sometimes, for instance as we get closer to urban areas, other reasons such as the "accessibility" of the area to urban populations can make it valuable, even if its pure ecological value is not high (Beaumont, 1994). The London Ecology Unit (LEU, 1985) suggested adding the notion of "presence in an area of deficiency", and this criterion is sometimes adopted for this exercise in urban areas.

Another similar set of criteria (listed in Morris and Emberton, 2001) has been suggested more recently as part of the so-called "New Approach to Appraisal" (DETR, 1998) and involves a shorter list of indicators, simpler than but not too different from Ratcliffe's:

- site category indicating the area's conservation importance, ranging from the various levels of legal designations to the lowest categories of sites with little or no biodiversity and ecological interest;
- features present, habitats and species;
- scale of importance (local, regional, national, international);
- level of importance (i.e. from low to high) and reasons for it;
- rarity;
- substitution possibilities by re-creation or re-location.

All these indicators tend to be used more in a qualitative rather than quantitative way. Some can also be interconnected but, in pure ecological terms, Ratcliffe's "naturalness" and "rarity" (and, by implication, "recreatability") tend to dominate, as they tend to reflect the *irreversibility* of any impacts. For most of these indicators there are no absolute measurement scales (in that sense they could be said to be "comparative") but in terms of rarity and naturalness, what the expert is looking for is the presence of certain habitats/species which are considered of importance in the accepted literature or legislation, because of their nature or because of their rarity or endangered status. In this respect, the *level* of society at which this importance is recognised will determine the overall status of importance of the area (Beaumont, 1994): local, county, regional, national or international interest (Figure 8.7).

8.2.4 Impact assessment

Ecological impacts are very diverse and complex (Morris and Emberton, 2001) and, in practice, their study usually concentrates on assessing the importance of the habitats and species affected, by habitat loss or fragmentation, by the project. Having determined the general quality of the area and the level of importance of its habitats and species, the assessment of the impact of the project is done by relatively simple cartographic *superimposition* of the two (Beaumont, 1994) with particular reference to the *temporality* of the various project effects. By superimposing a map of the different parts of the project onto the map of the ecology of the area where all the habitats and species are recorded – which GIS' *overlay* functions can make automatic, fast and

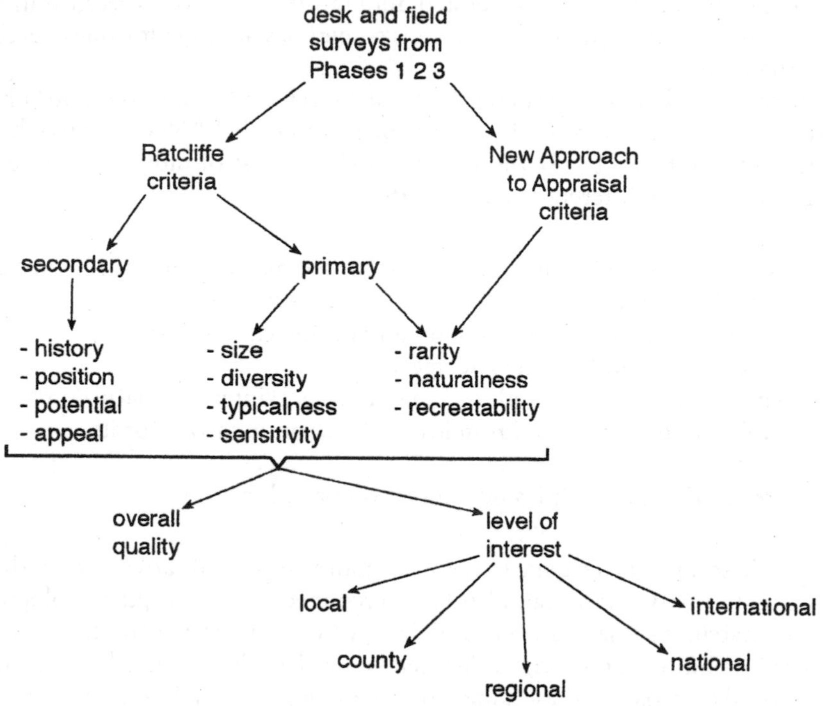

Figure 8.7 Ecological quality assessment.

accurate – a simple calculation of the "land take" of the different parts on the different habitats (and the fauna supported by them) will provide estimates of:

- the area of each type of habitat affected; and
- the number of different species (animal or vegetal) affected, by introducing the information about the densities of the various species observed in the surveys.

The importance of these estimated impacts will normally depend on (Beaumont, 1994): (i) their *magnitude*, as large concentrations of certain species (like some types of birds) can be by definition of considerable national or even international importance, and there are good guidelines on this, especially for birds; and (ii) the relationship between the size of a particular species or habitat affected and its rarity and preciousness. This is often estimated as the *percentage* that the population affected represents out of the existing population in a geographically defined area (at national level for instance), and disturbances exceeding 1 per cent would be considered significant.

On the other hand, even when an impact is important in terms of its magnitude or the proportion of the species affected, the temporal nature of the disturbance may make a difference depending on whether the environment disturbed is "recreatable" or not. If the part of the project creating the disturbance is temporary, its impact will be permanent if the environment it disturbed was non-recreatable. On the other hand, it will be possible to minimise the impacts on recreatable environments of temporary parts of the project (like areas occupied in the construction stage) by restoring the area affected to its former state once each temporary operation is over, and that will usually be done as part of the *mitigation* programme (Table 8.1 and Figure 8.8).

Table 8.1 Ecological impact assessment

Environment disturbed	*Project operation*	
	Temporary	*Permanent*
Recreatable	Non-significant (mitigation)	Significant
Non-recreatable	Significant	Significant

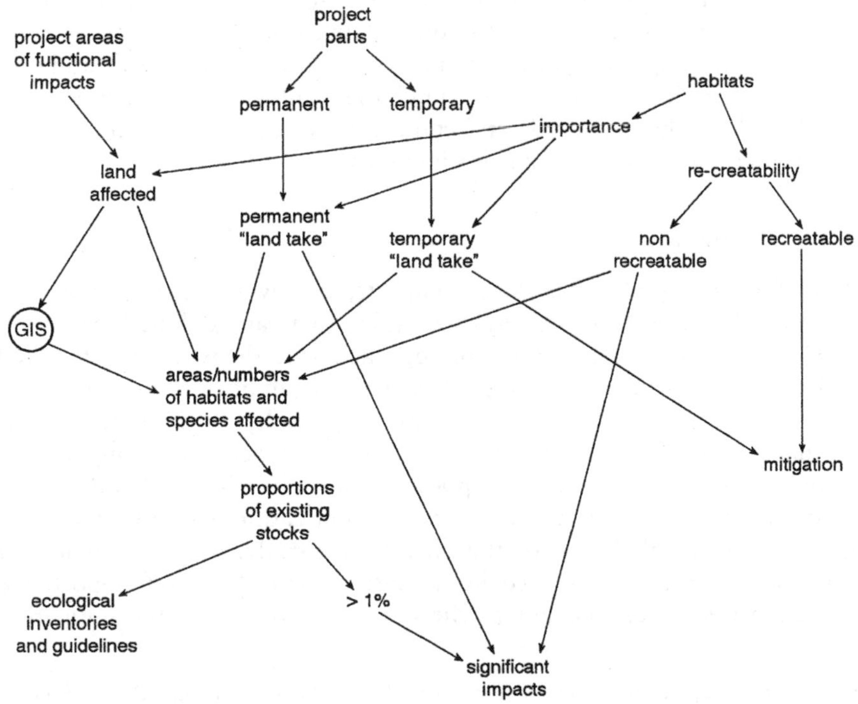

Figure 8.8 Impact significance in terrestrial ecology.

As a result, we end up with impacts of different levels of significance:

- unimportant impacts;
- important impacts which can be mitigated;
- significant impacts which *cannot* be mitigated (for any of the above reasons).

The New Approach to Appraisal (DETR, 1998) suggests a more qualitative scale of "combinations of impacts" (positive and negative) looking at: (i) *negative* impacts in terms of disturbance of the site's integrity making it unable to sustain the habitat, or if there is wildlife loss and its ecological objectives are affected; (ii) *positive* impacts – usually from mitigation measures – involving wildlife gains (for the full list, see Morris and Emberton, 2001). EEC (2002)[35] contains guidelines along similar lines for ecological impact assessment of projects and plans affecting "Natura 2000" sites in Europe.

Effects resulting from land being affected (being taken or being polluted) by the project are usually assessed in this way. Other impacts like the effect of noise or more complex ecological impacts are too complicated to quantify or simulate with models (practically non-existent), and tend to rely on "subjective" methods (Wathern, 1999) or on "analogy" with other similar experiences (Petts and Eduljee, 1994a). As Morris and Emberton (2001) explain, "in the absence of definitive quantitative evidence, impact prediction has to rely on judgements based on a knowledge of impact factors and ecological systems". The difficulty in simulating ecological impact prediction simply reflects the magnitude and complexity of the problem and, hence, the diversity of scientific arguments involved.

8.2.5 Mitigation

When there are significant ecological impacts, there will *always* be mitigation measures, because there is always something that can be done (Beaumont, 1994), but there is a basic choice of approach as to the way it is integrated with the impact study: (i) the impact assessment can be done first, and then mitigation measures are recommended as a result; or (ii) mitigation is "assumed" and the assessment of impacts is carried out taking it into account, although this second approach runs the risk of "hiding" the impacts, which may appear unexpectedly if mitigation does not work. The approach taken will depend on the client for the study, and it is also normal that *both* approaches are used in succession: first, in a draft report, the impacts are assessed and put to the client, together with the mitigation

35 Originally prepared by the Impact Assessment Unit (School of Planning, Oxford Brookes University) for the European Comission, and accessible on page *http://europa.eu.int/comm/environment/nature/natura_2000_assess.pdf*.

measures which would be required to make it acceptable; in a second stage, if the client accepts the recommendations, the impact study is done again but this time assuming the mitigation measures. When the impact assessment is submitted, the client will be held to the commitment to the proposed mitigations, which will be incorporated into the decision by the Planning Authority as "planning conditions". This is the most effective approach to mitigation, using impact assessment to influence the redesign of some aspects of the project. The best mitigation is achieved through trying to avoid the impacts in the first place, for example – as already mentioned in Section 8.2 of this chapter – by a judicious process of site selection for the project. However, this is uncommon, and the impact study usually has to consider possible mitigation measures which could be applied to the project as it stands at its different stages (Figure 8.9).

For the *pre-construction and construction* stages – usually involving many temporary operations – detailed lists of the most common mitigation measures can be found in Petts and Eduljee (1994a) and in Morris and Emberton (2001). Such measures tend to concentrate on "impact avoidance" (Beaumont, 1994):

- education of site-workers to make them habitat-sensitive;
- fencing sensitive areas nearby to stop accidental encroachment;

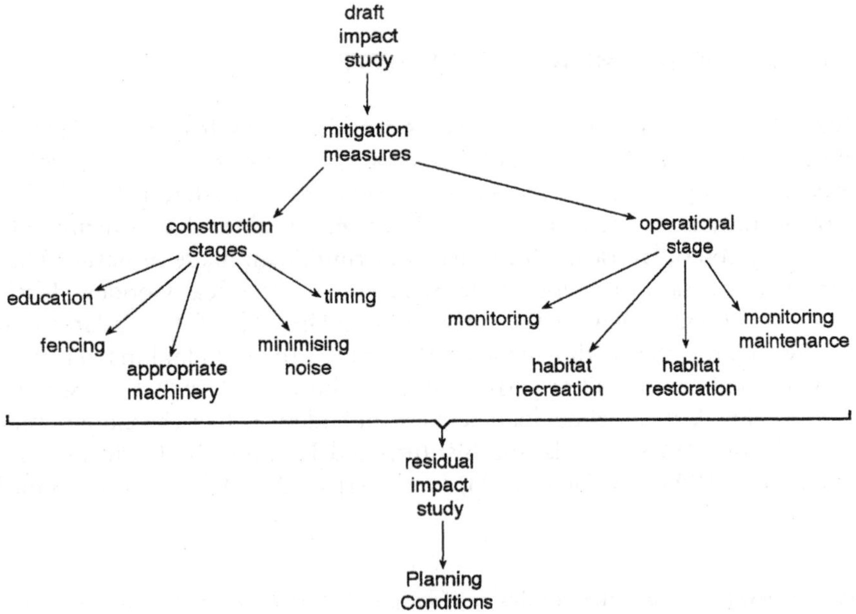

Figure 8.9 Mitigation in terrestrial ecology.

- using appropriate machinery, like the extreme case of using hovercrafts instead of wheeled equipment to move over sensitive bog or heath habitats;
- minimising noise;
- timing operations, for instance, waiting until the summer and avoiding the breeding season.

For the *operation* stage, when most effects are not temporary, mitigation is basically about containing the impacts and/or compensating for them:

- monitoring the environmental situation in areas nearby and not directly affected, as part of a "monitoring plan" recommended as a mitigation measure;
- similarly, as part of a "restoration plan", recreating lost habitats somewhere else, sometimes involving the "translocation" of species, or the recreation of communication between parts of habitats which have become fragmented;
- restoration of the areas left, trying to improve their interest and compensate for their losses by undertaking "creative conservation" (Beaumont, 1994) and landscaping with appropriate planting mixes in landscaping and plants;
- an element of "impact avoidance" can also be present at this stage in the form of monitoring and controlling any maintenance work that can cause damage.

8.3 LANDSCAPE IMPACT ASSESSMENT

The difficulty of this area of impact assessment derives not from the complexity of the science behind it – as it did with ecology – but rather from the *lack of science* to support it, as it is eminently *subjective* (Giesler, 1994).[36] This area of study is not new, and the study of the variables that contribute to the perception of landscape (be it urban or rural, man-made or natural) has attracted attention for a long time, especially in America. Goodey (1995) provides a good review of the early studies, Therivel (2001) updates that review and, together with Petts and Eduljee (1994b) and Hankinson (1999) provide good "manuals" of good practice in this area of impact assessment, supported by key guideline documents from bodies such as the Countryside Commission (1993), Landscape Institute and Institute for Environmental Assessment (1995) and DoT (1993) (see Therivel (2001) for a more detailed list).

36 The knowledge acquisition for this part was greatly helped by conversations with Nick Giesler, of Environmental Resources Management Ltd (London), and Andrew Bloore helped with the compilation and structuring of the material for this part. However, only the author should be held responsible for any inaccuracies or misrepresentations of views.

In this context, landscape is considered in two aspects, which are really "two sides of the same coin": on the one hand, landscape as a physical reality, a resource whose quality can be assessed in itself; on the other, landscape as the object of perception (mostly visual), to be assessed by reference to "receptors". This distinction gives rise to a basic division into two areas of impact assessment: impacts on *landscape quality* itself (the loss of valuable landscape features), and impacts on what is usually known as "*visual amenity*" – losses in the visual perception of the landscape. In practice, however, the latter derives from the former, and the two are just different stages in the process of assessment. The relative importance of these two aspects for assessing the significance of impacts may depend on the location: for example, in an urban setting, where the surrounding landscape is of relatively poor quality, the visual aspect may be the most important to establish the significance of the impacts; on the other hand, in a rural setting, the reverse may be true – as in the case of wind farms (Giesler, 1994). In between these two extremes, the "balance" of the assessment varies for the whole range of types of development, in terms of situation and size.

As with the assessment of most impacts, landscape and visual impacts are usually studied after the project in question has been clearly defined, which makes the whole process less effective. An earlier consideration of landscape impacts can improve the quality of the final product and the efficiency of the design process, as was the case with the project for the Forth Estuary (Scotland) crossing (Giesler, 1994). Such a study can be undertaken at the site selection stage – when the general area has been selected but the specific site has not – or after the location has been decided but before the design has started, not necessarily to "guide" the design in specific details, but to influence general considerations such as the general orientation of buildings or the location of access roads.

The assessment process flows from the determination of certain features of the project into the assessment of the landscape quality and then its visual amenity. However, before any fieldwork for the determination of the landscape quality can be done, the area of study must be determined and a "pre-study" of visibility is needed (Giesler, 1994), as only the area from where the project is likely to be visible will require investigation (Figure 8.10).

8.3.1 Project characteristics

As landscape impacts derive from the size and appearance of the project, it follows that those will be the main sets of characteristics to investigate (Figure 8.11):

1 Buildings and structures:

- location and dimensions, how much area they occupy;
- materials and colours to be used in their construction;

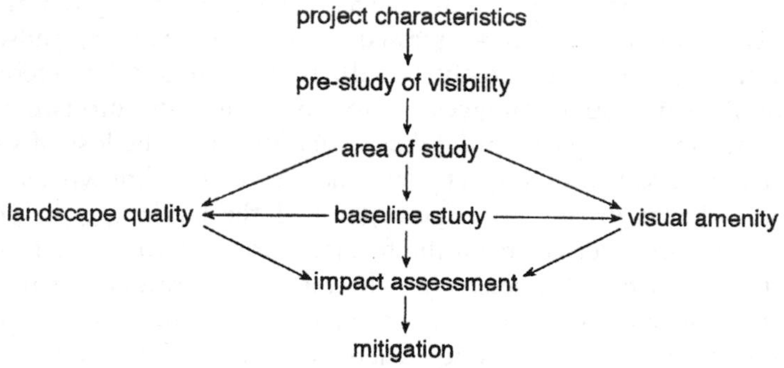

Figure 8.10 The logic of landscape impact assessment.

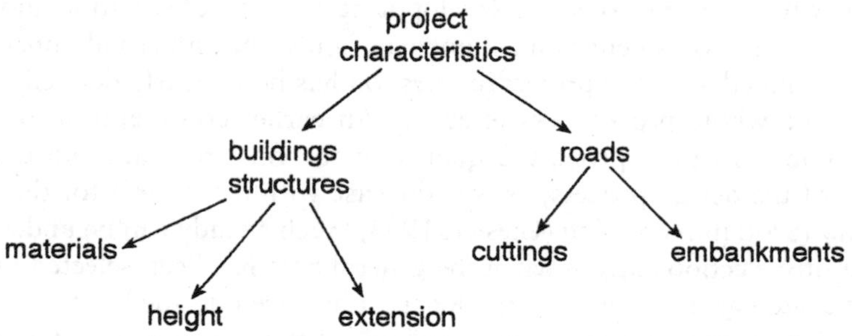

Figure 8.11 Project characteristics for landscape impact assessment.

- "extension" and width on various sides, as this will determine much of their visual impact;
- height, probably the most important dimension.

2 Roads and accesses:

- location and where any earth-movements will be;
- where the "cuttings" (if any) are going to be;
- where the "embankments" (if any) are going to be.

8.3.2 Area of study

The extent of the study area can vary from 1 km around the project (minimum) to as far as its area of visibility (Giesler, 1994). This means that, unlike other areas of impact assessment, in this case we cannot completely pre-determine *a priori* the "radius" of the area of study, because it requires carrying out part of the visibility analysis – to be done later as part of the analysis of visual impact – *before* we start the baseline work. A simplified version of this can be done using GIS technology, if we have the right information, which may be difficult or expensive to acquire. If we have topographic information (altitude) for a sufficient sample of points covering the vicinity of the project (extending several miles in all directions), we can use GIS to construct a "Digital Terrain Model" (DTM). A standard GIS "viewshed" function will produce a reasonably good first approximation to the area of visibility of the project, at least as far as the naked terrain goes, without the influence of vegetation, buildings, or other features which may act as "barriers". On the other hand, if this technology or the data necessary are not available, the area of study has to be defined "by eye" on the basis of an Ordnance Survey map (1:10,000 or 1:25,000) containing the contours of the terrain, as a substitute for the DTM. The inspection of such a map, together with experience which may be available – for example, looking at cases *similar or worse* than the project in the same area – should produce an idea of the likely extent of the area of visibility. This area can extend over considerable distances (10–20 miles) but the study does not have to go that far, because any visibility at such distance would be minimal, and the features of the project would appear very small on the horizon. In cases like this, the study area will extend no more than 5 km (Figure 8.12).

8.3.3 Preliminary landscape quality assessment

First of all, cartographic and photographic information about the site and the area should be collected. In addition to the OS maps (essential) already required for the visibility pre-study, aerial photographs can be even more useful, as they show shapes and features better. On the basis of this information, the first step in the assessment of the landscape – following the methodology suggested by the Countryside Commission (1993) – consists of breaking down the area into *landscape units*, broadly homogeneous landscape areas identified in terms of:

- *landform*: slope, valleys, ridges, etc.
- *landcover*: vegetation, land uses, woodland, etc.
- landscape *features*: buildings, fences, footpaths, rivers, ponds, etc.

On the basis of these factors, the size of the different landscape units is determined, making sure to cover three specific areas (Figure 8.13):

- the site to be occupied by the project;
- the immediate area around it;
- the estimated area of visibility of the project.

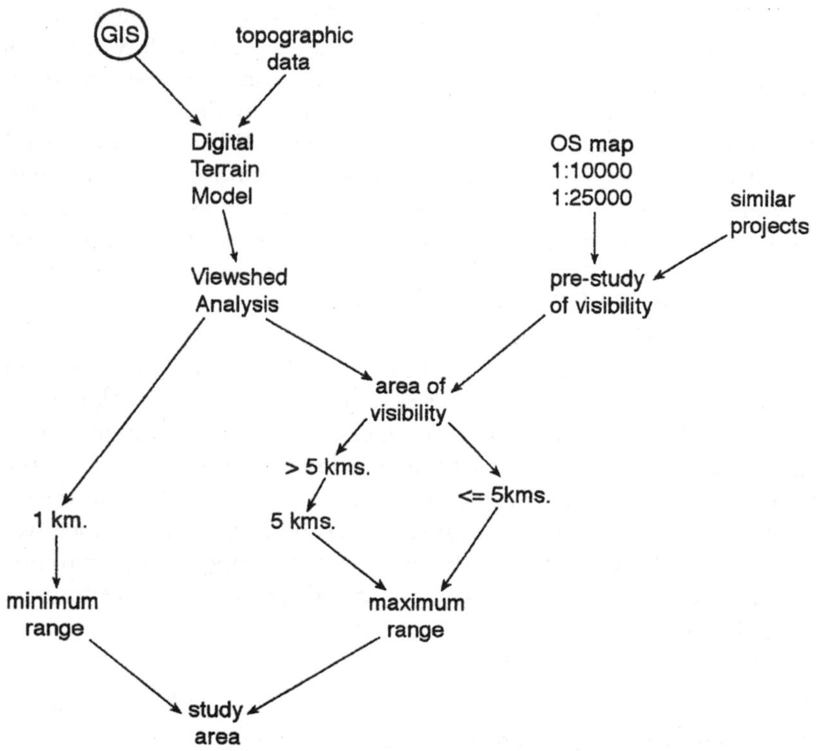

Figure 8.12 Study area definition for landscape impact assessment.

In this stage we also identify the main *receptors* (individuals and groups) with a view of the site, usually including one or more of:

- *home* receptors in individual residential properties or in residential areas;
- *recreational* receptors, users of footpaths, cyclepaths and/or leisure areas;
- *road* users;
- *workers* in local jobs, normally less important.

With respect to these receptors, it is important to identify both receptor *location*, which will guide the field visits, and *numbers*, which are essential to gauge the importance of the impacts: it can be argued that this is not the same for a project affecting a small community than one on the edge of a large city, seen by thousands of people (Giesler, 1994).

GIS can help with the identification, counting and/or measurement of the receptors because they are just features on maps (roads, cycleroads, footpaths, residential buildings, leisure areas), but the identification of landscape units involves judgmental decisions about the coherence – or lack of – between

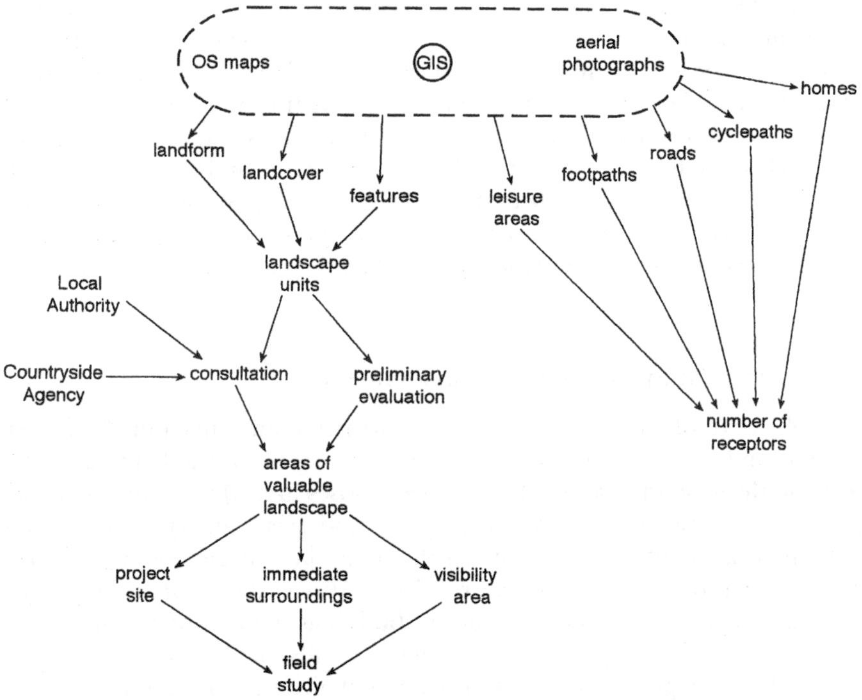

Figure 8.13 Preliminary landscape quality assessment.

characteristics which go beyond feature detection. GIS can help identify landuses and features in an area, and quantitative analysis of that could give some form of "average" characterisation of that area, but not at the level of detail necessary to define specific landscape units of varying extensions and boundaries.

The second step is the estimation of the levels of *quality* of the different units, identifying the areas of potentially valuable landscape, which will be the ones to be investigated in the field study. For this purpose, it is important to find out as much as possible about the site and the area around it by consulting two main sources:

- the County and/or District(s), who will be aware of any designations of areas of landscape interest (AONBs, etc.);
- statutory bodies, in particular the Countryside Agency for England, the Countryside Council for Wales and Scottish Natural Heritage for Scotland.

Some of these bodies undertake their own landscape evaluations. The Countryside Agency has its own landscape evaluation of the whole country by

1 km squares and, even if these km-square "average" evaluations are insufficient for the detail required in impact assessment, they can provide a good starting point, helping to characterise the area in landscape terms (like the ecological "area characterisation" referred to earlier in this chapter). If no areas of valuable landscape are found and no significant impact is expected, then the study does not need to proceed further. This is quite rare, because it is likely than in cases such as this the decision that a landscape impact study was unnecessary would have been made at the "scoping" stage and there would not even have been any preparatory work. If, as is normally the case, some "units" are found with some landscape quality, then a field study is necessary.

8.3.4 Field study and baseline assessment

The objective of the field survey is first to make an "inventory" of what there is and, to a certain extent, gain an impression of the landscape as it feels on the ground (Giesler, 1994). Its purposes are: (i) to complement the preliminary assessment of landscape units and their quality, to determine with first-hand information the worth of the landscape resource on and around the project site; (ii) to "correct" the estimated area of visibility with the features on the ground (vegetation, buildings, fences) by imagining the view of the project site – or using mechanical aids like balloons to simulate the height of the project (Hankinson, 1999) – which normally has the effect of *reducing* the visibility area; (iii) to assess the quality of the landscape *from the visibility area*, in particular from the point of view of the "receptors" identified before; and (iv) to assess the quality of the landscape *in* the visibility area, in the immediate surroundings of the receptors, as their perception of the landscape in the project area will be affected by the landscape around them.

The field study consists of a series of visits: (i) the *first* visits (Hankinson, 1999) should be to the project site and the area around it; (ii) then, visits to the locations where the *receptors* identified are likely to be. In both sets of visits, the landscape in the area should be assessed and also, when visiting the receptor areas, the landscape in and around the project site should be assessed *from the point of view* of the receptor area, as this will constitute the basis on which to assess the impact that the project is likely to have on the "visual amenity" of the landscape. The timing and duration of the overall impact study can dictate sometimes that there is very little choice of when the field visit for the landscape assessment is to take place. If this choice exists, however, it may be a good idea to make more than one field visit to the receptors in order to assess the visibility of the site with and without seasonal vegetation, which may change the extent of the visibility area considerably (Hankinson, 1999).

The approach to the field study is two-pronged (Figure 8.14): on the one hand, it involves a relatively objective approach, more a description than an assessment of the landscape as a resource, based on aspects similar to

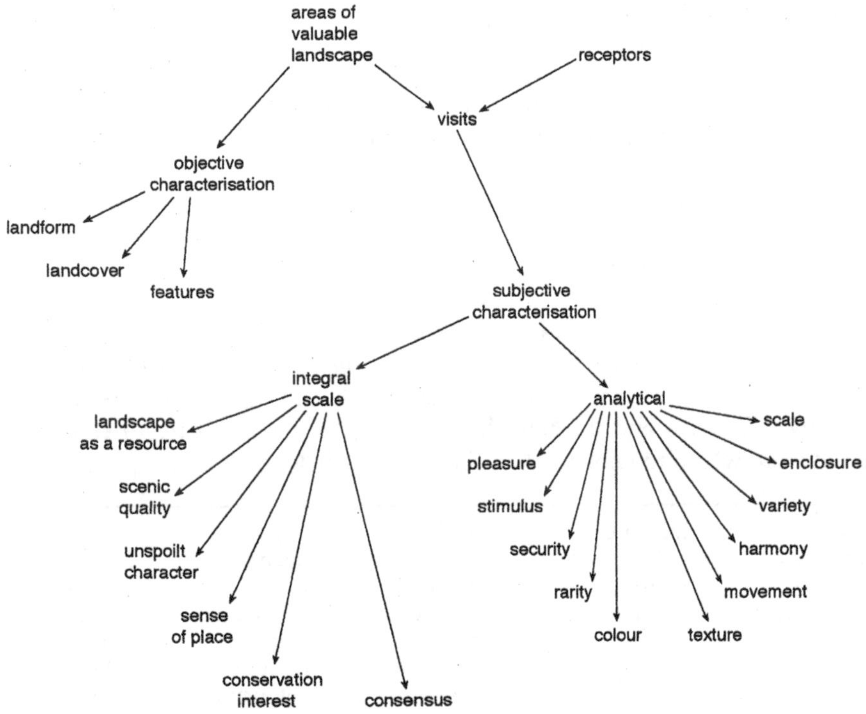

Figure 8.14 Landscape baseline study.

those used in the preliminary desk study: *landform*, *landcover* and landscape *features*. On the other hand, a more subjective approach is used to determine the "character" and quality of the landscape – and this is why it is invariably challenged at public inquiries (Giesler, 1994). This subjective assessment can be applied with an "integral" approach using a broad classification of the landscape into *one* quality scale using broad categories that encapsulate several aspects. The one proposed by Hankinson (1999) concentrates on the improvability/recoverability of the landscape as a resource:

- *irreplaceable*: pristine natural landscapes with their original features;
- *above average*: well-managed landscapes;
- *renewable, average*: ordinary pleasant countryside with human influence;
- *improvable*: degraded by abandoned human land uses;
- *seriously degraded*: derelict or polluted landscape unlikely to be recoverable.

In this approach, subjectivity is applied only once in an "aggregate" way. A more analytical approach can also be used which breaks down the problem (landscape quality) into sub-problems, using a "checklist" of *variables* that comprise landscape quality. The Countryside Commission (1993) suggests a relatively simple checklist, even if the variables themselves are complex:

- importance of the landscape as a resource (local, regional, national, etc.);
- scenic quality and combination of landscapes;
- unspoilt character;
- sense of place, with distinctive and common character and visual unity;
- conservation interest;
- consensus between professionals and the public.

Giesler (1994) suggests a list of more detailed variables: *scale, enclosure, variety, harmony, movement, texture, colour, rarity, security, stimulus, pleasure*. These are qualitative variables; hence their determination is mostly subjective and, to help the assessor, "scales" of categories can be used from which to choose, as in the example below:

Scale	intimate	small	large	vast
Enclosure	tight	enclosed	open	exposed
Variety	uniform	simple	varied	complex
Harmony	harmonious	balanced	discordant	chaotic
Movement	dead	calm	busy	frantic
Texture	smooth	managed	rough	wild
Colour	monochrome	muted	colourful	garish
Rarity	ordinary	unusual	rare	unique
Security	comfortable	safe	unsettling	threatening
Stimulus	boring	bland	interesting	invigorating
Pleasure	offensive	unpleasant	pleasant	beautiful

Source: ERM.

Such scales help to clarify the meaning of the variables they represent, to the extent that they can be used by members of the receptor groups also. These qualitative variables are intended to characterise the landscape more than to assess it, but an element of assessment is always present. The difficulty is that the scales contain varying mixtures of description and assessment that cannot be separated: while "enclosure" or "scale" may have a dominant descriptive character, "harmony" or "security" represents mostly value judgements. Another difficulty with such variables is that, even if some intuitive "worth" may be attached to each category, it is impossible to quantify it in objective terms and there is a danger of giving a false impression of objectivity (Giesler, 1994; Hankinson, 1999). Also, this implies that the different variables cannot be combined – for instance to work out an "overall index" of landscape quality as in the "integral" approach – and have to be used independently from each other, to make some kind of "cumulative impression" on an assessor loaded with subjectivity, based on some aesthetic notion of "harmonious balance" (Giesler, 1994). Probably the best final

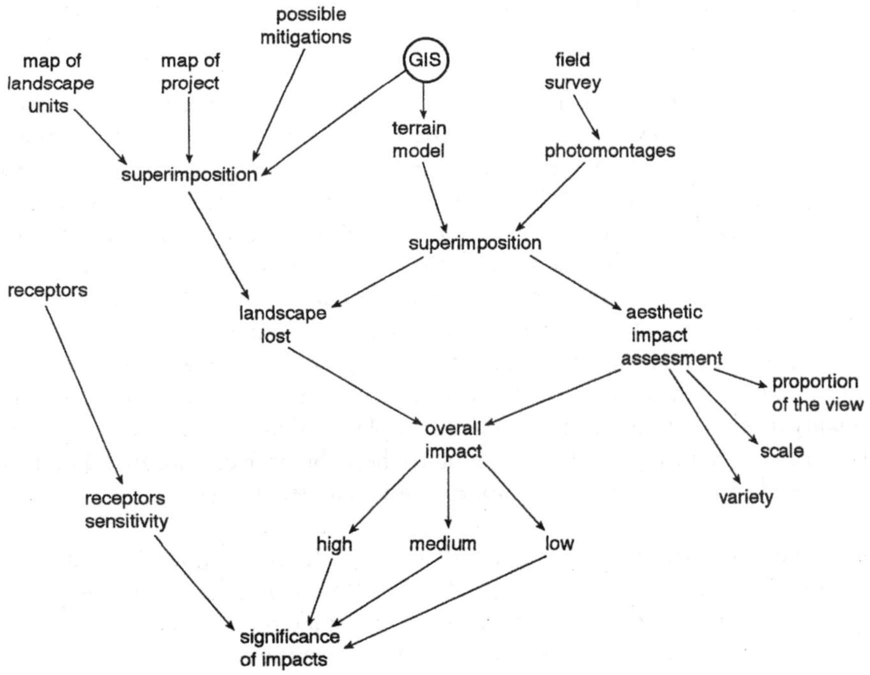

Figure 8.15 Landscape impact assessment.

assessment comes from the interaction between the two approaches, the "integral" and the "analytical".

The baseline study should be extended into a forecast of the future situation (Giesler, 1994), as the landscape is likely to change and as any mitigation by planting can take several years to mature. If such mitigation is going to be used, the study should look 15 years ahead. However, in practice the forecast of the changing baseline environment is rarely done, as landscape change is too difficult to predict.

8.3.5 Impact assessment

To assess the impact on the landscape baseline we have to go back to the basic distinction between landscape as a resource and landscape as visual amenity, as the two are assessed in different ways. The assessment of the impact on the landscape itself is not too different from the assessment of ecological impacts, and is based on identifying *how much* of the landscape will be affected (Figure 8.15):

- first, determining the landscape which will be *lost*: land, vegetation, features (ponds, etc.);

- second, determining the landscape that will remain untouched but be affected by the project, which will change the character of the surrounding area.

The first type of assessment can be done by simple "superimposition" of a map of the project on a map of the existing landscape units and their features, and the impact can be "measured" quite precisely, similar to measuring the loss of ecological areas. As with ecological impacts, the availability of GIS can make considerable difference, as the superposition and measurement can be done automatically, and more quickly and accurately.

The second type of assessment, goes back again to the subjective element that pervades this whole area of impact assessment, and which is more part of the impact on "visual amenity" than of the impact on the landscape itself. The visual amenity impact assessment presents the same difficulties as the baseline quality assessment, and has the added complexity that it is based not only on the intrinsic quality of the landscape unit where the project is located, but also on its scale and its surroundings, for example (Giesler, 1994):

- The *proportion of the view* affected: if there is a large landscape resource and only a fraction of it is affected by the project *without* disturbing the rest, the impact can be considered less important.
- A particular case of the previous type is when the *scale* of the landscape is vast and the project is seen from some distance, only spoiling a small part of the view.
- If there is "urban encroachment" and the landscape unit is next to or surrounded by urban land, the landscape becomes more precious and the impact on it becomes magnified.
- If a variety of landscapes exist and the project impacts on more than one type, then the impact is probably greatest.
- A variation to the problem of encroachment is that of the impact on a landscape already affected by other developments with only minor impacts, which raises the question of *cumulative impacts*: each of those projects may produce insignificant impacts, but a point can be reached when the *next* addition can make the whole area cross the threshold of significance.

An interesting "addition" to the sources of landscape impacts – which suggests the need for some sort of "cyclical" process of communication between the various areas of impact assessment – can be *the mitigation of other impacts*:

- raising the height of a stack can be recommended to mitigate air-pollution impacts, increasing the project's visibility;
- erecting barriers to mitigate noise impacts can also increase the visual impact;

- changing the position of access-roads can be used to mitigate traffic impacts, changing the impingement on the landscape.

After all these "partial" assessments, an impact study must move towards reaching an overall assessment of the landscape impacts. Although the problems of "reconstructing" an overall index of impact from a series of qualitative variables are well known, it is common for manuals to suggest such an approach. Petts and Eduljee (1994b) use a four-level scale *slight–moderate–substantial–severe*; Hankinson (1999) uses a three-level scale *low–moderate–significant* and Therivel (2001) uses a three-level scale *low–medium–high*. All these scales are qualitative and "impressionistic" as we should expect given the type of variables involved but they are mainly used to "summarise" the impact assessment.

The significance of the visual impact is determined by a combination of the magnitude of the impact and assumes receptor sensitivity. Receptors have varying degrees of sensitivity (ERM, 1993):

- *workers* in local jobs are considered to have *low* sensitivity;
- *home* receptors have *very high* sensitivity;
- *recreational* receptors and users of footpaths and cyclepaths have *high* sensitivity;
- *road* users have *mixed* levels of sensitivity, depending on the reason for travel (for example, recreational travellers will have higher sensitivity).

Finally, Table 8.2 (ERM, 1993) combines impacts and sensitivities into resulting degrees of significance.

To help support the assessor's views on visual amenity impact, it is a common practice to prepare *photomontages* showing the views of the project site from the "receptor" areas and, on them, superimpose some image giving an impression of the visual effect of the project:

- it can be a photograph of a model of the project, superimposed on the montage;
- or it can be simply a "wire-line" profile of the project as it will appear to the viewer, showing a 3D impression or just its skyline.

Table 8.2 Landscape impact significance

Observer sensitivity	*Impact magnitude*		
	High	*Moderate*	*Low*
Very high	Very high	High	Moderate
High	High	High	Moderate
Moderate	High	Moderate	Low
Low	Moderate	Low	Low

The photomontage itself can be enhanced by computer technology, or a virtual version can be prepared with a GIS terrain model, adding to it the various vegetations and features, with a degree of realism allowed by this technology that is increasing all the time. In addition, a simulation of the project, which can be made quite accurately using CAD or GIS, can be superimposed on the enhanced photomontage or on the GIS terrain model.

Landscape impact assessment concentrates mainly on the *operational* phase of the project, as it is mainly interested in the long-term effects, and the visual impacts of the construction stage are usually temporary. Impact assessment only concentrates on the construction stage if it is going to be long (more than 18 months or two years) and the final project is not going to have practically any visual impact. It is a question of considering "what proportion of the overall visual impact is going to come from construction and how much from operation" (Giesler, 1994), and the assessment concentrates on the worst of the two.

The landscape impact assessment study may take about five days (Giesler, 1994), undertaken usually within one working week, except when several visits are organised to cover seasonal variations. The field work is carried out by expert assessors, and the public is asked to participate – with surveys or public meetings – only when there is a budget for it and time permits. Difficult cases can arise when the public hold conflicting views about the visual impact of certain developments – as in the case of wind farms or afforestation projects.

8.3.6 Mitigation

As with other impacts, it is best to incorporate the concept of mitigation into the project design process or even at the site-selection stage using "primary" mitigation (Hankinson, 1999) – in which case the project does not require "secondary" mitigation – but this is very rare. Primary mitigation measures consist mainly of "compacting" the buildings and structures to minimise their intrusion (physical and visual):

- minimising the areas of existing landscape affected by paved and built-up areas.
- minimising built structures, for example by grouping structures together;
- minimising heights, for example by lowering the height of stacks.

Secondary mitigation of the impact on landscape *as a resource* may involve "compensation" as the only possibility left, which requires the evaluation of the landscape and, most importantly, the *trade-offs* between the landscape resource lost and other possible resources added as compensation (Hankinson, 1999), although this may be difficult in some locations. Typical secondary mitigation of *visual amenity* impacts involves (Giesler, 1994):

- soft landscaping on site
  (a) planting,
  (b) bunding.
- architectural treatment
  (a) camouflaging structures or buildings,
  (b) disguising the scale of the project.

There is nothing unexpected about this list, but the last two types are good examples of the difficulties of this type of impact assessment (Giesler, 1994) – involving aesthetic and subconscious processes – as the term "camouflaging" can refer literally to "playing tricks" on human perception. For example, concerning structures and/or buildings:

- using non-reflective finishes for buildings and structures;
- colouring different parts differently;
- if there is a lot of repetition in the structures (like an oil depot with many tanks), painting *one* tank in bright colours to attract attention, which can detract attention from the rest and make the whole project "look" smaller.

Scale is also one of the biggest contributors to visual impact (Giesler, 1994), and making the viewer "lose the sense of scale" can produce some mitigation:

- colouring the lower parts of large buildings/structures in darker colours and the upper parts in lighter colours;
- screening from view the lower parts of the project, where familiar points of reference (cars, people, houses nearby) – when visible – give the viewer a sense of scale (Figure 8.16).

8.4 CONCLUSIONS: THE LIMITS OF EXPERT SYSTEMS

As in the last chapter, we have seen that the published advice and expertise on impact assessment in terrestrial ecology and landscape can also be reduced to a logic which could be formalised into a "knowledge base" for an expert system. On the other hand, in the two areas discussed in this chapter we have encountered some of the difficulties involved in this "reduction". In terrestrial ecology, difficulty is associated mainly with "scientific" difficulty, the sheer complexity of the science behind it, and sometimes the problems of finding an expert with the right expertise, or sufficient published information, especially in countries new to impact assessment. Another more "external" difficulty – a mistake that "novices"

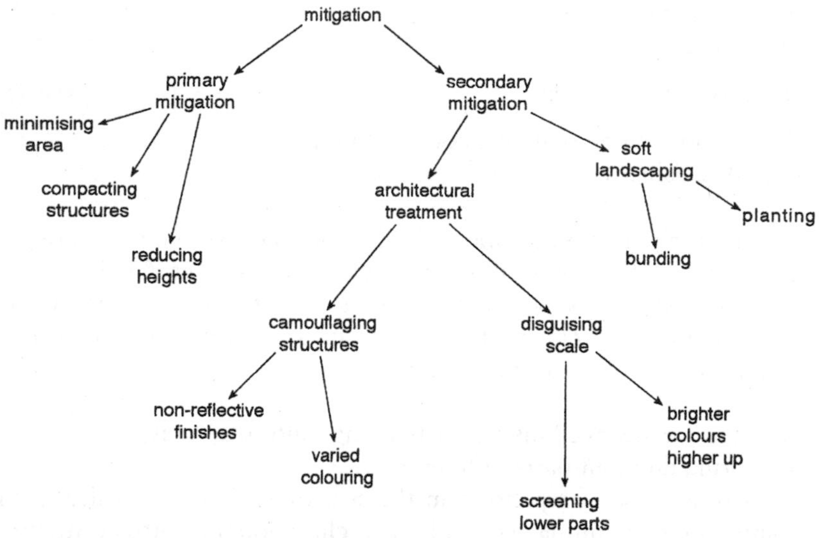

Figure 8.16 Landscape impact mitigation.

sometimes make – is to underestimate public feelings about their local sites and about local impacts. In landscape assessment, difficulty is also associated in the eyes of the experts with public feelings: a case is considered "difficult" when it is known that there are many different feelings about the project. The other great difficulty in landscape impact assessment, of course, is the subjective nature of practically the whole process (Figure 8.17).

In the context of the overall objectives of this book, these problems are useful to illustrate some of the limitations of expert systems, as discussed in Chapter 2 when introducing expert systems for the first time:

- Perceptual difficulty, the problem of "recognising" visually in the field the presence of certain elements. In the case of ecology this is about recognising species and habitats, involving expertise sometimes difficult to find. In the case of landscape it is about recognising landscape forms, with a "qualitative" aspect added, in that the expert is also expected to *assess* visually those landscape forms and the impacts on them, and to *anticipate* the visual effectiveness of possible mitigation measures.
- Scientific difficulty, the complexity and difficulty of the science itself (as in terrestrial ecology), where representing the full extension of the field – much of it still subject to research and debate – would probably prove too much for an expert system's knowledge base.
- Political difficulty, putting across the results of a very open-ended scientific field (in the case of ecology) and of a very subjective field (in the case of landscape) to a public passionate about their local environment and often with divided opinions.

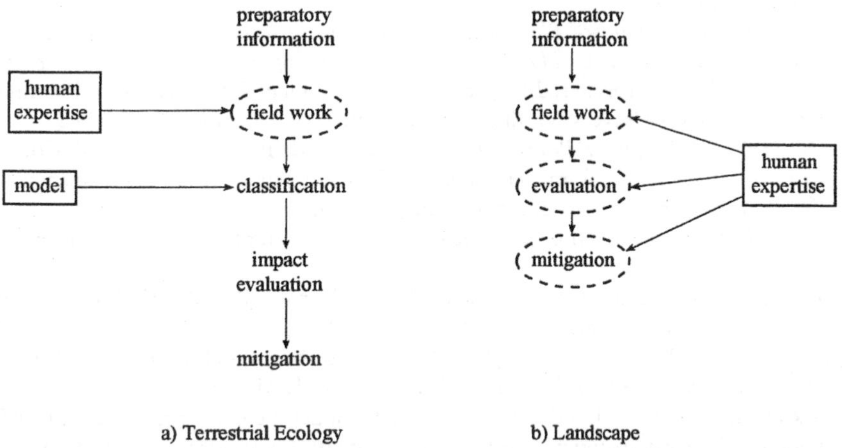

Figure 8.17 The role of expertise in landscape and terrestrial ecology impact assessment.

The first problem is a special one, derived from the difficulties of simulating vision: even if the technology to simulate visual sensors has progressed enormously, the capacity to *interpret* what is "seen" is far from perfect and is the subject of much research in areas akin to expert systems such as Neural Networks. This applies to the interpretation of sensor input and, in the case of landscape assessment, the whole problem of "qualitative" assessment would have to be added as well. The other problems are more to be expected, as noted in Chapter 2 about the limitations of expert systems when dealing with too large and complex problems, with expertise still subject to debate, and with "common-sense" issues (Waterman, 1986). In practical terms, expert systems dealing with these issues will probably have to include "gaps" in their logic, where the user will be required to consult a human expert for certain steps in the logical sequence.

REFERENCES

Beaumont, N. (1994) *Personal Communication*, Environmental Resources Management Ltd, Oxford.

Countryside Commission (1993) *Landscape Assessment Guidance*, CCP423, Countryside Commission, Cheltenham.

DETR (1998) *Guidance on the New Approach to Appraisal*, Department of the Environment, Transport and the Regions.

DoT (1993) *Design Manual for Roads and Bridges*, Vol. II: "Environmental Assessment", Department of Transport, HMSO, London.

EEC (1992) *Directive on the Conservation of Natural Habitats and Wild Fauna and Flora*, 92/43/EEC.

EEC (2002) *Assessment of Plans and Projects Significantly Affecting Natura 2000 Sites. Methodological Guidance on the Provisions of Article 6 (3) and (4) of the Habitat Directive* (prepared by the Impact Assessment Unit, School of Planning, Oxford Brookes University), European Commission, EU Environment DG.

ERM (1993) *Billingham Power Station (process plant part)*, Cleveland County Council, Environmental Impact Statement, Environmental Resources Management (for Northumbrian Environmental Management).

Giesler, N. (1994) *Personal Communication*, Environmental Resources Management Ltd, London.

Goodey, B. (1995) Landscape, in Morris, P. and Therivel, R. (eds) *Methods of Environmental Impact Assessment*, UCL London (Ch. 6).

Hanes, T. (1980) Vegetation and Wildlife Impact Assessment, in Rau, J.G. and Wooten, D.C. (eds) *Environmental Impact Analysis Handbook*, McGraw-Hill (Ch. 7).

Hankinson, M. (1999) Landscape and Visual Impact Assessment, in Petts, J. (ed.) *Handbook of Environmental Impact Assessment*, Blackwell Science Ltd, Oxford (Vol. 1, Ch. 16).

JNCC (1993) *Handbook for Phase 1 Habitat Survey – A Technique for Environmental Audit* (separate *Field Manual* also available), JNCC, Peterborough.

Landscape Institute for Environmental Assessment (1995) *Guidelines for Landscape and Visual Impact Assessment*, E & FN Spon, London.

LEU (1985) *Nature Conservation Guidelines for London*, London Ecology Unit.

Malloc, A.J.C. (2000) *MATCH II: A Computer Program to Aid the Assignment of Vegetation Data to the Communities and Subcommunities of the National Vegetation Classification*, Version 2.15 for Windows NT/95/98, Unit of Vegetation Science, University of Leicester.

Morris, P. (1995) Ecology – Overview, in Morris, P. and Therivel, R. (eds) *Methods of Environmental Impact Assessment*, UCL Press, London, 1st edition (Ch. 11).

Morris, P. and Emberton, R. (2001) Ecology – Overview and Terrestrial Systems, in Morris, P. and Therivel, R. (eds) *Methods of Environmental Impact Assessment*, Spon Press, London, 2nd edition (Ch. 11).

Morris, P. and Therivel, R. (2001) (eds) *Methods of Environmental Impact Assessment*, UCL Press, London (2nd edition).

Morris, P. and Thurling, D. (2001) Phase 2–3 Ecological Sampling Methods, in Morris, P. and Therivel, R. (eds) *Methods of Environmental Impact Assessment*, Spon Press, London, 2nd edition (Appendix G).

Petts, J. and Eduljee, G. (1994a) "Flora and Fauna", in *Environmental Impact Assessment for Waste Treatment and Disposal Facilities*, John Wiley & Sons, Chichester (Ch. 8).

Petts, J. and Eduljee, G. (1994b) "Landscape and Visual Amenity", in *Environmental Impact Assessment for Waste Treatment and Disposal Facilities*, John Wiley & Sons, Chichester (Ch. 13).

Ratcliffe, D.A. (1977) (ed.) *A Nature Conservation Review* (2 Vols), Cambridge University Press, Cambridge.

Rodwell, J.E. (ed.) *British Plant Communities*: (1991a) Vol. 1: *Woodlands and Scrub*; (1991b) Vol. 2: *Mires and Heaths*; (1992) Vol. 3: *Grasslands and Montane Communities*; (1995) Vol. 4: *Aquatic Communities, Swamp and Tall-herb Fens*;

(2000) Vol. 5: *Maritime Communities and Vegetation of Open Habitats*. Cambridge University Press, Cambridge.

Therivel, R. (2001) Landscape, in Morris, P. and Therivel, R. (eds) *Methods of Environmental Impact Assessment*, Spon Press, London, 2nd edition (Ch. 6).

UKBG (UK Biodiversity Group) *Tranche 2 Action Plans*: (1998) Vol. I: *Vertebrates and Vascular Plants*; Vol. II: *Terrestrial and Freshwater Habitats*; (1999) Vol. III *Plants and Fungi*; Vol. IV: *Invertebrates*; Vol. V: *Maritime Species and Habitats*; Vol. VI: *Terrestrial Species and Habitats*. Peterborough.

Waterman, D.A. (1986) *A Guide to Expert Systems*, Addison Wesley.

Wathern, P. (1999) Ecological Impact Assessment, in Petts, J. (ed.) *Handbook of Environmental Impact Assessment*, Blackwell Science Ltd, Oxford (Vol. 1, Ch. 15).

Westman, W.E. (1985) *Ecology, Impact Assessment and Environmental Planning*, John Wiley & Sons.

9 Socio-economic and traffic impacts

9.1 INTRODUCTION

With the exception of ecological impacts, most impacts are assessed by the repercussions they have on humans (noise, air pollution, landscape, etc.) and to that extent they all could be considered social in nature. However, impacts usually referred to as "socio-economic" have the characteristic that they are transmitted through the workings of society itself, its economy and the behaviour of its population as a result of the project. In this respect, traffic impacts can also be considered under the same heading, as they also result directly from social behaviour – with vehicles as "instruments". This view of socio-economic impacts suggests the need to consider how society works in order to assess any impacts on it, and that can face us with a problem similar to what we found when dealing with ecology, i.e. the extreme complexity of the science that studies the field, in this case, social behaviour. It can be argued (Vanclay, 1999) that social impacts have always been the central concern of the *social sciences*, and that to analyse these impacts we have to use the rigour of such sciences. In this sense, the usual approach to the study of these impacts can be said to only "scratch the surface" of social impacts, concentrating on relatively superficial indicators of impact but without getting into their deeper social repercussions in terms of *social change*, the true measure of social impact. On the other hand, in practical terms it might prove difficult to engage in deep social research involving wide-ranging surveys for every project requiring this type of impact assessment. This is one of the dilemmas of socio-economic impact assessment – and one that impact studies address in varying degrees – especially since this area of impact assessment is relatively new and still has to become fully established as part of the standard collection of impacts to consider.

9.2 SOCIO-ECONOMIC IMPACTS

These types of impacts are relative newcomers to impact assessment, as the initial emphasis of this growing area of interest and legislation was placed more on "environmental" impacts, probably on the assumption that the socio-economic side was already being covered by the town planning system (Glasson, 2001). Only in the 1990s did socio-economic impact studies become a standard component – albeit sometimes rather "thin" (Glasson, 1994)[37] – of a growing number of environmental statements, following the good-practice literature which has accompanied this "coming of age" (Petts and Eduljee, 1994b; Glasson, 1995, 2001; Chadwick, 1995, 2001; Vanclay, 1999; Chadwick, 2001 also contains a very good bibliographical compilation). There has been some debate about the nature of, and what to include in, socio-economic impacts. Our definition of these "people impacts" includes direct economic impacts, which normally lead to indirect wider economic/expenditure impacts, demographic, housing, other social services (such as education, health, police) and socio-cultural impacts (including lifestyle, community integration, cohesion and alienation). The general logic advocated for these studies is similar to that of other impacts (Figure 9.1).

Although *economic* and *social* impacts can be studied separately – partly because economic impacts tend to be positive while social impacts tend to be negative – the logic they follow is similar, and usually starts from a common base, and it is only after "scoping" the impacts that the two lines of enquiry separate.

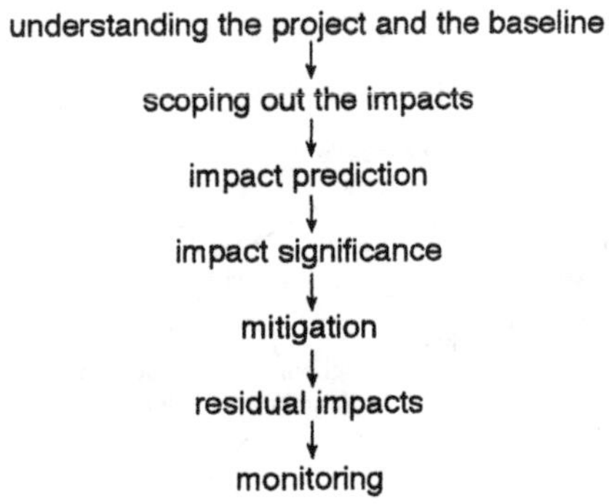

Figure 9.1 The logic of socio-economic impact assessment.

37 The face-to-face part of the knowledge elicitation for this area of impact was approached in a way similar to the other areas of impact, i.e. by holding structured conversations between Agustin Rodriguez-Bachiller and an expert in the field, even if in this case the expert (John Glasson, of the Impact Assessment Unit in the school of planning, Oxford Brookes University) was part of the authorship of this book, and references to those conversations will be made in the usual manner. Duma Langdon helped with the compilation and structuring of the material for this part.

9.2.1 Understanding the project

In socio-economic terms, what matters about the project is its *capital investment* and its *human-resources* (labour and users/customers) plans for the *construction* and *operation* stages, the study of the latter often extending up to 2–3 years into full operation. This involves first of all the detailed quantification of the socio-economic components of the project, but also it concerns more qualitative social/employment policies associated with it (Figure 9.2). Starting with the *quantitative* information, concerning the expenditure in *physical* factors first, we need to know the magnitude and nature of the project:

1 For the *construction* stage, the investment over time in:

- infrastructure,
- equipment,
- buildings,
- non-labour services.

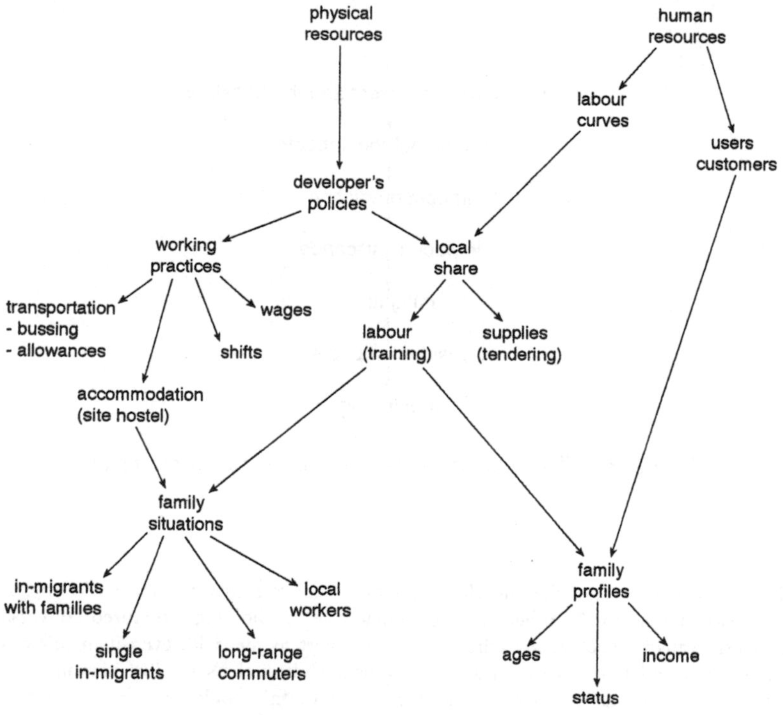

Figure 9.2 Information about the project for socio-economic impact assessment.

2 For the *operation* stage, the expenditure over time on:

- goods,
- raw materials,
- non-labour services,
- maintenance.

On the *human resources* side, we need to know:

1 The "labour curves" over time for *construction* and *operation* (see an example in Glasson, 2001):

- number of workers,
- occupational categories/skills.

Differences in the labour force between construction and operation can be important, as some infrastructure/utilities projects (like power stations, roads) involve much more labour during construction than operation, while manufacturing and especially service projects (business parks, new settlements) tend to the opposite. On the other hand, when the latter happens it tends to be because of a high number of visitors/ users, and not because of a high number of workers operating the project, as most types of projects tend to be more and more capital-intensive.

2 Visiting users/customers over time (only for the *operation* stage):

- numbers,
- socio-economic profile.

In the construction stage it is unlikely that there will be significant numbers of visitors, users or customers, and in some types of projects (like energy projects) this will also be the case for the operation stage. Other projects (like leisure facilities, retail parks, new settlements) depend on large numbers of visitors/users, whose impacts must be considered.

On the *qualitative* side, it is crucial to identify the developer's policies concerning labour practices on the one hand, and the expected level of local sharing in all the activities, on the other. On the *working practices*, it is important to know:

1 wage levels;
2 shifts to be used (e.g. two or three);
3 accommodation policies (like provision of an on-site hostel);
4 transportation policies:

- bussing workers (especially for the construction stage),
- providing travel allowances up to a certain distance.

Also, it is most important to find out if the developer has any specific policies about the expected *local share* of each part of the project:

1. Expected proportion of local/non-local labour, usually decreasing as the skill level increases; Glasson (2001) gives a typical profile of the *proportions of local labour* expected in major projects:

   - site-services, security and clerical: 90 per cent,
   - civil engineering operatives: 55 per cent,
   - mechanical and electrical operatives: 40 per cent,
   - professional, supervisory and managerial: 15 per cent.

   Sometimes developers are less inclined to employ local labour when the area has a reputation for labour problems.

2. Training policies: including training in the employment package can be useful to overcome any prejudice against taking on local unemployed people. As a general rule, the higher the occupational category of the staff the longer will be the training needed and the *less* likely workers are to come from the locality.
3. Policy on local suppliers and putting contracts out to tender: in the construction stage, during normal operation.
4. Purchasing agreements that the firm running the project (often a national firm) may have with non-local firms.

As a result of some of these policies, a profile will emerge of the proportion of workers at different occupational levels likely to be in different family/housing situations (during construction and operation):

- workers in-migrating to the area with their families: in the construction stage – if it lasts for several years – it will be of the order of 10 per cent or 20 per cent of the external workforce, during operation it is likely to be the vast majority (90 per cent) of the in-migrating workforce;
- workers in-migrating to the area but without their families;
- long-distance commuters;
- local workers.

Although all this information about the project is necessary to carry out a detailed impact study, developers cannot always provide it. Decisions on some aspects of the project (like staffing) may be at an early stage and we can either use aggregate figures for labour or investment (and carry out the analysis at an *aggregate level*) or we can use other similar projects as sources of comparative information to "flesh out" the project, when estimating the likely composition of the labour force, or the likely proportions to be in-migrants, commuters, or locals.

9.2.2 Understanding the baseline

The next step is to understand the host society which the project is likely to impact. As with the project, the study of the socio-economic baseline involves on the one hand finding out about the social situation from *data* and, on the other, finding out what the social *attitudes* and sensitivities are, which give social meaning to the data (Figure 9.3). Studying the facts alone may allow us to calculate the quantitative value of some of the impacts, but it will reduce the study of their significance to the kind of technocratic study of indicators (the "checklist approach") which Vanclay (1999) critically refers to, and only the study of the local *culture* will give us sufficient information to assess the significance of those impacts.

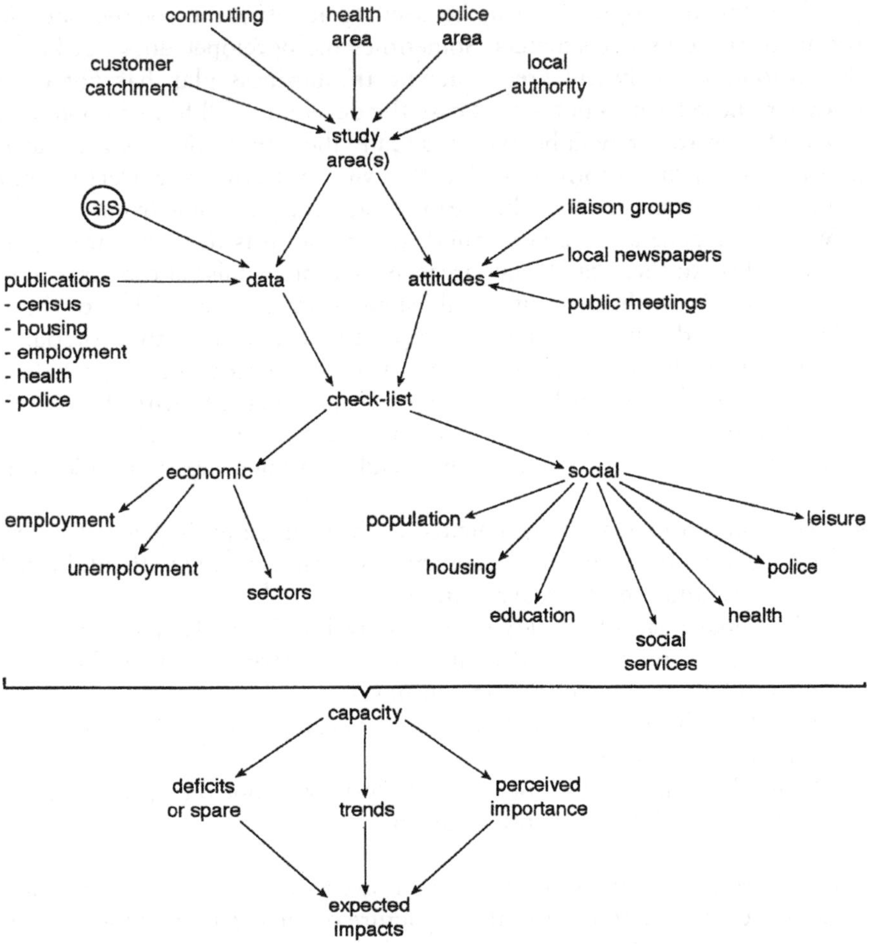

Figure 9.3 Baseline study for socio-economic impact assessment.

The first step is to define the *area(s) of study*, trying to match as much as possible the "areas of influence" of the project. The most important of these areas of influence is the *commuting area* for the project workers:

- For the construction stage, it can be substantial and, for some workers, up to the 90-min isochrone or beyond, as short-term construction workers are prepared to travel longer distances.
- For the operation stage, the catchment distance is usually considered closer, with workers usually living near to a project at which they may work for many years.

When dealing with projects that involve visiting users/customers, a different type of travel area can come into the picture, the *market area* of the project. When such catchment area is known – maybe as part of the "business plan" of the developer – it can be used to identify the socio-economic profile of those users/customers. Sometimes the developer does not know the customers' catchment area – maybe the business plan has not been drawn in those terms – but in that case the developer will have a good idea of *who* the customers will be (which is really the information we are after), and we can get that information *directly*, without having to extract it from published information about the area they are likely to come from.

With these general criteria in mind, the question is to define area(s) of study as close to these catchment areas as possible, whilst at the same time trying to maximise the amount of published information available for those areas; the final decision is usually a compromise between the two criteria. It is common for the study to use several sets of study areas – each providing their own set of data – as long as they overlap sufficiently with the "core" area of influence, and as long as they do not differ too much from each other. The final data-collection area may end up being a superimposition of:

- Local authorities, well documented in the Census: in the UK, a County can be a good starting point, sometimes complemented with additional Districts (and even Wards) around it.
- The Department of Employment's "Travel to Work Areas", which are quite large and can be adequate for the construction stage, but they tend to be excessive for the operation stage.
- Health Authorities are too large, but they can provide good data on the health-care situation.
- Similarly, Police Authorities are also too large, but they provide good data on the crime-prevention situation.

For the respective areas of influence – however defined – the information to be collected helps to put together a picture of the *capacity* of the area (in economic terms and in social terms) and the existence of any *surplus or deficit* in any of these aspects, which will help determine the extent of any impacts. But we also need to find out the *perceptions and attitudes* of the various sectors of the local population about what are the problems (if any) in the area, as it is these perceptions that will ultimately shape the "meaning" of the new project for the local population and the significance of its impacts. With this double objective in mind, the "information sweep" should be carried out at several levels through:

1 Desk-based data collection from published statistics and local studies if they exist.
2 Assessment of social perceptions and feelings in the area:

- establishing *liaison groups* between the study team, the developer and the community;
- browsing through the local press;
- talking to employment and planning officers in the local authority to check if something is "going on" such as problems developing, other competing projects coming to the area, or local anxieties;
- talks with the Department of Employment's manpower sections about local labour markets, and their policies and opinions about incoming change;
- interviews with key-individuals in the community;
- investigating general *public opinion* directly, either *informally*, in casual conversation with locals while doing other parts of the field work, or *formally*, by more organised public information-gathering: (i) by systematic surveys on specific issues identified informally; (ii) in public meetings organised to increase public awareness of and participation in the impact assessment exercise; such meetings normally refer to *all aspects* of the project (and not just to its socio-economic side) and can represent one of the few points in the impact assessment process where all areas of impact assessment come together. This type of systematic investigation of public opinion presents the usual problems discussed before about public participation: although impact assessment experts invariably think it a good idea, developers tend to be reticent about it, as it can raise awareness about the proposed development and generate a reaction against it from quite early in the process. This is a typical example of what Vanclay (1999) refers to when saying that one of the problems of social research is that the investigation itself can change the social reality it is investigating.

The "information sweep" can be summarised in the following checklist (for a fuller discussion see Glasson, 2001 and Chadwick, 2001):

For the *economic* side of the study:

1 The situation of those in *employment* in local firms:

- age,
- gender,
- economic sector,
- occupational category.

The best source for this type of data in the UK is the National Online Manpower Information System (NOMIS), which can be accessed by subscription.

2 The *unemployment* situation:

- numbers unemployed,
- how long unemployed,
- occupational category.

The best source for this information in the UK is the Department of Employment Data Sources (e.g. Labour Market Trends) that update and publish unemployment, vacancy and redundancy data on a monthly basis, and with a regional disaggregation.

For the *social* side of the study:

1 *Population*

(a) latest figures by *age groups* from the Census (sometimes going down to Ward level with the Small Area Statistics)
(b) population *trends*:

(i) from the mid-year estimates;
(ii) population projections for Regions and Counties produced by the Office of National Statistics;
(iii) Planning Local Authorities usually have working figures about population trends at County and District levels as part of the Structure and Local planning activity.

2 *Housing*

(a) the latest stock (from the Census or from surveys by the local authority): deficits, surpluses (e.g. under-occupation), vacancy rates, second homes;
(b) housing prices/rents (from local estate agents and newspapers, also from some local Building Societies);

(c) housing construction/renovation *trends* (from "Local Housing Statistics" in England and Wales and "Housing Trends" in Scotland);
(d) availability of *temporary accommodation* (normally for tourism) as a possible accommodation alternative, especially for construction workers: Bed and Breakfast, guest houses, caravan sites
(in the UK, the Regional Tourist Boards have good information about local capacity and occupancy rates; local Tourist Information Centres can often provide more "on the ground" information);
(e) with respect to *trends* in the supply of tourist accommodation, local authorities will have information from the inflow of planning applications.

3 *Education*

In the UK, Local Education Authorities have good information on education, which can be complemented with data from the Department of Education and Employment:

(a) current *supply* (schools and Colleges of Further Education): capacity, numbers of pupils, pupil/teacher ratios;
(b) *trends* and planned changes: trends in local demand can be calculated by "rolling on" the data collected about people of school age, although with the increased freedom of choice of school, the level of use of schools is influenced not only by local demographics, but also by how each school compares with others.

4 *Health*

In the UK, the following kind of information can be provided by the Family Health Service Authorities and by the Regional Health Authorities:

(a) General Practitioners in the area;
(b) size of doctors' lists;
(c) turnover of doctors;
(d) spare capacity in local hospitals (if any).

5 *Social services*

From the Department of Health and Social Security, information can be gained on:

(a) homes for the elderly: places, spare capacity;
(b) children's homes: places, spare capacity.

6 *Police and emergency services*

From the Police Authorities, data can be obtained on crime/arrests and on general feelings about the crime-prevention situation. This can be extended to other emergency services if it is perceived that there are problems of capacity or dissatisfaction in the area concerning those services.

7 *Social facilities*

As with other services, what interests us here is the existing *capacity* and whether it is considered sufficient, if there is spare capacity, if any of these facilities (or the lack of facilities) create *problems* for the community or for the authorities, such as the police: leisure, sports, pubs, clubs.

The socio-economic field is one of the very few areas of impact assessment where *trends in the baseline* (without the project) are central to the assessment. Population projections (10–15 years ahead) for the local area are crucial, and from them other projections are made of demand for housing, schooling, health care and other services. Geographic information systems can be used as a storage and "synthesizer" of large amount of information (from the Census and many other sources) and, to that extent, an existing GIS with all or part of the information needed for the baseline could be used at this stage as an important source. In this context, GIS would not really be used in its analytical capabilities, but *only as a database* with the ability to display *maps of the information*, with the advantages this can add to the understanding of the area.

The ultimate objective of the information sweep is twofold: (i) to determine the capacity (present and future) of the system for extra jobs and extra demand for services; and (ii) to understand how the local population feel about the situation and the incoming change. This should give an idea of the aspects of society where the new project is likely to produce its impacts, which will need to be investigated further.

9.2.3 Economic impact prediction

It is at this point that the economic and social lines of enquiry part company, not because their objectives differ but because the approaches they use diverge. Economic impacts could be interpreted in a wider sense, to mean all the economic effects of the project and the transformation – quantitative and qualitative – of the local economy that could result. In practice, however, the study of economic impacts focuses on the likely *overall quantitative growth* that a project can generate, and this growth is usually studied focussing on two areas: (i) changes in Local Authority finances, and (ii) growth in the local economy. First, the financial situation of the Local Authorities affected are likely to change in various ways (Chadwick, 2001):

1 On the *income* side:

   (a) there will be increases in council tax, as new people buy property in the area;
   (b) population increases will mean changes in the Local Authorities' position in the calculation of the "Standard Spending Assessment" contribution by central government (which are proportional to the resident population), although short-term temporary workers will not make a difference;
   (c) similarly, there should be an improvement coming from non-domestic rates, which are paid to a central pool and then re-allocated to Local Authorities by population levels.
   (information on this can be found in "Finance and General Ratings Statistics" ["Rating Review" for Scotland] from the Chartered Institute of Public Finance and Accountancy Statistical Information Service)

2 On the *expenditure* side, the effects of growth can be more difficult to calculate, as the published figures on the various costs allow the calculation of average costs, which in reality "hide" two types of costs: *fixed* costs which do not change with growth and *variable* costs (per head) that do, and the growth in expenditure would only affect the latter.

With respect to the growth of the local economy, it can be quantified in terms of employment or income but the basic reasoning is the same: an injection of new demand for workers and/or goods will make the local economy grow, and the question is to forecast *by how much*. It is known from economic theory that the economic effect of an expenditure in an economic system is *greater* than the original amount because of the "cascade effects" it generates, as if the original injection had been "multiplied" by a factor greater than one. Hence, the calculation of this type of economic effect focuses on calculating the two elements involved (Figure 9.4): the magnitude of the economic injection, the "multiplicand", and the greater-than-one multiplying factor, the "multiplier".

9.2.3.1 *The multiplicand*

A development project usually generates several injections into the economy – some are one-off and some are permanent during the life of the project. For the purposes of multiplier analysis, these injections can be grouped under two main headings: *investment* and *jobs*, both during construction and operation. If the information available about the project is limited to overall figures and we are carrying out the study at an *aggregate level*, these two project injections will constitute our main multiplicands.

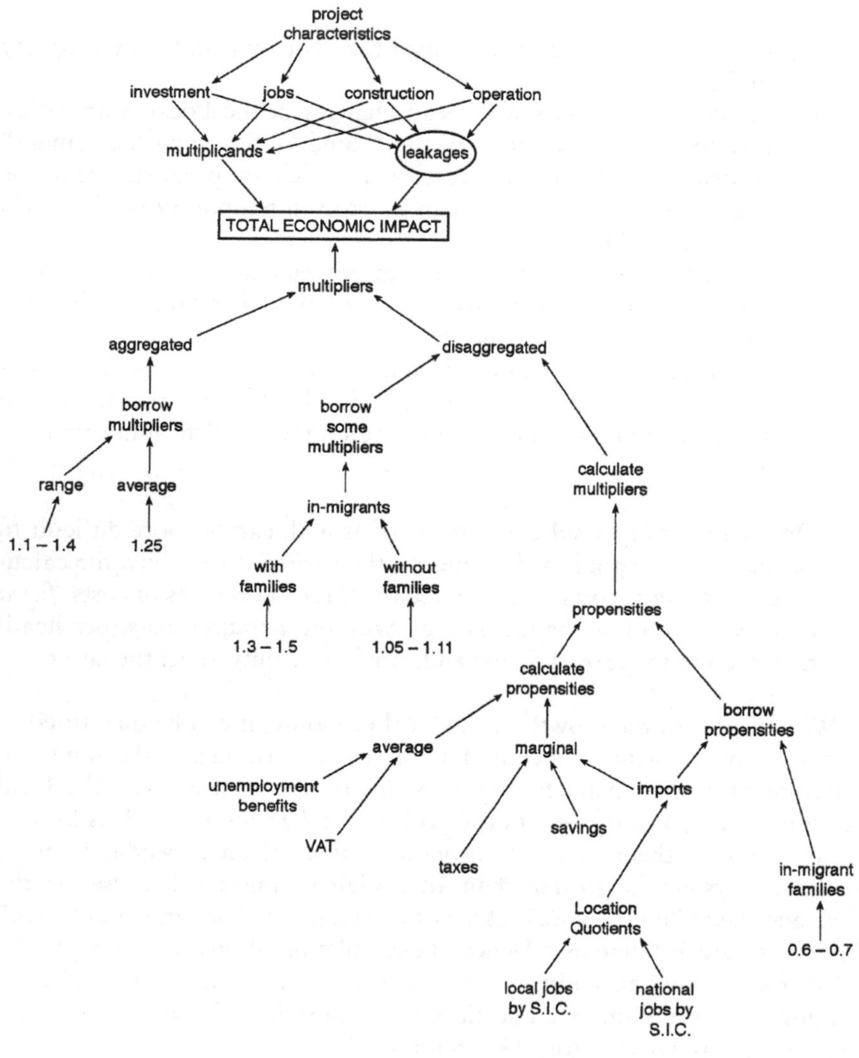

Figure 9.4 The multiplier.

If a more *disaggregated* approach is attempted, these injections are broken down into a more detailed list of multiplicands:

1 During the *construction* stage, assumed to happen *once* (if there are expansions/modifications to the project later, for the purposes of impact assessment they are considered in most cases as new projects and their impacts need to be assessed afresh):

   (a) the initial investment involved in the creation of the project (infrastructure, buildings, equipment);

(b) labour (and their wages) to work in the construction of the project:

  (i) coming from outside the area: single temporary in-migrants, temporary in-migrants with their families, temporary long-range commuters;
  (ii) local labour.

2 During the *operation* of the project (these multiplicands apply during the *life of the project*):

(a) regular demand for inputs (goods, raw materials, services, rented floorspace);
(b) stable labour (and wages) to work in the project:

  (i) coming from outside the area: single permanent in-migrants, permanent in-migrants with their families, permanent long-range commuters;
  (ii) local labour;

(c) expected users/visitors and their expenditure in the local area (*not* in the project, e.g. entrance fees).

Not all these categories will be present in all projects, and some will be negligible and not worth calculating (like the number of permanent long-range commuters for the life of the project), although some of these categories "evolve" into others: for instance, it is common for long-range commuters to become in-migrants, or for single in-migrants to bring their families later if the labour situation stabilises.

If we are following a disaggregated approach to our multiplier analysis, it is useful to consider the different multiplicands separately, not because they are conceptually different – we can add apples and pears if we are only interested in their cost – but because they work their way into the system differently. In particular, these multiplicands do not apply "in full" to the local economy because they suffer "leakages" to the outside. Typical *multiplicand leakages* are:

1 Leakages from the initial investment (the *construction* stage) which can be:

(a) the equipment – and its installation – which the firm undertaking the project brings with it, maybe because it involves specialised technology not available locally, like a nuclear reactor or a waste incinerator;
(b) goods or services likely to be imported during construction, maybe due to prior arrangements with other outside firms.

2 Similar leakages can happen during *operation*:

(a) raw materials, goods and services for the running of the project imported from outside the area, sometimes due to prior purchasing agreements with other firms;
(b) property rents going to landlord's resident outside the area;
(c) profits going to shareholder's resident outside the area.

From the earlier investigation of the project, where we quantified all the investments and jobs involved (if the information was available), these leakages must be deducted and the residual amounts spent *locally* can be calculated, for use in the next stage in combination with the economic multiplier.

9.2.3.2 *The multiplier*

Although there are various types of multipliers,[38] it is the Keynesian version that is normally used for this type of study. We can expect these multipliers to be greater than one. In fact, the so-called "income multiplier" would be infinite were it not for *multiplier leakages* (in addition to the multiplicand leakages discussed before). Multipliers are usually expressed by a formula of the type $1/(1-f_{\text{leakages}})$ where the function f depends on the particular way in which the leakages are calculated. In the UK, the standard approach derives from early discussions of "regional" multipliers (Brown, 1967; Steele, 1969), and starts from an adaptation of the classic Keynesian way of expressing a change Y in the Gross Domestic Product of an economic system (national, regional or local) at factor costs in terms of its components:

$$Y=J-T_{\mathrm{d}}-U+C-M-T_{\mathrm{i}}$$

J expenditure on value added in the area, this is the "autonomous" part of the equation, usually taken to mean the "injection" of resources from outside the system which, in our context, can be used also to represent public or private investment on development projects

38 The three most commonly known approaches to the definition of multipliers (see Glasson, 2001) differ in the level of disaggregation they use to look at the economy and its interactions: the Input–Output approach breaks down the economy into (many) economic branches, the Economic Base approach breaks down the economy into basic and non-basic activities, and the Keynesian approach considers the economy as a whole. Partly because of this, the first two become quite difficult to use at local level: with respect to the Input–Output approach, it is virtually impossible to find a reliable local I–O table; with respect to the Economic Base approach, there are conceptual difficulties in defining what is basic and non-basic at the local level.

T_d direct taxes (like Income Tax), a leakage which can be expressed as $t_d \times Y$ (where t_d is the *marginal propensity* to pay taxes with rising Y)

U change (decline) in transfer payments (unemployment benefits for example) from Government with rising income and employment, a leakage which can be expressed as $u \times Y$ (where u is the *propensity* to lose transfer payments with rising Y)

C change in consumer expenditure at market prices, which can be expressed as a function of the income left after deducting the previous leakages (direct taxes and loss of transfer payments): $c \times (Y - T_d - U)$ where c is the marginal propensity to consume part of the disposable income left

M imports for consumption, a leakage that can be expressed as a function of consumption $m \times C$ (where m is the *marginal propensity* to import with rising consumption) which, substituting the expanded expression for C, becomes: $m \times c \times (Y - T_d - U)$

T_i indirect taxes (like Value-Added Tax), another leakage, which can be expressed as a function of "local" consumption (after discounting the imports) $t_i \times (C - M)$ where t_i is the *propensity* to pay indirect taxes with rising consumption which, substituting the expanded expressions for consumption and imports, becomes: $t_i \times (Y - T_d - U) \times (1 - m)$.

Substituting T_d and U by their expressions ($t_d Y$ and uY) and substituting all these expressions into the master equation for Y, we get:

$$Y = J + Y \times c \times (1 - t_d - u) \times (1 - m) \times (1 - t_i)$$

It has also become standard practice (Steele, 1969) to assume consumption and saving as complementary, and C can be substituted by $Y - T_d - U - S$ in the master equation, where S (savings) is another leakage which can be represented as $s \times (Y - T_d - U)$ where s is the *marginal propensity to save*, and it is assumed that $c = 1 - s$. Substituting in the last equation and simplifying, we derive the standard formula for the multiplier (Glasson, 2001):

$$Y = J \times \frac{1}{1 - (1 - s) \times (1 - t_d - u) \times (1 - m) \times (1 - t_i)}$$

We can see that the increase in value-added Y would be equal to the "autonomous investment" J multiplied by a factor *greater than one* (as the denominator is less than one). The succession of expressions in brackets expresses how leakages "accumulate", each one applying to what is left after the others. The main problem with calculating these leakages is the difficulty of knowing *marginal* propensities – representing the proportions of the *next* income increase to be used in various ways – and the usual compromise is to use *average* propensities instead, which represent the proportions of the *whole* income. Unless fresh survey data is available to estimate the likely proportions of extra income to be used in different ways, published information usually shows *overall* figures, and proportions calculated from them will only represent *average* behaviour and not marginal behaviour. This is not a major problem in some cases (unemployment benefits or VAT, for instance) when the proportion lost will be the same independent of the level, but in most other cases (direct taxation, savings, imports) it is well known that the proportions tend to increase with income.

Having calculated the multiplicands – coarse or disaggregated – derived from the project (see previous section), what we have to do now is to:

- *calculate the multipliers which apply to each multiplicand*;
- *multiply each multiplier by its multiplicand*;
- *add up all the multiplications*, and this sum will be the total economic impact.

Sometimes the disaggregation of the multiplicand can introduce complications that require modifications of the way we calculate the multiplier. For instance, Brownrigg (1971) modified the standard calculation of the multiplier to account for in-migration of some of the labour force, breaking down the calculation of the multiplier into two stages:

- First, in-migrant workers inject some of their demand for goods and services into the local economy, with their own propensities to leak (ignoring the loss of transfer benefits, and using *average* propensities, as all their income is used for the calculation) and their *first-round multiplier* (M_1) can be calculated.
- Second, this "multiplied" injection into the local economy generates its own *subsequent-rounds multiplier* (M_2) for the whole local population, calculated using the normal procedures and propensities (marginal if possible).
- Finally, the overall multiplier for this particular labour group can be calculated as $1+M_2\times(M_1-1)$.

Local area multipliers normally vary between 1.1 and 1.4 (Glasson, 2001) meaning that for each pound brought directly by the project, an extra 10–40 p is generated indirectly. The range of values is relatively narrow, and if we are carrying out an aggregate multiplier analysis (maybe because the budget for the project is not high) it is possible just to "borrow" these values and assume that they will apply to our project, expressed as a range (1.1–1.4) or as an average (1.25).

Even if we are carrying out a disaggregated study of the various multiplicands, and given the difficulties of calculating propensities, we can:

1 Borrow the multiplier values for some of the multplicands from other studies of similar projects; for example, power-station impact studies have produced consistent multiplier values for typical labour groups (Glasson, 2001):

   - for in-migrant workers without families 1.05–1.11 (between 5 p and 11 p extra);
   - for in-migrant workers with families 1.3–1.5 (between 30 p and 50 p extra).

2 Or we can calculate the propensities (to leak) and the multipliers for the disaggregated multiplicands from scratch.

Calculating the various *propensities* associated with each type of multiplicand we can sometimes use some simplifications:

- Some propensities can be ignored (assumed zero) with some multiplicands: for example, when calculating the multiplier for outside labour, we can ignore changes in transfer payments like unemployment benefits, as incoming labour may prevent a fall in local unemployment.
- Some propensities will be common to all multiplicands (like Value Added Tax).
- Some propensities will be common to several multiplicands (like the propensity to save or to pay taxes) likely to be similar for all labour of the same occupational standard irrespective of whether they are local or not.[39]

The single most important propensity, which is likely to show the greatest range of variation and the greatest influence on the final value of the multiplier, is the propensity to *import*. It is also one of the most difficult to calculate for sub-national economic systems, given the difficulty to find published data on imports–exports between regions, let alone smaller areas like the ones normally used in impact assessment. We can try to get around this problem by:

- "Borrowing" import propensities from studies which have used a similar breakdown of multiplicands; for instance Glasson *et al.* (1988) found when studying power stations in fairly remote locations that the propensity to import for in-migrant workers with their families could be as high as 0.6–0.7 (60–70 p).
- "Approximating" the quantification of imports–exports with indicators, a typical example of which is the use of Location Quotients, a classic tool of spatial economic analysis (Florence *et al.*, 1943) which can be

39 But transient staff may be more likely to save than permanent staff.

adapted to estimate the likelihood of a local area needing to import from outside.

9.2.3.2.1 *Location Quotients*

Location Quotients (LQs) calculate the level of concentration in a local area of a particular branch of the economy by comparing the local situation with the situation in a wider area – the whole country or the region(s) around the local area – and the Location Quotient of an industry gives a quantitative measure of that level of concentration. It works industry by industry (often based on the categories in the Standard Industrial Classification (SIC): construction, manufacturing, etc.), and the LQ of an SIC category in an area is calculated by dividing the proportion which that category represents in the local area (measured usually in proportion of jobs), divided by the proportion which that same category represents in the larger area:

$$LQ(X) = \frac{\text{local jobs in industry}X/\text{all jobs in the local area}}{\text{jobs in industry}X\text{in the parent area}/\text{all jobs in the parent area}}$$

If $LQ(X) \geq 1$, it means that the concentration of industry X in the local area is the same or more than in the parent area, therefore it is unlikely that the local area will be requiring any imports of X. On the other hand, if $LQ(X) < 1$, it means that the concentration of industry X in the area is less than in the wider economic system, and this can be taken to mean that the local area is likely to need to import some of its requirements of X from the parent area, on the assumption that all areas ultimately require similar proportions of everything. The *proportion of imports of X required* can be estimated as $1 - LQ(X)$, the extra proportion needed to bring its LQ value up to one. If we make this calculation for all the relevant SIC categories, the weighted average of the proportion of imports needed for all the categories can give an approximation to the overall propensity to import in that local area.

9.2.4 Social impact prediction

The estimation of the *magnitude* (we shall discuss significance in the next section) of the social impact is based on comparisons between the likely *extra demands* on local services and housing derived from the project and the local situation in the area. These demands will derive from the population changes generated by the project. Hence the first step in the calculation of social impacts is a *demographic study* of the likely population changes in the area with respect to the baseline (see Section 9.2.2) focussing on changes directly derived from the labour curves of the project and the family situations likely to be generated (see Section 9.2.1) by increases in in-migrant households (year by year):

- temporary (mostly during construction): single persons, whole families;
- permanent (mostly during operation): single persons, whole families;
- day workers (mostly during construction).

Some of these categories include very small number and are unlikely to create any problems. The two main categories usually requiring attention are (i) temporary single workers during construction; and (ii) whole families during operation. We are particularly interested in:

1 numbers of households (one per worker, with or without family);
2 family sizes for different ages of the heads (from the Census);
3 total number of persons;
4 demographic characteristics:

    - proportion of persons in education age by broad age groups: 0–4 years of age (for nursery education) and 5–18 years of age (for school education);
    - proportion of young people (under 30).

In addition, the "local share" of the new jobs can generate some demographic changes in the local community:

- some would-be "economic" out-migrants (part of the baseline trend) may find jobs in the project and decide not to emigrate;
- some local workers may decide to move jobs and start working in the project, leaving behind vacant jobs which may generate further in-migration.

The first type of impact that can be estimated from this population study is *demographic*:

- overall size of the incoming population compared with the size of the local population;
- proportions of new/old populations by broad age groups.

The demographic impacts can be calculated in terms of the proportional increases in the various age groups that the new population represents with respect to the old.

In areas of service where needs can be predicted accurately and there is a recorded "capacity" in the system, impact analysis consists of comparing the new needs with that capacity. For example, the calculation of *housing/accommodation impacts* on the local area follows from a combination of the accommodation needs of the incoming population, the provisions for on-site accommodation made by the developer, and the local accommodation situation:

1 From the overall incoming population, deductions must be made to account for any plans for on-site accommodation (hostels, etc.) especially during the construction stage.
2 Single-person households (those not to be accomodated on-site) are a special category because they can share:

   (a) with other outside workers,
   (b) in "digs" with local families.

3 Some families will only require temporary accommodation:

   (a) in caravan parks,
   (b) in Bed and Breakfast accommodation.

   These temporary needs must be compared with the local provision of this type of accommodation.

4 Most families of two or more persons will require permanent/semi-permanent accommodation, and their numbers must be compared with:

   (a) the local level of vacancies (over and above the level needed for normal operation of the market):

      (i) for sale, suitable for permanent workers and even sometimes for workers in a long construction phase (several years) 10–20 per cent of whom might buy property for that period (Chadwick, 2001);
      (ii) for rental, for temporary workers, usually in the construction phase
      (Concerning vacancies, it must be remembered that a 4–6 per cent is always present in a "healthy" housing market, and when vacancies fall below those levels it is usually accompanied by an undesirable rise in prices);

   (b) the local rates of housing renovation/completion.

5 To these needs for the incoming population must be added the local housing needs derived from their own situation, which will in fact be *in competition* with the needs arising from the project:

   (a) a local housing deficit may exist due to overcrowding or poor standards;

(b) additional future housing needs are likely to arise from the dynamics of the local population itself.

Similarly, *education impacts* are calculated by comparing the education needs of the incoming population with any spare capacity in the local education system:

- We can multiply the number of children in the incoming population calculated in the demographic study by the expected rates of school participation (national figures can be found in the Department of Education and Employment Statistical Bulletin) for the various age groups noted earlier.
- The impacts of the project on the education system can be calculated by comparing these expected *demands* for education for the various age groups with the existing spare capacity (if any) in the local schools and colleges.
- As with housing, to these needs will have to be added the additional future local needs arising from the dynamics of the local population.

In the case of *health and social services* impacts, we *can* identify the spare capacity of the system through data such as:

- General Practitioners' lists;
- beds in hospitals;
- places in old persons' homes;
- places in children's homes;
- foster-children places.

What we *cannot* predict so precisely are the levels of demand to be generated by the incoming population. The best we can do in this situation is:

- To identify the typical age groups which tend to be the main "customers" of such services and quantify them in the new incoming population: infants (0–4), school-age children (5–18), old-age pensioners.
- A good measure of the likely impact of these new "potential customers" (the increased pressure on the services) can be the *proportional increase* they represent with respect to their respective numbers in the local community without the project.
- We can now compare these percentage increases with the spare capacity (also expressed as a percentage) and with the expected endogenous growth in demand from the existing population.

In the case of some services, the notion of "capacity" cannot be clearly defined, and the approach has to be adapted accordingly. For example, in the case of *police* services:

- We can quantify the numbers in the incoming population belonging to age groups which usually show relatively high crime rates, what can be loosely described as "young itinerant males" (Glasson, 1994).
- As with health, we can quantify the *proportional increase* they will represent with respect to the local population, which we can take as an approximation to the magnitude of this impact.

This is the case also with *leisure* services, where we can identify the customers but we *cannot* clearly define the capacity of the system, and we can approach the question in a similar way:

- quantify numbers in the incoming population in the age groups likely to participate in these activities (pubs, leisure and sports centres, etc.);
- calculate the proportional increases they represent with respect to those groups already present in the area and take these as a measure of the likely impact.

9.2.5 Impact significance

The discussion of the "significance" of impacts merits a separate section because the socio-economic area of impact assessment is one where the determination of the significance of various impacts is a problem in itself, as there are no precise standards to meet for most of them. In many other areas of impact assessment it is the measurement of impacts that is the greatest problem and, once measured, they only have to be compared with the accepted standards to establish their significance. The concept of significant socio-economic impacts on the other hand is a typical example of *fuzziness* where the frontier between "belonging" and "not belonging" to a category (being and not-being significant) is not a sharp dividing line identified by a certain value, but a grey area extending over a *range* of values for which there are varying proportions of people interpreting the situation in one way or the other. As we emphasised from the beginning of the discussion of socio-economic impacts, what gives meaning to the effect of the project on the local society is how they are perceived by the members of that society, and an important part of the impact assessment job is to determine "who wins and who loses" (Glasson, 2001) as a result of the project, giving special attention to the most vulnerable sections of society.[40] When discussing the study of the baseline (see Section 9.2.2) we already covered the different aspects of the local society we should consider. What is left for our discussion here is to identify sets of criteria (mostly qualitative) to help us determine if the impacts measured in the previous section are significant, and we can do this following the same order in which we discussed both the baseline and the impacts.

40 This is a point also made by Vanclay (1999).

Economic impacts are normally assumed to be *positive* (just as social impacts are normally assumed to be negative) because they represent economic growth, especially if there is high local unemployment and the local authority is anxious to boost the local economy. However, the magnitude and speed of that growth can have negative effects:

1 It is important how the local business community thinks they will be affected by rapid growth:
   (a) if there is "spare capacity" in the system, so that increasing demand can be met without additional capital investment;[41]
   (b) will growth be seen as an opportunity for expansion;
   (c) or will growth increase competition with others.
2 Also, the injection of local jobs can raise some anxieties, for example:
   (a) if the "local share" of the new jobs is below approximately 1/3 there is likely to be public resentment against the project;
   (b) if, on the other hand, the local share of jobs goes above 2/3 this can increase the dependence of the local economy on the project and the danger of a "boom and bust" local cycle when the project closes down (especially when the project is large compared with the size of the local economy).

Social impacts on the other hand are often assumed to be negative, which is not always the case. One type of *positive* social impact can be that the population growth derived from the project can *make viable* facilities which could not be sustained before and are either non-existent or struggling to survive – as is the case in many small communities in rural areas – making it possible to keep them if they were already there, or to open them anew:

- post offices;
- police stations;
- some particularly vulnerable health facilities like community hospitals or nursing homes;
- leisure facilities;
- local shops.

On the *negative* side of social impacts, *demographic impacts* can generate potential anxieties (Figure 9.5):

41 This is one of the assumptions of income multipliers and if new investment is likely to follow from the increased demand a different type of multiplier applies.

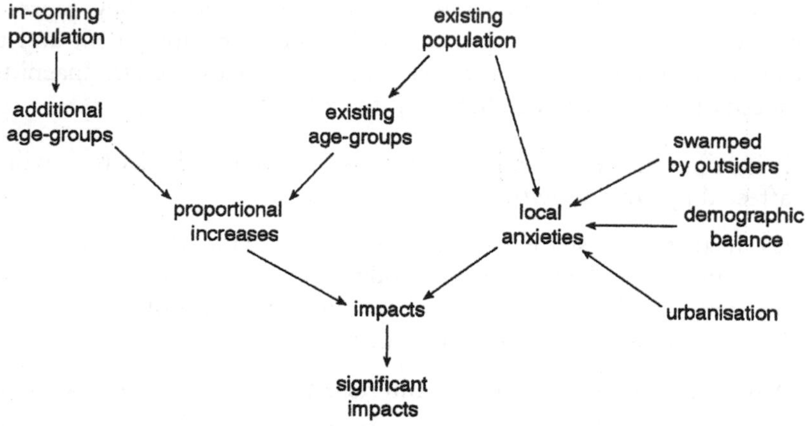

Figure 9.5 Demographic impacts.

- the local community being "swamped" by outsiders;
- the composition of the new social influx being very different from the local community, for example, a middle-aged middle-class local population as opposed to a young working-class incoming population;
- the nature of the area changing from "rural" to "urban".

Housing is an area where something equivalent to "standards" are available in the form of *capacity levels*. It is with respect to those levels that the impact is assessed (see previous section), and its significance should be indicated by the extent to which capacity is exceeded (not forgetting to leave a "healthy" level of vacancies) (Figure 9.6). The two housing aspects that normally remain the source of local anxiety and are difficult to quantify are:

- Prospects of rising house prices with increasing demand above what the local population can afford, resulting from the differences in income levels between the incoming and local populations.
- Location conflicts between housing for the new population (maybe in housing estates) and established neighbourhoods when the two groups are very different in age or composition.

Education is another area of impact where we have "capacity standards" and we can judge the significance of any impacts by how much that capacity is eroded (Figure 9.7):

- When capacity is exceeded impact significance will be high, proportional to the extent of the excess.
- Even when capacity is not exceeded the impacts can have some significance, as class sizes increase.

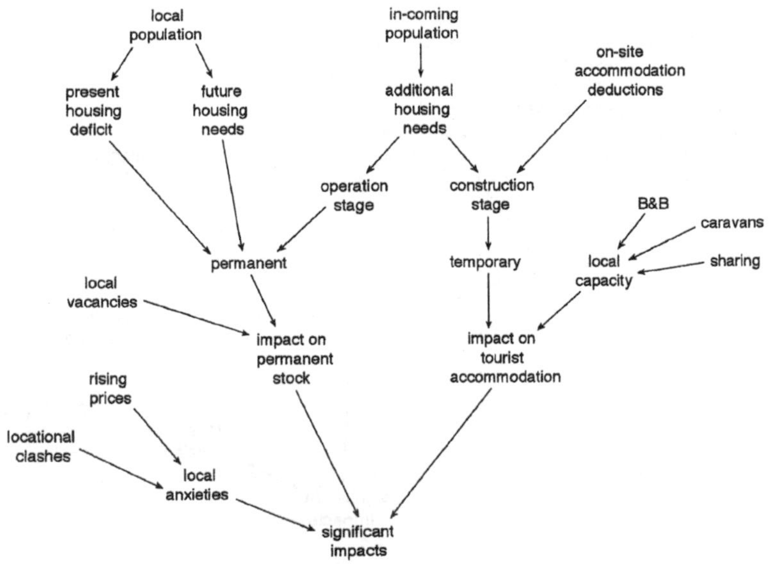

Figure 9.6 Housing impacts.

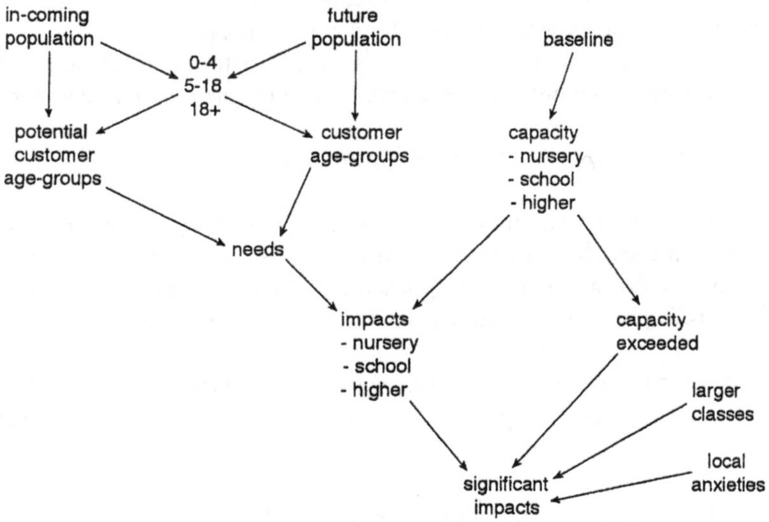

Figure 9.7 Education impacts.

In the case of *health and social services*, the significance of the impacts will derive partly from their measurement (see the previous section), comparing the percentage spare capacity in different parts of services with the percentage increases in the social groups who are likely users of those services (mainly the young and the old) (Figure 9.8):

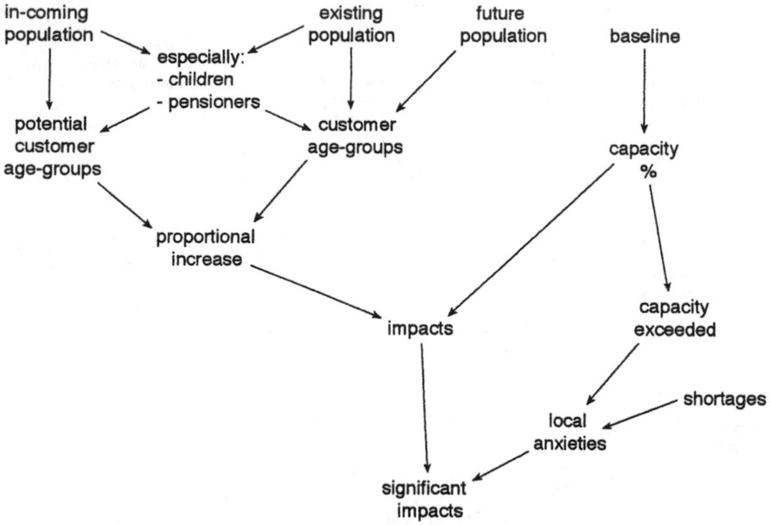

Figure 9.8 Impacts on social facilities.

- In objective terms, impact significance will be proportional to any excesses of the expected demand over the capacity.
- In socio-cultural terms, what can also exacerbate the significance of any possible impacts is any *existing anxiety* about the provision of these services.

The same applies to *police* services (Figure 9.9):

- In objective terms, impact significance will be proportional to the expected growth in those population groups most prone to crime.
- In socio-cultural terms, the significance of these trends will be qualified by *existing anxieties* about the crime situation in the area.

Finally, in the case of *leisure services* (where impacts are predicted on the basis of the expected growth in the customer population) we can establish the significance of any impacts:

- proportional to the expected growth in customer population;
- exacerbated by any local anxiety about the provision (or lack of) these services in the area.

9.2.6 Mitigation

As we argued at the start, most impacts have ultimately a social dimension, so it can be argued in the same way that *all* the mitigations suggested for other types of impacts (noise, pollution, etc.) are also socio-economic

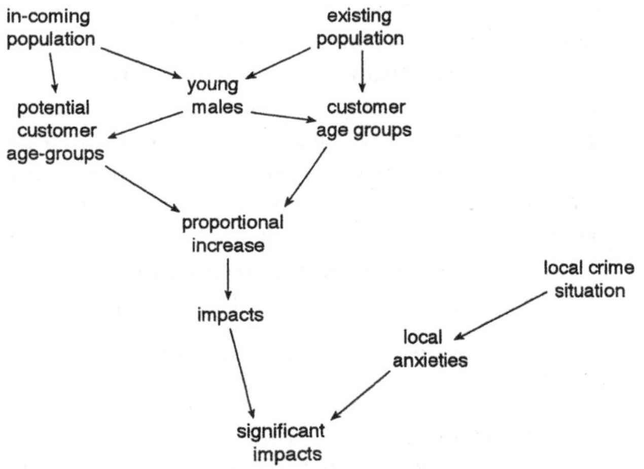

Figure 9.9 Negative impacts on police services.

mitigations. However, we shall discuss here only those mitigation measures directed towards socio-economic impacts as discussed in the previous sections. The general checklist of measures for the mitigation of socio-economic impacts can be sketched out as follows:

1 *Facilitating/maximising the positive* impacts:

   (a) putting in place training programmes for local labour;
   (b) facilitating the contractual benefits to local firms:

      (i) organising liaison committees to maximise the level of information;
      (ii) organising seminars to advertise what is available on the development site and to tell the developer about local firms;
      (iii) involving representatives of the Chamber of Commerce to alert local businesses to project opportunities;
      (iv) making contractual arrangements so that, where possible for equal prices, local firms are given priority by the developer.

2 *Minimising/stopping the negative* impacts:

   (a) some negative impacts can be stopped altogether: for example, changing the shifts and labour arrangements to eliminate night work at the construction and operation stages;
   (b) reducing the impacts by *internalising* them to the project site or to resources internal to the project:

      (i) minimising the temporary relocation of workers (for instance, during the construction stage) by facilitating commuting: travel/petrol allowances, bussing-in workers;

(ii) building alternative accommodation on site to accommodate temporary workers: hostels, caravan parks, site camps;
(iii) building bars and leisure facilities on site;
(iv) having an on-site medical centre and health facilities (an additional advantage of these measures is that not only do they mitigate negative impacts, but they can become themselves a *positive* impact for the benefit of the local community, if they remain in place after the temporary workers have gone);
(v) providing minibuses to ferry back to their accommodation workers who go out drinking, to minimise the possible impact of disorderly behaviour and/or drunk driving.

3 *Spreading* the impacts among the local facilities with spare capacity:

(a) encouraging project workers to use unoccupied (or under-occupied) local accommodation;
(b) compiling a directory of local capacity (B&B, rooms for rent, etc.).

4 *Compensating* for impacts you cannot stop or internalise:

(a) specific compensation to individuals, for instance buying their property if it is badly affected and cannot be adequately protected;
(b) community compensation (similar to the notion of Planning Gain)[42] also known as "amelioration package" (Glasson, 1994) by which the developer can provide additional facilities for the area (for example: a new bypass, a swimming pool, leisure or sports facilities).

Even more than with other types of impacts, *monitoring* any residual impacts after mitigation is crucial in socio-economic impact assessment. In the first place, for reasons common to all impacts: (i) as a check on the implementation of the mitigation measures by the developer; (ii) to gather useful information that can be used in other impact studies. But second, in the case of socio-economic impacts, another reason makes monitoring even more important: the fact that the monitoring itself – for instance by the continuation of the liaison committees and other working groups during the study – can *act as mitigation* also, as the changing perceptions and anxieties of the local population can be detected and quickly responded to as they arise.

9.3 TRAFFIC IMPACTS

Traffic has the peculiar characteristic that it is at the same time an impact in itself and the source of other indirect impacts like noise and pollution.

42 Town planning legislation in the UK allows for the negotiation with developers of additional works to be included in projects in exchange for planning permission, and these "extras" (usually for the benefit of the local community: an additional road link, or a leisure facility) are referred to as "Planning Gain".

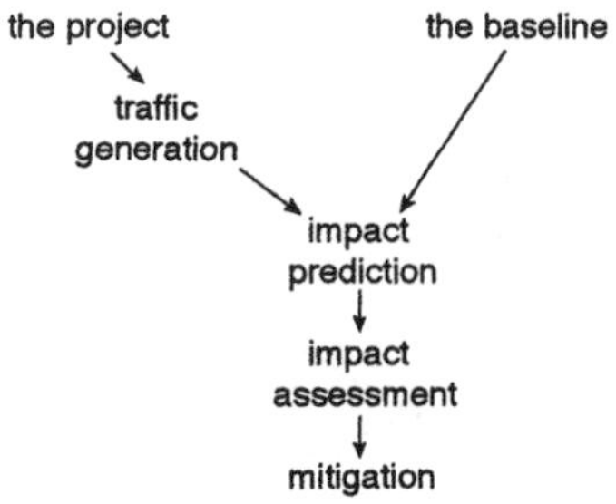

Figure 9.10 The logic of traffic impact assessment.

The study of all these impacts can again be sketched out using the standard logical sequence (Figure 9.10).

This logical sequence is common to all the traffic impacts (direct and indirect) as far as the prediction of future traffic generation, but beyond this point the different impacts are studied following their own logic. Traffic is one of the most developed areas of urban and spatial studies and, as with other types of impacts already discussed, best practice advice on traffic impact assessment has been appearing mostly since the mid-1990s (Petts and Eduljee, 1994a; Hughes, 1995; Richardson and Callaghan, 2001) in turn referring the reader to the wide range of official guidelines which has also appeared, like IEA (1993), DoT (1993), IHT (1994), DETR (1998).

9.3.1 The development project

Concerning traffic, there are two main types of projects: (i) transport-infrastructure projects (roads, rail) which affect traffic generated by others (usually by diverting it); and (ii) development projects that generate traffic themselves. For both types we need similar information about the project (Ferrary, 1994)[43] – this is relatively little compared to what we need to know for the assessment of other impacts (Figure 9.11):

1 For the *construction* stage:

   (a) hours of work;
   (b) vehicles used by construction workers (numbers, access points);
   (c) heavy vehicles (numbers, routes and access points):

43 The knowledge acquisition for this part was greatly helped by conversations with Chris Ferrary, of Environmental Resources Management Ltd (London), however, only the author should be held responsible for, any inaccuracies or misrepresentations of her views. Julia Reynolds helped with the compilation and structuring of the material for this part.

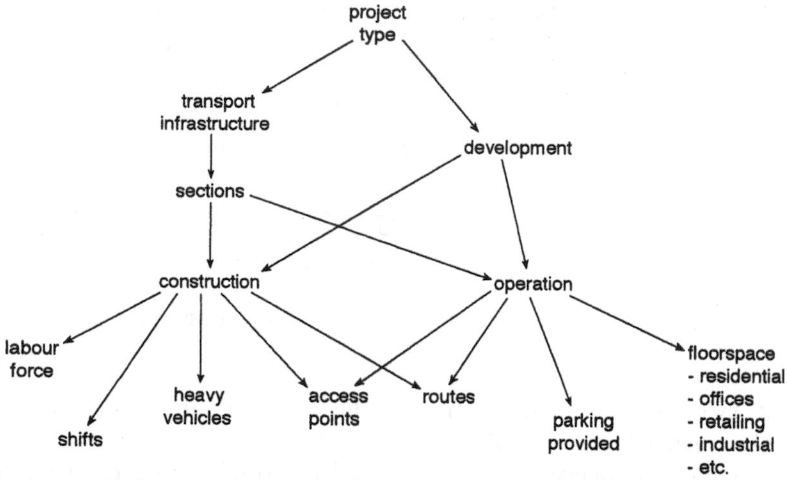

Figure 9.11 Project information for traffic impact assessment.

(i) removing earth and rubble when clearing the site;
(ii) supplying construction materials;
(iii) heavy plant and equipment.

2 For the *operation* stage:

(a) when the project is of the "development" type: *floorspace* – in area or in units of use – of different landuses (offices, retailing, residential, etc.); in traditional traffic-modelling style, we could ask for detailed knowledge of all the traffic generators in the project in order to simulate their trip-making behaviour and aggregate the results later:

(i) vehicles to be used by employees and visitors/customers;
(ii) goods vehicles needed for the operation of the project.

However, in practice, all we are after is how many trips they are likely to make and there are good sources available today which can estimate trip generation with sufficient accuracy using case-based or average figures by unit of floorspace

(b) routes and access points for the traffic;
(c) parking spaces provided in the project (if the project is of the "development" type).

If the project is a linear transport-infrastructure project (like a new road), we break it up into *route sections* and study each section separately.

9.3.2 Baseline study

The *area of study* can vary depending on the size of the project and the likely importance of the traffic effects:

- if the project is relatively small and the area is not particularly congested, we only look at the access roads to the project;
- if the project is large and/or the area is congested, we look at a wider area around the project, wide enough to include road *junctions* (usually starting no further than 500 m).

In most cases, the study of the baseline is quite focused (Figure 9.12), concentrating on: (i) the access arrangement between the project and the local network; (ii) identifying potentially sensitive receptors of traffic impacts (receptors in the area and network users); and (iii) determining the existing traffic conditions in the local network.

Figure 9.12 Baseline study for traffic impact assessment.

1 The *access* situation can be studied from Ordnance Survey maps (or a GIS) usually concentrating on the area immediate to the project:

(a) main route(s) to and from the project;
(b) alternative routes;
(c) planned modifications (if any) to the network, from the County Council.

2 The recommended lists of potential *receptors* of traffic impacts (Hughes, 1995, and Richardson and Callaghan, 2001, after IEA, 1993 and IHT, 1994) can be summarised as:

(a) motorised road users in public and/or private transport;
(b) cyclists;
(c) pedestrians;
(d) people at home and at work;
(e) sensitive groups and/or locations (the elderly, hospitals, etc.);
(f) open spaces, leisure and shopping areas;
(g) ecologically sensitive areas.

3 Concerning the *traffic situation* in the network, the Department of Transport compiles data from regular nationwide traffic counts on major roads, and County Councils usually have local traffic flow information:

(a) it is common to have traffic *daily totals* for most roads derived from 12 h or 24 h surveys and certain formulae (or simple fractions) can be applied to such averages to derive from them the likely traffic levels at peak times, for the purposes of comparison with other relevant information like road capacity;
(b) if possible, it is also useful to get an idea of the *traffic composition* (percentage of heavy vehicles in particular);
(c) if a wider area of study is used, *junctions* should also be studied:

(i) their layout;
(ii) traffic situation in them (level of congestion and queues).

(d) if the project is of the *transport infrastructure* type (especially if the project is likely to *increase* the traffic in other parts of the network), in addition to local data on traffic we shall need to know more about the *generators* of that traffic:

(i) the distribution of trip *origins*, mainly population;
(ii) the distribution of trip *destinations*, mainly: places of work, shopping areas, schools.

This information will be needed if we intend to carry out some form of *simulation* of the traffic in the area with and without the project (probably a function of the time and the budget we have for the study).

The whole methodology is heavily focussed on private motorised traffic, but other aspects can be studied also (Richardson and Callaghan, 2001) if they have been identified as potential receptors, even if in many cases they are relatively peripheral:

- The availability and use of *public transport* is included in the baseline studies when the level of provision is sufficiently high for it to constitute a real alternative to private transport.
- A similar consideration can be made with respect to *bicycle traffic.*
- Similarly, *pedestrian flows* are usually only considered if their levels are substantial.

This introduces an element of "scoping" in the study of the baseline: a general impression of the area is formed and possible traffic impacts are identified (congestion, pedestrians, cyclists, etc.), leading to the collection of the baseline information likely to be relevant for the study of those impacts. Occasionally, when the information is not available and the resources allow it, *traffic surveys* can be carried out. This can happen in remote locations where there are no reasons for the local authority to count the traffic (Ferrary, 1994). Such surveys should concentrate on counting traffic *at the worst times* although sometimes this is not possible: for example, some areas have seasonal traffic but the time of the study may not coincide with the high season.

The baseline should be *forecast* over a 10–15 year period. Any future changes to the access situation (future roads or junctions) or the relocation of receptors should be in the local authority's planning documents. Concerning the forecast of traffic data in the UK:

- It is normal for County Councils to have traffic forecasts for their own Local Transport Plans.
- The Department of Transport publishes figures for expected national growth in traffic, and we can use these figures to multiply present flows (the so-called "growth-factor" approach to traffic prediction).

Before using official traffic forecasts it should be clarified if they are based on existing landuses only (forecasting changes in trip-generation rates) or if they assume some new developments as well. If the latter is the case, using those forecasts as baseline could introduce an element of double counting, as the project being assessed might be part of those new developments already included in the forecasts (Hughes, 1995).

9.3.3 Traffic generation

Two totally different approaches are used to estimate the traffic generated during construction and operation. For the *construction* stage, no standard pattern is assumed, as the organisation of this phase can be quite specific to each project, and the estimated traffic derives directly from the description of the project (see Section 9.3.1 above). For the *operation* stage on the other hand, a pattern of traffic generation is usually assumed based on the type of project it is, and the approach varies considerably depending on whether the project is a transport infrastructure or another type of development.

If it is a *transport-infrastructure* project (apart from the construction stage), the project does not generate traffic itself but it affects the traffic in the area, usually by attracting or diverting some. Hence, the traffic generation we must study is that of the area, and how it is affected by the project. We can simulate trip distribution with a traffic model like SATURN (Richardson and Callaghan, 2001) and run the model *with and without* the project to see how the traffic in different parts of the network is likely to be affected by the presence of the project. However, in practice this is not often done as part of impact assessment, partly because a simulation like this adds a very expensive element (in time and money) to the study, and partly because the accuracy of such simulations at the local scale is not guaranteed. Also, it is common that this type of project is intended to *alleviate* the local traffic situation (a bypass, a new road or rail link, a motorway) and it is usually expected that its traffic impact – leaving aside indirect impacts like noise or pollution – is likely to be *positive* on the whole. Although the redistribution of flows resulting from the project can produce both positive and negative impacts, it is not considered essential for the purposes of protecting the environment and the community (which are after all the aims of impact assessment) to have an accurate simulation.

When the project is of the *development* type that generates traffic itself, the standard approach is to apply to the landuses present in the project rates of trip generation derived from a database like TRICS (Trip Rate Information Computer System).[44] This is a computer database well established in the UK (Hughes, 1995; Richardson and Callaghan, 2001) containing *case-based* trip generation information for projects classified by Use Class (according to the Use Classes Order), type of locations (central, edge of town, suburban) and size. Sometimes case-based data is not detailed enough for our project, and other published sources (like some traffic engineering manuals such as Slinn *et al.*, 1988) can be used, containing *average* trip generation information for various landuses (Table 9.1).

Tables like this one containing daily totals and peak flow levels can also be used to calculate the *conversion fractions* between daily averages and any of the peak hours for any specific landuse. For example, if the total daily trip generation for offices is 4.8 (per 100 m^2) and the incoming peak morning flow is 1.5, we can calculate the proportion of total daily traffic to be found in the morning peak as $1.5/4.8=0.3125$ and we can use this fraction

44 Produced by JPM Consultants Ltd, London.

Table 9.1 Trips per 100 m² of floorspace

Land uses	*Peak a.m.*		*Peak p.m.*		*Total daily each way*
	Arrivals	*Departures*	*Arrivals*	*Departures*	
Offices	1.5	0.1	0.1	1.1	4.8
Business parks	1.2	0.1	0.2	0.9	4.0
Warehousing	0.3	0.1	0.1	0.3	2.1
Industrial	0.7	0.2	0.2	0.6	4.2
Retail parks	0.5	0.2	0.8	1.0	12.2
Supermarkets	2.4	0.7	6.2	6.4	68.0
Residential (*trips × household*)	0.2	0.5	0.5	0.2	3.9
Hotels (*trips × bedroom*)	0.2	0.2	0.2	0.2	3.2

if we need to convert one flow into the other for this particular type of floorspace.

When the categories covered by either of these approaches are not detailed enough for the types of landuses present in the project, or when public transport is prevalent and "modal-split" is an intervening factor,[45] other approaches can be used to supplement the above (Richardson and Callaghan, 2001):

- comparisons with similar developments (in the same area if possible), using existing information or carrying out traffic surveys of those developments;
- using simulation models like SATURN or TRIPS (for public transport).

9.3.4 Impact assessment

Traffic produces a wide range of impacts (Petts and Eduljee, 1994a), some are direct effects of traffic and some are indirect effects which we can consider under other impact-assessment headings (Figure 9.13):

1 *Direct* traffic effects:

(a) increased traffic and congestion, time delays;
(b) severance, increased problems to pedestrians wanting to cross the road;
(c) nuisance and fear, danger to pedestrians;
(d) danger to cyclists.

45 The term "modal-split" refers to the relative rates of take-up of different modes of transport when making trips between certain origins and destinations, and it is often narrowed down to a two-way split between "private" and "public" transport.

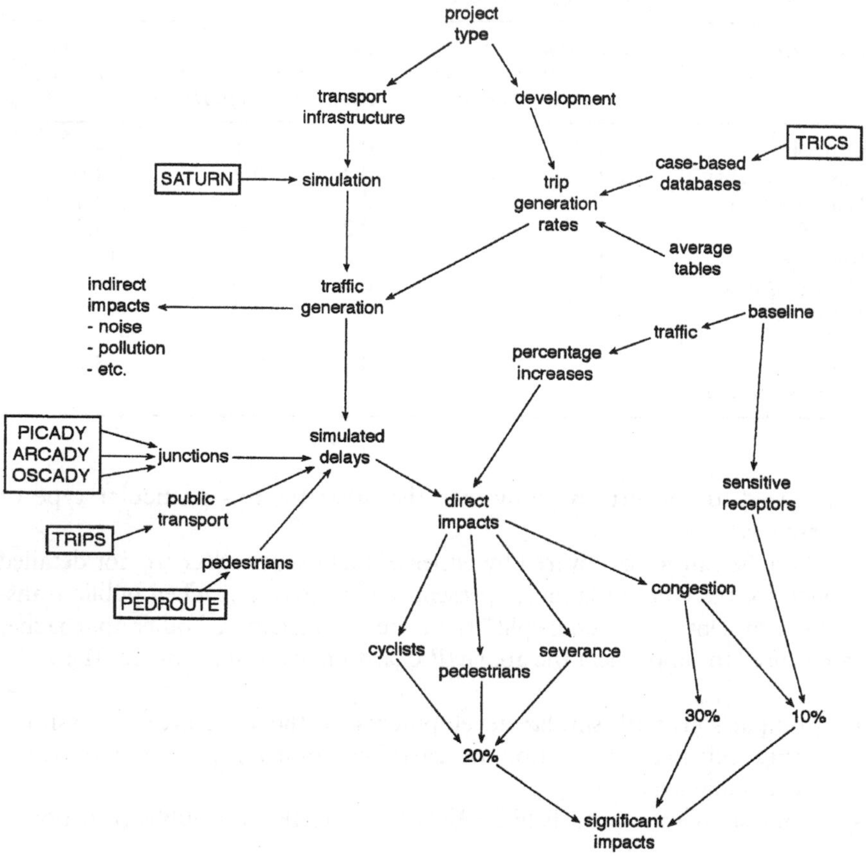

Figure 9.13 Traffic impact assessment.

2 *Indirect* effects that can be considered under other headings:

(a) noise and vibration (considered under "noise");
(b) air pollution (considered under the heading of the same name);
(c) visual intrusion (considered under "landscape");
(d) ecological effects (considered under "ecology").

It is rare for impact assessment studies to cover all these impacts, but it is common practice for good-quality impact assessment reports to look at what can be seen as the "standard package" of traffic impacts: increases in congestion, noise and air pollution (ERL, 1991). Here, however, we are not going to discuss any of the indirect impacts – we considered traffic as one of the potential sources when discussing each of those impacts in the previous chapters – and we shall discuss only the direct traffic impacts.

The direct traffic impacts are "measured" first by calculating the *percentage increase* in various types of traffic (usually overall traffic and heavy vehicles are used as indicators) and, second, their effects on the receptors have to be worked out. Some impacts – for instance the effects on other traffic on the roads – are assumed to be proportional to those traffic increases, and those percentages themselves are taken to represent the impacts. For other receptors, the effects are more complex and must be *simulated* using specially designed models (Richardson and Callaghan, 2001):

1 For *junctions*, models can simulate the length of any queues appearing and the time delays involved:

   (a) PICADY for priority junctions;
   (b) ARCADY for roundabouts;
   (c) OSCADY or LINSIG for signalled junctions.

2 For *public transport*, models like TRIPS can simulate the impact on service and access times.
3 For *pedestrians*, specialised models like PEDROUTE can also be used.

With such simulated information, the increases in delays produced by the new traffic can be calculated, and the impact of such delays can again be measured as percentage increases.

To establish the *significance* of such percentage increases is more difficult, and judgement forms an important part of it (Hughes, 1995), helped by some general *standards*:

1 Concerning general impacts *on traffic*, the IEA (1993) guidelines suggest that for the two standard measures of traffic impact (percentage increases in overall traffic flow and in heavy vehicles), the thresholds of significance should depend on the type of area:

   (a) in a normal area, 30 per cent or more;
   (b) in a sensitive area (or concerning sensitive receptors), 10 per cent or more.

   Similar thresholds can also be used for increases in time delays at junctions (although the IEA guidelines say nothing about this)

2 Concerning impacts on other types of road users like *cyclists* or *pedestrians*, research suggests (Crompton, 1981, quoted in ERL, 1991) that increases of 20 per cent or more will make such impacts significant.

On the other hand, we may want to qualify these thresholds with the baseline level of traffic: if the baseline traffic is very low, an increase of over 30 per cent may still not represent a significant impact, while if the traffic in a road or a junction is near its capacity, a smaller increase would make it significant. Introducing *capacity* considerations adds certain difficulties: first, we must have information on the capacities of the roads and junctions and, second, capacities are usually expressed in vehicles-per-hour while the traffic figures used in impact assessment come usually as daily totals, and to use them we have to make some adjustments:

1. Convert baseline daily flows into peak hour flows using conversion formulae or fractions (from the data).
2. Compare these peak flows with the capacity of the road or junction (for roads for example, about 1000 vehicles per hour per lane, for junctions it varies with the type of junction) and calculate what percentage of the capacity is being used at peak time.
3. Add the traffic increases and recalculate the percentage of the capacity being used:

   (a) the impact is considered significant if the increases make the road or junction reach or exceed its capacity;
   (b) if the increased flows are still well within the capacity, the impacts are not significant.

It is usually at *junctions* that capacity-related impacts show up best, as the roads are only likely to reach their capacity *after* their junctions (Ferrary, 1994).

In general, development-type projects tend to have negative traffic impacts with additional "indirect" impacts (noise, pollution, visual intrusion) also negative, while transport-infrastructure projects (except during the construction stage) tend to have mainly "positive" traffic impacts and, again, mainly negative indirect impacts.

9.3.4.1 Loop-back

In the assessment of traffic impacts there is an interesting iterative process between impact assessment and the reconsideration of the study area discussed in Section 9.2.2 (Ferrary, 1994):

- To start with, the study focuses on the area immediate to the project, and the traffic impacts we look for are just the traffic in and out of the project.
- If the project is large in traffic terms and/or the area is sensitive to traffic, we reconsider the area of study, *extending* it to the next set of junctions in the network, and we assess the traffic impacts on those junctions.
- If the traffic impacts are still significant, we reconsider the area of study again, *extending it further* to the next set of junctions, and we assess the impacts again.

- Each time we extend the study further, the traffic becomes "diluted" further over a wider area, and we repeat this process until the extra traffic at all the junctions does not generate any significant effects.

This progressive widening of the geographical extent of the area of study (and focusing on the "next ring" of junctions) could be *automated* with a GIS of sufficient sophistication in its handling of networks. However, there is no clear way of doing it, with the current generation of off-the-shelf GIS, without having to engage in potentially complex "macro" programming.

9.3.5 Mitigation

Many of the mitigation measures for traffic impacts discussed in the literature refer to the *indirect* impacts of traffic (noise, pollution, visual intrusion) and have in common with socio-economic mitigation the emphasis on "internalising" the effects to the site, for example to minimise dust or noise. With respect to the mitigation of *direct* traffic impacts, mitigation measures tend to be of a very different nature depending on the stage of the project they apply to (Ferrary, 1994; Hughes, 1995; Richardson and Callaghan, 2001):

Measures that can be used during *construction* relate mainly to the traffic to and from the site:

- using more local materials and suppliers to reduce long-distance trips by construction lorries;
- using a range of modes of transport to spread the load;
- using designated specific routes agreed with the Local Authority;
- using particular times (e.g. overnight) to move highly disruptive, slow-moving heavy abnormal loads.

For the *operation* stage, the development is "fixed" (Ferrary, 1994) and mitigation measures tend to focus on the network around the project rather than the project itself:

- improving specific junctions to reduce congestion;
- small-scale road widening around the access points; large-scale improvements are seldom suggested because, if they are necessary it usually means that the project is "in the wrong place" and it would not get planning permission (Ferrary, 1994) anyway;
- traffic calming measures;
- pedestrian facilities like crossings, central islands and signals;
- measures and facilities to support cycling;
- public transport support (bus lanes, etc.).

While most developers usually can accept the first group of measures related to the construction stage, they are more reluctant to accept the second related to the operation stage, as they tend to feel that their responsibility only extends "within the site" (Ferrary, 1994), and that mitigations outside are for the Local Authority to apply, although this is now being overtaken by the increasing focus on "green commuter" planning. In this sense, it can be said that, when the second group of measures are adopted by developers, they are perceived more like Planning Gain operations.

9.4 EXPERT SYSTEMS AND MODELS, PROBLEMS AND CHOICES

In traffic impacts we find again (as with noise and air pollution) an area where the application of expertise *can* go through stages of heavy modelling:

- to simulate the traffic generated;
- to simulate the impacts on junctions, public transport, pedestrians.

Modelling in this area is quite developed and it can be simply a question of knowing which model to choose. However, the modelling experience is far from "seamless" and when asked about what makes a project difficult to assess (Ferrary, 1994), the case of "big developments with a lot of modelling to do" was quoted. Also, the interpretation of the output from models is one of the typical mistakes that "novices" make (as opposed to experts). Sometimes models are very sensitive to small variations in some of the data input into them, and small errors in these inputs can make the models produce abnormal results, which only an expert can detect (and double-check the inputs accordingly). These are some of the reasons why hard modelling is not always used; in fact, simulations are one of the first things (traffic surveys are the other) to be "simplified out" of traffic impact studies when the time or budget is more limited. As an alternative to simulation, the values we need to proceed can be calculated from databases (as in the case of trip generation) or using the raw data with simple formulae (like the calculation of percentage increases in traffic) (Figure 9.14).

In the case of socio-economic impacts, the modelling is "softer" but still can be present, mainly at two points in the process (Figure 9.15):

- when calculating the economic multiplier;
- when making demographic projections on which to base the social impacts;
- there is also considerable "low-level" modelling going on throughout the whole study, involving the calculation of percentages, ratios, capacities, etc., although these do not require "models" as such, as the normal functionality of any normal expert system or GIS can handle them. The problem with such calculations is not in their modelling, but in handling the information they use, and the wide range of formats they use.

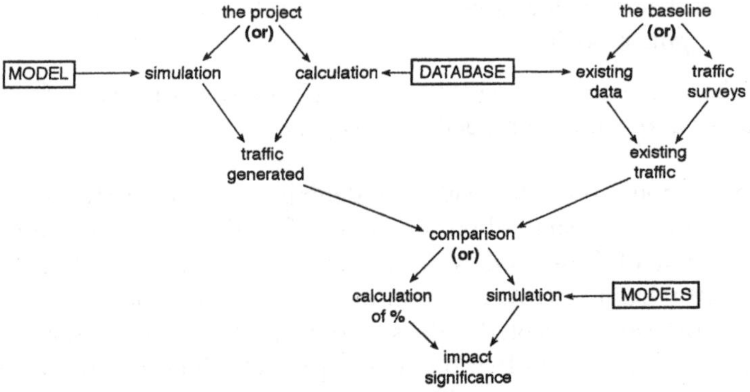

Figure 9.14 Simulation and calculation as alternatives in traffic impact assessment.

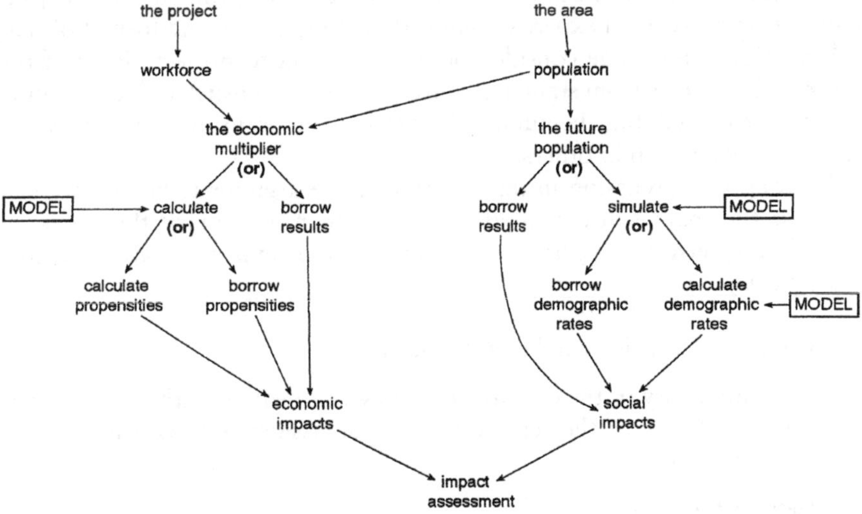

Figure 9.15 Simulation and calculation as alternatives in socio-economic impact assessment.

As with traffic, there is a choice of the *level of modelling* adopted, that in the case of socio-economic impacts presents quite clear intermediate levels:

1 At one extreme, the case of modelling everything from raw data, using a "black-box" style of modelling.
2 As an intermediate case, simulating the core of the calculations but "borrowing" some of the parameters:

(a) some or all of the propensities for the multiplier;
(b) the demographic rates (fertility, mortality) necessary for the demographic modelling.

3 At the opposite extreme, "borrowing" the overall results from other sources and avoiding modelling altogether.

These decisions about the level of modelling are based partly on budgetary considerations – as usual – but also on the fact that local variations in the results of some of these models can be relatively small (as with multipliers) and the error from borrowing existing values can be compensated by the simplicity of using accepted values (from similar projects for instance). At Public Inquiries, it is often the actual project which is more challenged than the socio-economic calculations made for the EIA (Glasson, 1994), although there is sometimes concern when results are presented as "ranges" rather than precise, unique figures. The main difficulty with socio-economic impact assessment is getting all the information needed – about the project and about the area – and one of the major problems of encapsulating their study in an expert system is that the quantity and format of that information can change considerably from project to project. Some of the information comes from standard sources (like the Census) and its handling can be automated, but the number of *ad hoc* sources in this type of impact study is greater than in others.

Relating the discussion in this chapter to the arguments before, we can see that the main sources of *difficulty* – which also mark the potential boundaries between expert systems and human experts – are emerging quite clearly:

1 Complexity, which can be of two kinds:

(a) complexity of the case in hand, as when dealing with big projects;
(b) complexity of the science behind the assessment, as was the case with ecology.

2 Getting the data:

(a) sometimes from developers, about aspects of the project which are sensitive or still undecided;
(b) from surveys, consultees or published sources.

3 Politics (!), expressed by a hot climate of opinion surrounding the project:

(a) arguments between the Local Authority and the developer;
(b) public concern and debate, in the press or even in the streets.

The study of traffic impacts threw up also an interesting variation in the standard logic of impact assessment, which derives from the nature of traffic itself: an *iterative loop* in the logic (Figure 9.16): The study starts assessing

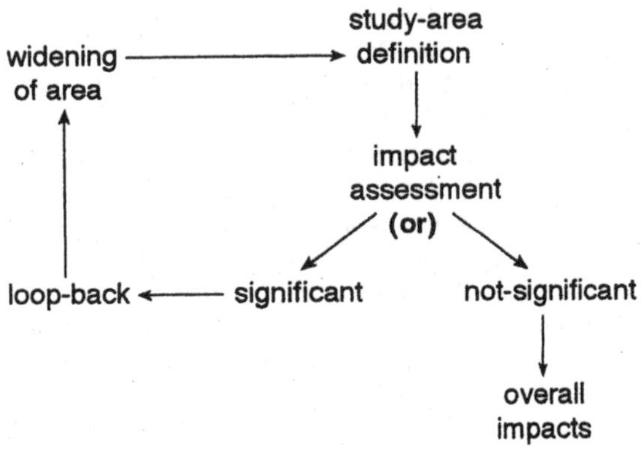

Figure 9.16 Iterative loop of area definition and assessment of impact significance.

the impacts at one geographical level and this level is widened if the impacts are found to be significant. Because traffic gets "thinner" as it spreads over a wider network, significant impacts become less likely as the area is widened, and the iterative process continues until no further significant traffic impacts are found. This also applies to the determination of socio-economic commuting zones at the construction stage.

REFERENCES

Brown, A.J. (1967) The "Green Paper" on the development areas, *National Institute Economic Review*, NIESR, Vol. 40 (May).

Brownrigg, M. (1971) The regional income multiplier: an attempt to complete the model, *Scottish Journal of Political Economy*, Vol. 18.

Chadwick, A. (1995) "Socio-economic impacts 2: social impacts", in Morris, P. and Therivel, R. (eds) *Methods of Environmental Impact Assessment*, UCL Press, London, 1st edition (Ch. 3).

Chadwick, A. (2001) "Socio-economic impacts 2: social impacts", in Morris, P. and Therivel, R. (eds) *Methods of Environmental Impact Assessment*, Spon Press, London, 2nd edition (Ch. 3).

Crompton, D.H. (1981) *Pedestrian Delay, Annoyance and Risk*, Imperial College, London.

DETR (1998) *Guidance on the New Approach to Appraisal*, Department of the Environment, Transport and the Regions, HMSO, London.

DoT (1993) *Design Manual for Roads and Bridges*, Vol. 11: "Environmental assessment", Department of Transport, HMSO, London.

ERL (1991) *Clinical Waste Incinerator at Knostrop*, Environmental Impact Assessment Report, Yorks. Water Enterprises Ltd and Environmental Resources Limited.

Ferrary, C. (1994) *Personal Communication*, Environmental Resources and Management, London.

Florence, P.S., Fritz, W.G. and Gilles, R.C. (1943) Measures of Industrial Distribution, in National Resources Planning Board, *Industrial Location and National Resources*, Government Printing Office, Washington, Ch. 5, pp. 105–24.

Hughes, A. (1995) Traffic, in Morris, P. and Therivel, R. (eds) *Methods of Environmental Impact Assessment*, UCL Press, London, 1st edition (Ch. 5).

Glasson, J., van der Wee, D. and Barrett, B. (1988) A local income and employment multiplier analysis of a proposed nuclear power station development at Hinkley Point in Somerset, *Urban Studies*, Vol. 25, pp. 248–61.

Glasson, J. (1994) *Personal Communication*, Impact Assessment Unit, School of Planning, Oxford Brookes University.

Glasson, J. (1995) Socio-economic impacts 1: overview and economic impacts, in Morris, P. and Therivel, R. (eds) *Methods of Environmental Impact Assessment*, UCL Press, London, 1st edition (Ch. 2).

Glasson, J. (2001) Socio-economic impacts 1: overview and economic impacts, in Morris, P. and Therivel, R. (eds) *Methods of Environmental Impact Assessment*, Spon Press, London, 2nd edition (Ch. 2).

IEA (1993) Guidance Note 1 *Guidelines for the Environmental Assessment of Road Traffic*, Institute of Environmental Assessment, Lincoln.

IHT (1994) *Guidelines for Traffic Impact Assessment*, Institution of Highways and Transportation, London.

Petts, J. and Eduljee, G. (1994a) Transport, in *Environmental Impact Assessment for Waste Treatment and Disposal Facilities*, John Wiley & Sons, Chichester (Ch. 15).

Petts, J. and Eduljee, G. (1994b) Social and Economic Impacts, in *Environmental Impact Assessment for Waste Treatment and Disposal Facilities*, John Wiley & Sons, Chichester (Ch. 16).

Richardson, J. and Callaghan, G. (2001) Transport, in Morris, P. and Therivel, R. (eds) *Methods of Environmental Impact Assessment*, Spon Press, London, 2nd edition (Ch. 5).

Slinn, M., Guest, P. and Mathews, P. (1998) *Traffic Engineering Design – Principles and Practice*, Arnold, London (Ch. 13).

Steele, D.B. (1969) Regional multipliers in Great Britain, *Oxford Economic Papers*, Vol. XXI, pp. 268–92.

Vanclay, F. (1999) Social Impact Assessment, in Petts, J. (ed.) *Handbook of Environmental Impact Assessment*, Blackwell Science Ltd, Oxford (Vol. 1, Ch. 14).

10 Water impacts

10.1 INTRODUCTION

We have discussed in some detail a wide range of types of impacts, reducing them to relatively simple logical processes with a potential for automation as expert systems. Although not all the standard areas of impact assessment have been covered, there has been enough variety to illustrate most of the problems and issues involved when "translating" expert behaviour and judgement into a simple logical process that a non-expert can follow. This can be illustrated by discussing one last area of impact that encompasses most of the issues raised in other areas: *water*, which really consists of a succession of *several* impact assessments. Water impact assessment is probably the most difficult, because of the extreme variety of impacts that can affect water, and because of the extreme variety of standards and legislation covering them (see Bourdillon, 1995 for an early list). It can be said that Environmental Impact Assessment is a by-product of the relative cultural sophistication normally associated in a society with a certain degree of development, but concerns with the quality and quantity of water have been central to all societies throughout history, and this makes it probably the most extensively documented – and regulated – area of impact assessment. Also, in terms of the line of argument we are following here, water impact assessment involves really a *chain* of several areas of impact, each of which can be looked at as we have been doing in previous chapters. These areas can be seen as "modules" which form part of water impact assessment, linking the original source of impacts – the project – to the ultimate impacts on humans or on the natural environment (Figure 10.1).

- The project can produce certain effects *directly* on a water system (discharges to it, abstractions from it) and it can also have certain effects on the groundwater.
- The behaviour of the groundwater will determine possible *indirect* effects on water systems, as well as other effects on the soils and on the usage

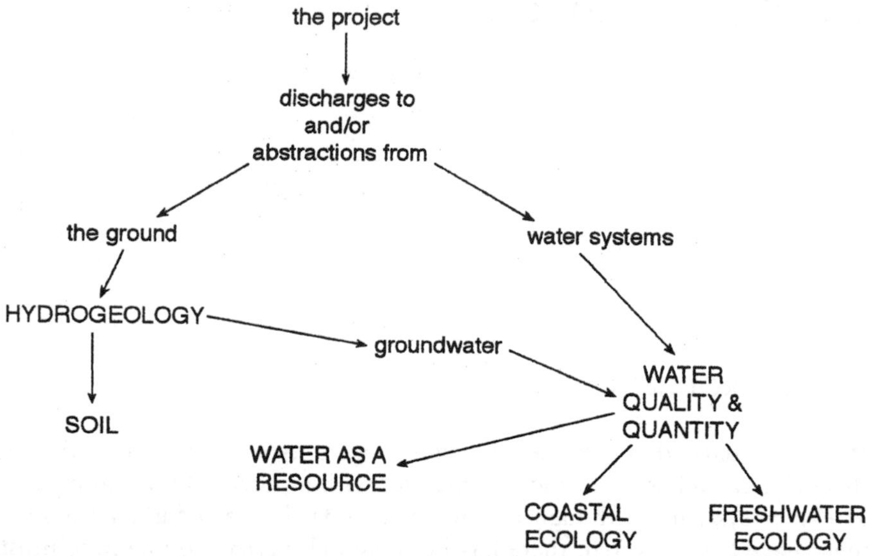

Figure 10.1 The interlinked logic of hydrogeology, water, and water-ecology impacts.

of water as a resource; in impact assessment, these impacts are usually covered under headings like "hydrogeology and soils".

- Be it directly or indirectly, the water system is affected in terms of "water quality/quantity": volume, flow and possible contamination.
- In turn, the water system affected has effects (impacts) on the usage of water as a resource (for drinking, leisure, etc.) and on the ecology of its environment.
- Usually, water ecology impacts are studied under one of two headings: "freshwater ecology" (rivers and lakes) and "coastal ecology" – with a third category of "estuarine ecology" used sometimes – depending on the type of water system.

We can treat the study of water impacts as a sequence of impact studies, from hydrogeology to water quality to ecology. As already mentioned, the literature on each of these areas is vast, and impact assessment manuals can be good summaries of the field (like the very detailed account in York and Speakman, 1980) and can also provide good "guides" to the literature (Westman, 1985; Petts and Eduljee, 1994a and 1994b; Atkinson, 1999; Biggs *et al.*, 2001; Hodson *et al.*, 2001; Morris *et al.*, 2001d; Thompson and Lee, 2001). It is also an area with much legislation and regulation, the latest of which being the EC's Water Framework Directive (EU,

2002).[46] Here, we are going to follow the same logic as before, discussing each of the main steps in the flow chart above, and treating in greater depth each of the specific areas of impact study.

10.2 THE PROJECT

Since we are looking for both "direct" effects and "indirect" effects (through hydrogeology) on all the water systems around the project (surface or underground), we are interested in those aspects of the construction or operation of the project likely to generate such effects. For the *construction* stage, the list concentrates on the type of project it is and its features on the one hand, and on construction practices on the other (for reasons that will become clear later, we are marking with an asterix * those aspects with a link to hydrogeology):

1 The type of project and the presence/absence of certain *features* may involve:

- tunnelling or mining (*);
- quarrying or deep excavations involving soil removal (*);
- site-levelling involving earth movements (*);
- foundations involving piling (*);
- temporary modification or manipulation of water systems, changes in the course of a river, erection of water-protection barriers;
- construction of drainage systems (*).

2 Concerning on-site working practices:

- number of workers;
- phasing of construction;
- materials used for construction;
- policy concerning the control of dust and particulates by vehicle and earth movements;
- vehicle movements, and the type of fuel to be used (especially diesel);
- on-site policies about storage of fuel and oil tanks and dealing with losses and leakages (*);
- policy about disposal of empty fuel and oil tanks (*).

46 Also available as a consultation document circulated by the UK's Environment Agency on the internet in page *http://www.environment-agency.gov.uk/yourenv/consultations/305276/?versione1&lang=_e* or by e-mail from *waterframeworkdirective@environment-agency.gov.uk*.

During the *operation* stage of the project, the presence of any features which could alter or contaminate the water systems may include (* indicates a link to hydrogeology):

1 project areas: area affected, area paved (*);
2 number of persons using the site: workers, customers/visitors, suppliers;
3 what facilities are included in the project: canteens, toilets, water-related facilities like swimming pools;
4 concerning the discharge of foul water from the project: connected to existing sewers, a new sewer (*);
5 storage tanks (*):
   (a) their contents,
   (b) their location: above ground, below ground;
6 pipelines and their location (*): above ground, below ground;
7 other discharges apart from foul water from toilets and kitchens:
   (a) composition of the discharges: materials, chemical composition, flow rate, temperature,
   (b) concentration: from a point source, diffuse,
   (c) location of the discharges:
      (i) to a water system, are there balancing facilities (like a pond) before the release outside? (*),
      (ii) to the ground (*): as run-off water, to soak-aways.
8 water abstractions:
   (a) from a surface water system,
   (b) from bore-holes from underground aquifers (*),
   (c) flow/volume required.

This list is really a combination of the individual lists we would require if we were studying hydrogeology, soils or ecology, which overlap considerably with each other. How to proceed next is dictated by the project features present. In the first place, the list of project features can show that the project will discharge to (or abstract from) a surface water system *directly* (and we can study these impacts on the water), or that it will discharge to the ground or affect the ground in other ways (earth-movements, etc.). If the latter is the case, these ground-related actions can produce two kinds of impacts to be studied separately: (i) impacts on the soil itself, which we would study as a separate area of impact assessment; (ii) impacts on the hydrogeology beneath it, which we also study as another area of impact assessment, and finally, the hydrogeological effects in turn are likely to impact on the surface water system (Figure 10.2).

This discussion does not cover Soil impact assessment, because the focus is on direct or indirect impacts *on water*. Soil impacts merit a whole section

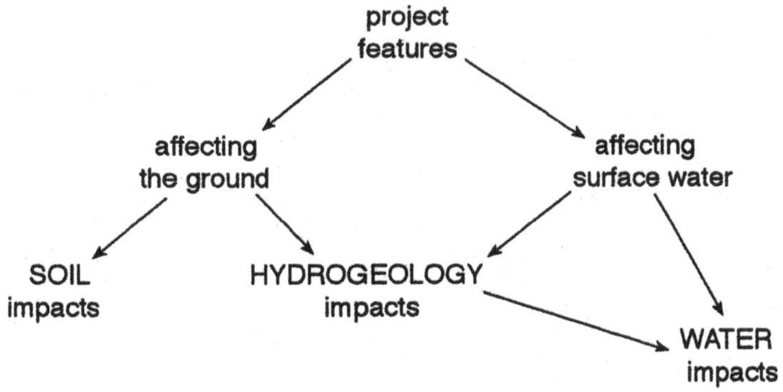

Figure 10.2 The chain of project effects on soil, hydrogeology and water.

of their own (they usually do in manuals and in Environmental Statements) involving baseline studies, impact identification, standards, mitigation, etc. Good-practice guidance on soil impact assessment can be found in Petts and Eduljee (1994a), Hodson (1995) and Hodson *et al.* (2001), which also contain references to government guidance and standards.

The focus of our hypothetical study of impacts here depends on what features are present in the project and whether they are likely to produce direct or indirect impacts on water: (i) if the project has features that suggest there are likely to be hydrogeological issues involved (indicated by the presence of project features marked *), we proceed to the study of hydrogeological issues; (ii) if the hydrological effects from the project are only *direct* ones to a surface water system, we can move on directly to study water quantity and quality in those systems (Section 10.4 below).

10.3 HYDROGEOLOGY: THE BASELINE

Hydrogeology – often studied together with "soil" – is another typical area of impact assessment, and it follows a logic similar to the others, from the baseline study to the determination and mitigation of impacts (Figure 10.3). The baseline study develops from a map-based desk study into an exercise in consultation with organisations that have the relevant information (Simonson, 1994),[47] and only rarely – for big projects with a big budget for the impact study – does it involve fieldwork to collect information. As this process evolves, the *area of study* also changes,

47 The knowledge acquisition for this part was greatly helped by conversations with John Simonson, of Environmental Resources Management Ltd (Oxford branch); Mathew Anderson helped with the compilation and structuring of the material for this part. However, only the author should be held responsible for any inaccuracies or misrepresentations of views.

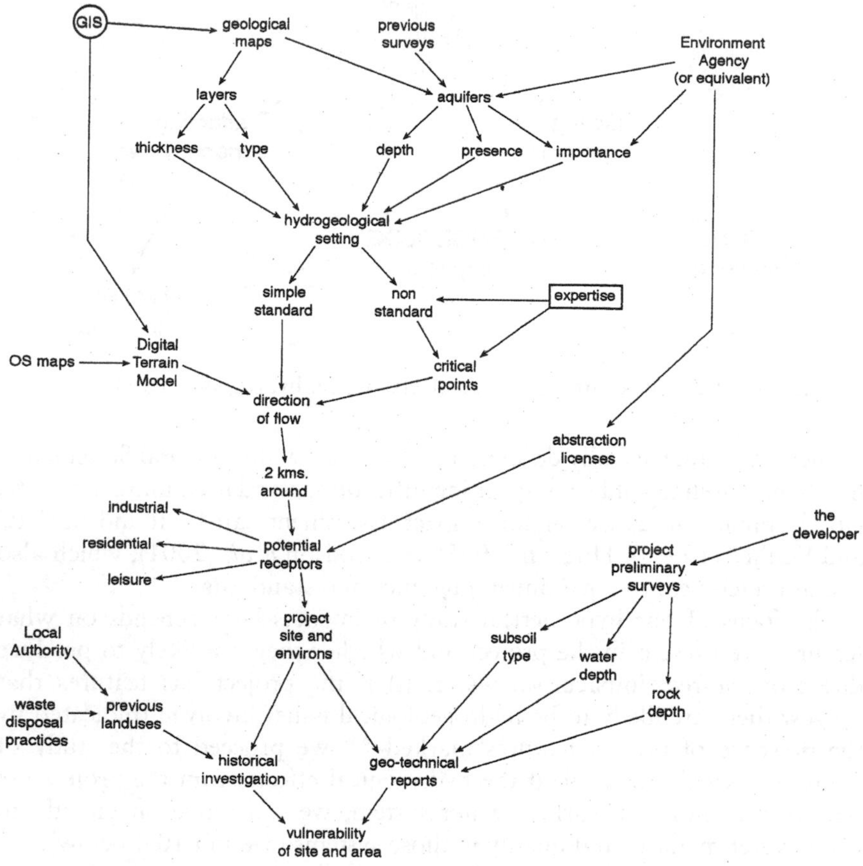

Figure 10.3 The logic of the hydrogeological baseline study.

focusing gradually on the site, so that the baseline study is really applied at several spatial scales (regional, intermediate, immediate), which also correspond to three *time* scales, as the diffusion of effects through groundwater is quite slow.

At a *regional scale*, the *geological setting* is studied with geological and hydrological maps (some of them also available in digital form)[48] from the British Geological Survey, at what could be seen as the scale of very long-term effects. The aim is to build a mental *model* of the geological structure

48 British Geological Survey operates an online "Geoscience Data Index" of all the data and maps they hold (Hodson *et al.*, 2001) in digital form, and the Centre for Ecology and Hydrology in Wallingford also has produced the National Groundwater Level Archive (CEH, 2000) with map-information about wells and major aquifers (see Morris *et al.*, 2001).

on which the project site will impact. Such model can be a variation of the simplest hydrogeological model (Morris *et al.*, 2001d) where there is a standard series of geological strata below the project:

- a thin layer of organic soil at the top;
- next, a thicker layer of more or less permeable sub-soil (clay, sand, gravel);
- then, an impervious bedrock layer below.

An aquifer may be present in the second layer, saturating it in water up to a certain level, the water table. When aquifers are present, they can be in direct contact with the rest of the upper layer ("unconfined" aquifers) or they can be "confined" by a layer of impervious material above them which can also protect them from penetration and contamination.

The main focus of this stage of the study is to identify the various layers – organic soil at the top, sub-soil, bedrock, and intermediate "confining" layer if it exists – under and around the site:

- type of sub-soil (clay, gravel, sand), thickness;
- type of bedrock beneath it: depth, other types of bedrock in the area;
- type and thickness of any confining layers;
- presence of special geological features: faults/fissures, foldings.

The depths and slopes of the different layers will give an idea of the direction in which the bedrock is "dipping", and whether this is an area of geological complexity or one which can be treated as a standard case. The latter would have the same layers superimposed all over the project area, no discontinuities (changes in the type of layer, or breaks in the structure), and the project not near any edge between layers where vulnerable points may appear. The ultimate aim of the study is to get a picture of how any groundwater present is likely to move and "carry" any possible contaminations:

1 Identifying the presence and type of *aquifers* (one or several) under the area of study:
   (a) the depth of the water-table;
   (b) the depth of the aquifer;
   (c) determining if the aquifer is *protected* by other hard layers (gravel for instance), especially from above.
2 Most importantly, anticipating the likely movement of groundwater:
   (a) the likely *direction of flow* of groundwater, derived from the way the bedrock underneath slopes;
   (b) where it is likely to discharge to a water system.

Geological data will normally come mainly from already-prepared *maps* (Simonson, 1994), but much of this information can also be obtained through *field surveys*. Such surveys can come from previous studies of the same area or can be commissioned by the project developer – if the time and budget for the study permits. Field surveys usually involve a combination of:

- *sampling* of the location of the survey-points;
- thin *bore-holes* to extract samples from all the layers which can be measured and analysed;
- *wells* for the analysis of aquifers, the water table and water quality.

Whatever the source of the data, the derivation from it of a mental "model" of the geology of the area draws on the considerable complexity of geology as a science, and can present the same type of difficulty for automation (for an expert system) that was encountered for ecology: the complexity of the science behind the expert's approach, which he/she will need to refer to when the case in hand is "non-standard". On the other hand, the prediction of directions of flow for standard situations can be made easier by automation using GIS if the geomorphology maps are in digital form and include depth and thickness of strata. Using such maps in conjunction with Ordnance Survey maps, standard GIS Digital Terrain Models can be built, from which slopes and directions of flow of potential streams can be derived for any of the geological strata included in the GIS maps.

At an *intermediate scale*, an area of possible medium-term impacts around the project (2 km around the site, which could take 10–20 years for groundwater to reach) is studied to identify *potential receptors* in the area, the types of users of groundwater in the area, including: industrial, residential, leisure, other uses. Ordnance Survey maps can be used, as well as surveys carried out by the Local Authority.

The general objective of the baseline study is to establish the site's *vulnerability* (Simonson, 1994). At one extreme, if the clay is very thick (50 m) the aquifer below is well protected; at the opposite extreme, if the water table is high, there is no protecting hard layer, and the aquifer is used for public water supply, then the situation is highly vulnerable. This vulnerability can also be established in the UK from *maps* from the Environment Agency: degrees of vulnerability are recorded in Groundwater Vulnerability maps (at 1:100,000) produced in paper as well as digital format, and if the site is in a Groundwater Source Protection Zones, on maps in digital form for GIS use.

After the desk-study of maps (GIS-based or not), the baseline study moves on to a consultation stage to complement the cartographic information: (i) with the Local Authority, to find out if it is a Regionally Important Geological site; and (ii) especially, with the UK's Environment Agency (or equivalent in other countries):

- to find out the degree of vulnerability of the area as a: major aquifer, minor aquifer, non-aquifer;
- to learn what abstraction licenses there are in the area: location, volume, type of use;
- and also, to ascertain the Agency's general views about the area and if they have any concerns about groundwater quality.

At *the next scale down*, potential receptors in the *immediate surroundings* of the site (100–500 m) are considered, where the effects from the site could permeate in a short time. Finally, the *site itself* is studied in its geological structure in the greatest detail possible with the available information, as it is there that the project will produce the *impacts* (rupturing a layer of clay, reaching an aquifer, etc.). Sometimes, the project has required a geological survey by the developer before assessing its feasibility and determining its structural requirements (foundations, etc.). When this is the case, the geo-technical reports for that survey of the site and adjacent areas are another invaluable source of baseline information.

An important part of the baseline study of the site and the surrounding area is a *historical investigation*, based on consultation with the Local Authority and/or the local library:

- study of *previous landuses*, in case there have been previous possible contaminations of the land/aquifers (if *industry* has been present on or around the site);
- historical waste-disposal practices in the area.

The aim is to estimate the likelihood of historical contamination of the area, which is expensive to measure directly with boreholes.

10.3.1 Hydrogeological impacts

A comprehensive list of all types of water impacts (not just hydrogeological) can be found in Morris *et al.* (2001d), and Hodson *et al.* (2001) discuss a short list of more specifically hydrogeological effects. In general, such effects fall into three main types, which in turn originate from different types of project actions:

1 Physical *disruption* of the geological setting derived from physical/structural features in the project (Figure 10.4):

   (a) alteration of some layers, in particular by *extraction*:

      (i) for foundations,
      (ii) for subterranean facilities.

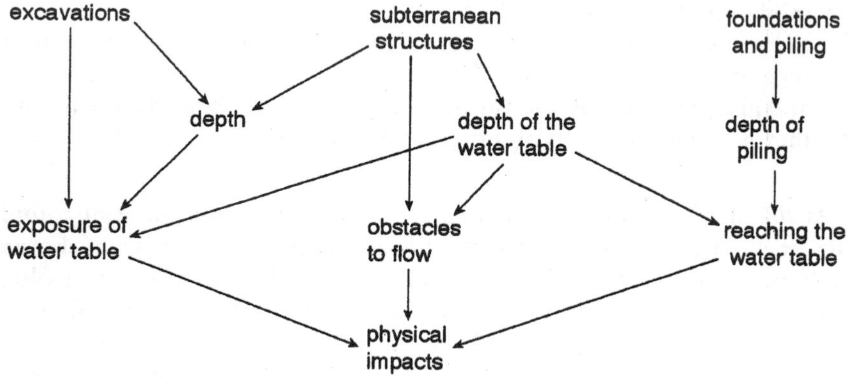

Figure 10.4 Physical disruptions from the project.

(b) affecting the water table:
   (i) exposing it after excavations,
   (ii) reaching the water table,
   (iii) *piercing* the "confining" layers.

(c) *obstacles* to groundwater flow, by subterranean facilities or foundations, resulting in:
   (i) obstructions,
   (ii) changes in direction.

2 Impacts on the *volume* of groundwater:

(a) by reducing aquifer-recharges, typically by paving (or putting tarmac) over land thus reducing the amount of water making its way into the ground;
(b) by abstractions of water for project use;
(c) by changing the flow-pressure by *emissions* of water (even if clean).

3 *Contamination* of aquifers:

(a) during *construction*:
   (i) accidental spillages of oils/lubricants or combustion fuels to the ground,
   (ii) non-accidental effluents derived from vehicle-washing,

(b) during *operation*:
   (i) accidental spillages of oils/lubricants or combustion fuels to the ground,
   (ii) accidental spillages from liquids stored on site,
   (iii) discharges of contaminated water from the project to "soakaways" from which it may percolate to reach aquifers if the geological setting allows it,
   (iv) gradual discharges of leachates/contaminants from underground projects like landfills.

The prediction of *physical-disruption* impacts is based on an initial *qualitative* identification of the presence in the project of features with such potential effects: excavations, subterranean structures, foundations and piling. In addition to the presence of these features, some simple calculations have to be made to establish if they are likely to constitute an impact. Practically all projects involve the disturbance of the land and a certain amount of excavation (and foundations work), but these operations will only impact on hydrogeology if they affect an aquifer by going deep enough to "reach" it. This can happen directly or by piercing other hard layers protecting it, or even by piercing through the bedrock under it (providing new "escape routes" for the groundwater). This possibility can be anticipated by comparing the *depth* of each subterranean operation (from the project specifications) with the depth of the aquifers, the water table and the bedrock layers from the baseline study.

The other two types of impacts are determined in a more *quantitative* way. Impacts on the *volume* of aquifers can be estimated *as percentages* of the existing aquifer (Figure 10.5):

- Concerning the loss of recharge land, the area lost can be compared to the area of recharge-land above that section of the aquifer, and the percentage loss can be calculated.

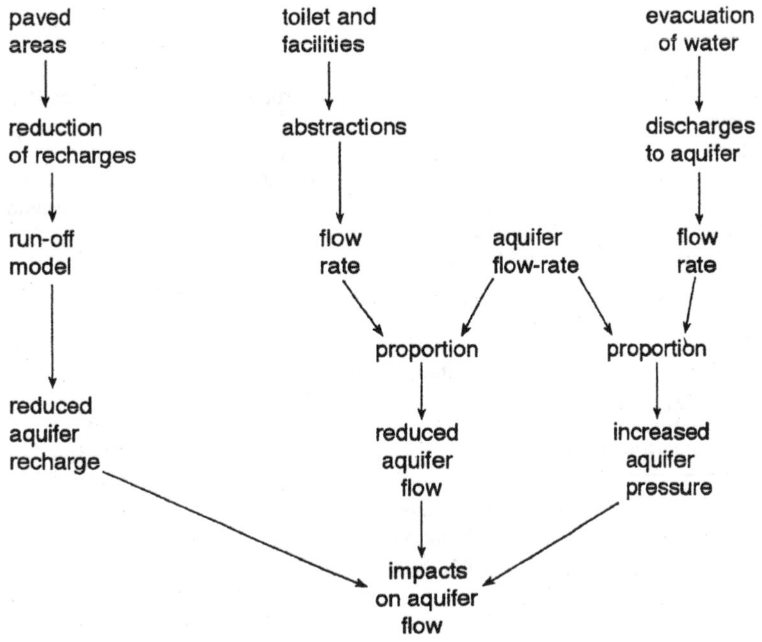

Figure 10.5 Impacts on the volume of aquifers.

- Concerning the impact of abstractions, the volume of flow required for the abstraction can be compared to the total flow in the aquifer underneath, and the percentage loss calculated in the same way.
- Similar calculations can be made with respect to any emissions of extra water into the aquifer, comparing its volume to the flow of the aquifer.

With respect to the *contamination* of aquifers, its determination is also predominantly quantitative, but it also can go through a *simulation* stage, which can involve some modelling (Figure 10.6). Once the likely discharges to the soil (intentional or accidental) have been determined, the question is how the polluted water will travel: (i) how much of it will reach aquifers through the soil (and eventually reach surface water systems)?; (ii) how much will reach the water systems directly through the surface?:

- With respect to discharges and spillages on the surface, *run-off* models can be used to separate the contaminated water likely to travel through the surface from the water likely to filter through the soil. This separation is expressed as the proportion of water likely to run-off (the "run-off coefficient") and Appendix C of Morris and Therivel (1995) show values for this coefficient for a range of types of soil and landuses.
- The behaviour of underground pollutants can be approached using transport models, mostly produced in the US (a good review can be

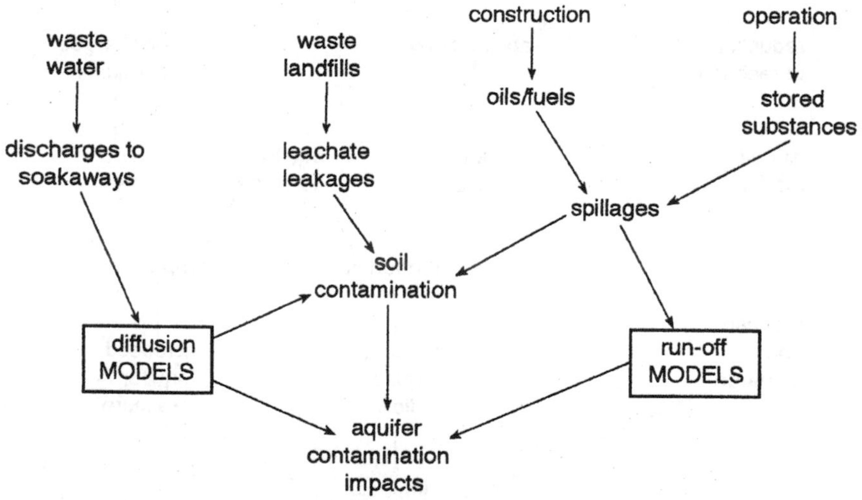

Figure 10.6 Contamination of aquifers.

found in USEPA, 1988) like MODFLOW or AQUA, or even adaptations of air-dispersion models like ISC3.[49]

- For the particular case of land-fills, underground leachate leakages and their behaviour can also be estimated using models which use the "release rates" of different linings used for landfill projects (Petts and Eduljee, 1994b), models like LANDSIM (Hodson *et al.*, 2001), or the Simplified Vertical and Horizontal Spreading model researched in Britain by the Department of the Environment (Eduljee, 1992).
- *Darcy-type* models can be used to simulate the movement of contaminated water through the soil. Darcy's Law applies to groundwater aquifers and it expresses groundwater velocity as proportional to the "groundwater slope" (difference in hydraulic heads between two points divided by their distance) and to the "hydraulic conductivity" of the soil, whose value for different types of soil can be found in published sources.

Such models can be used to estimate the flow of contaminants towards the aquifers, which in turn will determine the level and speed of the contamination and the resulting concentration of pollutants in the water at the point of discharge into the water system. In addition to the speed of diffusion of the contaminant, its density will also affect its behaviour (Simonson, 1994), as a contaminant denser than water will travel through the water table in a very different way to another less dense than water. A comprehensive list of water models can be found in Morris *et al.* (2001d), which also gives a cautionary note about their use, the difficulties they present and how relatively expensive they are, maybe explaining the reason why they are not used often (Simonson, 1994). In practice, modelling what happens two kilometres away is an exception rather than the rule, and hydrological impact studies tend to concentrate on the project-site and immediate area and just *comment* on the potential effects for any sensitive receptors in the area further away. Normally, a "generic" approach to the impact study (no field-surveys and no modelling) is used – taking about five person-days of work. Only if a particularly sensitive area is found, is the developer asked to undertake a more detailed study (Simonson, 1994), on the grounds that the Environment Agency would not accept it otherwise.

The standards for contamination of groundwater are quite fragmented for various types of projects and are not very clear (Simonson, 1994). As a result, the assessment of *significance* of impacts tends to be based on judgement based on the vulnerability of the site – of the nearest groundwater receptors – and the type of contaminant. Also, the Environment Agency has

49 Industrial Source Complex 3, by UK-Adams (see the section on air pollution in Chapter 7).

a framework that can be used for deciding on the acceptability of development from the point of view of groundwater (EA, 1998).

10.3.2 Hydrogeological mitigation and monitoring

Impacts are produced directly by certain project features, hence mitigating their effects involves modifying the project to varying degrees. At one end of the scale, we have impacts produced by the "presence" of certain features, and mitigating those impacts can only be done by avoiding/modifying those features, involving some level of *redesign* of the project:

1. avoid piling if it would reach the water table;
2. avoid subterranean work or reduce depth;
3. in any earthwork involved, compliance with:
   - (a) British Standard 6031 (1981) "Code of Practice for Earthworks"
   - (b) British Standard 8004 (1986) "Code of Practice for Foundations".

At an intermediate level, some mitigation measures involve *slight modifications* to the project – sometimes involving the choice of certain materials over others – without modifying its fundamental design:

1. Measures to contain polluting substances:
   - (a) locating storage tanks and pipelines above ground;
   - (b) bunding of tanks in bunds with capacity greater (110 per cent for instance) than the capacity of the tanks;
   - (c) placing storage tanks over impervious surfaces;
   - (d) placing diesel/petrol powered fixed plant on impervious drip-trays.
2. Measures to reduce run-off and increase underground recharge:
   - (a) removing as little vegetation as possible, and using replanting;
   - (b) using gentle gradients in the terrain rather than steep slopes;
   - (c) using porous surfaces when paving the ground.
3. Surface-flow retention measures:
   - (a) detention/balancing ponds
   - (b) grass swales
   - (c) vegetated channels.
4. Better drainage:
   - (a) routing all waste water to sewers or treatment works;
   - (b) site-drainage control to reduce polluted run-off;
   - (c) routing the drainage through siltation wells/lagoons to allow sedimentation of solids;
   - (d) using oil-interceptors before the drainage leaves the site, correctly dimensioned for the volumes of drainage and of rainfall expected (sufficient volume and sufficient retention time).

Finally, at the other end of the scale, there is mitigation that involves just *improved site management* measures, during both *construction* and *operation*, including:

1 lining excavated areas;
2 minimising leachate generation;
3 guarding against spillages/leaks or accidental discharges;
4 compliance with:

(a) British Standard 6031 (1981) "Code of Practice for Earthworks",
(b) British Standard 8004 (1986) "Code of Practice for Foundations".

Most of these mitigation measures do *not* eliminate the impacts but only *reduce* their effect in a "gradual" way, and this is why *monitoring* is an essential part of mitigation if potential impacts have been identified (Simonson, 1994). A groundwater-monitoring network should be installed to detect changes in groundwater flow or chemical composition. The monitoring stations should be located on the perimeter of the site: monitoring measurements should be taken at least yearly (preferably quarterly); and this monitoring should continue *indefinitely*, until monitoring reveals no leakages over several years.

10.4 WATER QUANTITY AND QUALITY: THE BASELINE

Whether the "route" from the project to surface water-systems is direct or indirect (through hydrogeology) the study of the effects on the water itself is an essential part of any impact study near a water system, in relation to both water *quantity* and water *quality* (Clarke, 1994).[50]

The first question (Figure 10.7) is to identify the water systems to be investigated in relation to the project: (i) first, if the project relates *directly* to a water system (by using it, discharging to it, or abstracting from it), that water system must be included; (ii) second, any water system within a distance such that the effects of the project may affect it (through the surface or underground) must also be investigated. The latter can be derived from a study of *catchment areas*:

- Water-catchment areas for surface water systems can be worked out with GIS using a Digital Terrain Model or a "surface" and basin-analysis functions available in sophisticated systems (like Arc-Info GRID).

50 The knowledge acquisition for this part and the "Freshwater ecology" part (see Section 10.5 below) was greatly helped by conversations with Sue Clarke, of Environmental Resources Management Ltd (Oxford branch); Owain Prosser helped with the compilation and structuring of the material for this part. However, only the author should be held responsible for any inaccuracies or misrepresentations of views.

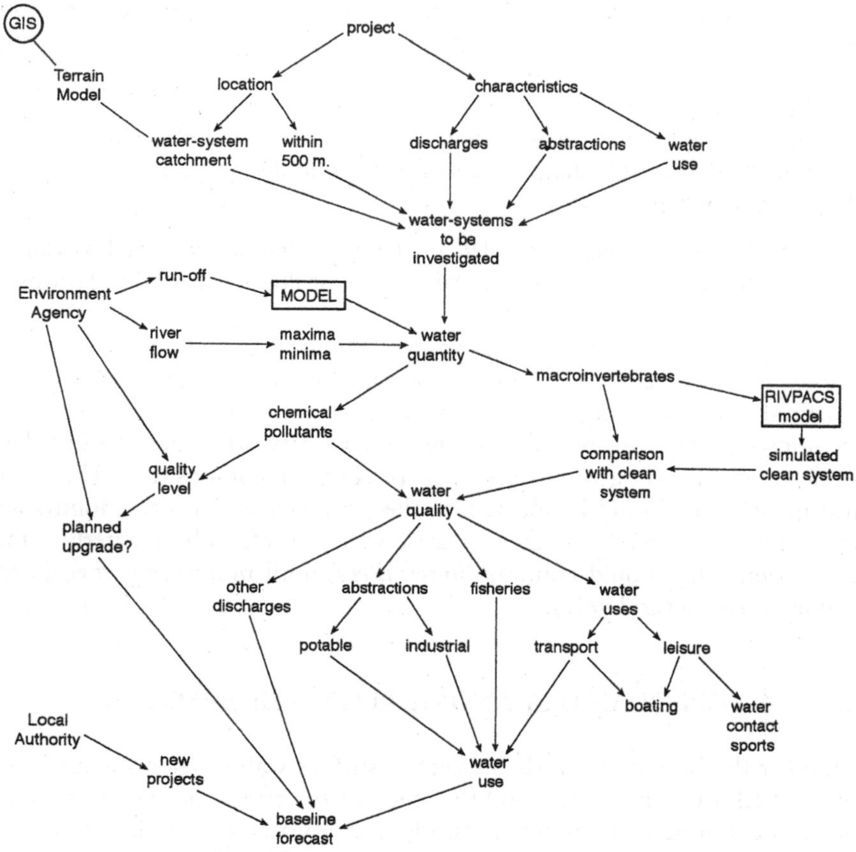

Figure 10.7 The water baseline study.

- The boundaries of all catchments in the UK greater than 0.5 km^2 are included in the CEH CD-ROM database (CEH, 1999b) described in Morris *et al.* (2001d).

If the project is inside the catchment area of a water system, that system should be included in the study. The area of study can be estimated by similar considerations to those for estimating areas around the project for the hydrogeological baseline:

- For short-term effects, any water system within 500 m of the project is worth investigating.
- For medium-term effects (several years), that distance can be extended to 2 km.

These limits can be extended if there are particularly sensitive systems a little further away, based on consultation with the Environment Agency.

Considering *each water system in turn*, the baseline study looks first at the hydrological situation in terms of *water quantity*, based on recent data (at least the last 3 years):

- river flow data: yearly maxima, yearly minima;
- rainfall in the area;
- soil type: on the banks of the water system, between the water system and the project;
- run-off flow.

The information is obtained mainly from the Environment Agency. Rainfall data can be obtained from the FEH CD-ROM database (CEH, 1999a) from the Centre for Ecology and Hydrology (their National River Flow Archive), or from the Meteorological Office. Run-off can be calculated by *modelling*, using run-off models like PC-IHACRES for the UK, from the centre for Ecology and Hydrology (Morris *et al.*, 2001d).

The second area of baseline study is *water quality*, which is normally monitored by the Environment Agency leading to a nationwide classification by the National Water Council of all water systems into different "classes" of quality. Water quality is normally studied based on physico-chemical and biological considerations. The *physico-chemical* approach is a quantification of the physical and chemical *contents* of the water system, and normally relies on the Environment Agency, which regularly monitors rivers and canals for their "General Quality Assessment", looking at levels of: Biochemical Oxygen Demand (BOD), Dissolved Oxygen (DO), ammonia, some nutrients. To complement this data, other sources can be used to measure indicators from an "ideal" checklist of substances and characteristics including (Morris *et al.*, 2001d):

- nutrients (phosphorus, nitrate, chlorophyll A);
- Dissolved Oxygen;
- organic matter (BOD);
- sediment;
- metals;
- oils;
- other pollutants (Ammonia, Hydrogen sulphide, Cyanide, Pathogen from faecal contamination);
- acidity and pH;
- alkalinity;
- electrical conductivity;
- temperature.

Surveys can also be used, although they can be expensive if the sampling is extensive. So-called "grab" samples at *one* point are simple enough, but to get good average values for the whole body of water it is better to use *composite* samples based on "3-D systematic" sampling that can require a substantial number of sampling stations.

The extent of pollutants covered in the baseline study should depend on the types of discharges anticipated (York and Speakman, 1980):

- if certain chemicals are to be discharged, those same chemicals should be studied;
- if nutrients are to be discharged, nitrogen, phosphorus and BOD should also be studied;
- if the project involves thermal discharges, the temperature must be studied;
- if organic waste discharges are anticipated, oxygen resources must be analysed (BOD, DO, temperature);
- if sanitary sewage is to be discharged, total coliforms and fecal coliforms must be measured;
- if dissolved solids will be discharged, such solids should be measured as well as conductivity, alkalinity, acidity, pH;
- if suspended solids will be discharged, such solids should be measured, as well as sediments.

The baseline study normally uses Environment Agency monitoring data for the previous three years – because this span fits well the normal life span of aquatic invertebrates – which are also used for the baseline assessment (see below). If the water system involves fisheries or the water is particularly pristine, the time span for data collection can be extended back to five (or even ten) years.

Because chemical methods of water quality assessment are limited and have problems, it is very common to use *biological indicators* instead, on the basis that looking at the species – and their concentration – that survive in the water can tell us most of what we want to know about its pollution levels. The most common biological indicator is the presence of *macroinvertebrate families* that are regularly monitored by the Environment Agency. It can be said that while water samples tell us about quality at a spot-point in time, invertebrates give a longer-term picture of what the environment is like (Clarke, 1994). If monitoring data is not available, *surveys* can be undertaken, using simple methods like "kick" sampling walking along the water course with a "sweep net" collecting the sample.[51] As discussed in Morris *et al.* (2001d) the concentrations of invertebrates and the various indices calculated from them are not sufficient indicators of water quality. To establish the baseline quality of the water we should *compare* its situation with a hypothetical "clean" case, and we can do this with a simulation model:

51 The Freshwater Biological Association has produced useful "identification keys" for aquatic invertebrates.

- Use the RIVPACS model (Morris *et al.*, 2001d) to simulate the various indices of invertebrate presence for the hypothetical "clean case" of a water system of similar characteristics.
- Compare such indices with those calculated from the baseline data for the system in question, this should give good indicators of baseline water quality.

In addition to studying the water quality in itself, to put the project in context it is useful to know of any *discharges by other activities* (like waterworks, or other developments) near the project or *upstream* from it, in consultation with:

- the Internal Drainage Board;
- the Local Authority;
- British Waterways (if it is a canal);
- the Water Authority.

After water quantity and quality, the baseline study should look at current *water use* (often referred to as "water as a resource") under three main headings:

1 *abstractions*:
  (a) volume,
  (b) type (potable, industrial).
2 *fisheries*:
  (a) species,
  (b) abundance,
  (c) whether it is for exploitation or for fishing.
3 *uses* of the water:
  (a) frequency/intensity of use,
  (b) type:
    (i) transport,
    (ii) leisure: boating, water-contact sports (bathing/swimming, water skiing, wind surfing), angling.

The water-use baseline can often be forecast by using trends. However, for water quality, it is important to anticipate changes by, for example, asking the Environment Agency about any plans to *upgrade* the quality level of the water system, and the Local Authority about other projects in the pipeline.

10.4.1 Water quantity impacts

The project can affect the quantity of water in various ways, more or less direct:

1 The most *direct* influences:

   (a) abstractions,
   (b) discharges.

2 At an *intermediate* degree of directness, *run-off* flow can be affected by:

   (a) paving,
   (b) planting,
   (c) changes in the shape of the ground surface.

3 *Indirectly*, the flow of groundwater can also be affected by:

   (a) soil water (soakaways, etc.),
   (b) changes to aquifer flow: abstractions, discharges.

The developer will have already agreed with the Environment Agency the maximum discharges/abstractions permitted by law. For our calculations of impacts, we assume that the quantities discharged/abstracted will be the maximum permitted, even if they will be lower most of the time, and this difference is usually mentioned in the impact assessment report (Clarke, 1994).

The effects of direct and indirect actions on the aquifer can be quantified from the project characteristics. *Run-off* effects, on the other hand, usually have to be *simulated* with a run-off model like TF-55[52] (see Morris *et al.*, 2001d, for a discussion). The total *extent* of the impact will be the total variation in flow resulting from adding together all the positive additions resulting from the various project features and subtracting all the negative values like abstractions and reductions in run-off. The resulting flow change can be *positive* (the most common) or *negative*, and the *significance* of the impact is determined by comparison with the natural oscillations in flow as measured in the data (Clarke, 1994):

1 If the flow change brings the total flow (normal average flow plus flow-change) to *within* the natural range of maxima–minima experienced in the past, the impact will *not* be significant:

   - The proportion of past-measurements within which the impact falls (50, 95, 99 per cent) can be used as a measure of the *degree of confidence* in the non-significance of the impact.

2 If the resulting total flow exceeds the maxima–minima range from past data, the impact will be considered *significant*:

   - The proportion of past-measurements the impact exceeds (50, 95, 99 per cent) can be used as a measure of the *degree of confidence* in the significance of the impact.

52 "TF-55 Urban Hydrology for Small Watersheds".

In this approach, *the baseline provides the standard of impact significance*; i.e. an impact will be considered significant only when it exceeds the "normal" oscillations of the baseline.

10.4.2 Water quality impacts

As we mentioned at the beginning of the chapter, water quality is one of the most regulated aspects in the UK and in Europe, and Morris *et al.* (2001d) provide quite comprehensive lists of European Directives and UK legislation relating to water quality. They also note two broad types of standards and regulations concerning water: (i) controls *at source* defining the limits of what a project can discharge to surface water systems, to the ground, to drainage systems, etc.; (ii) controls at the *reception point* in the water, defining the maximum concentrations of various substances allowed in waters of different types, the temperature ranges allowed, etc. It is likely that the controls of the first type will have been cleared with the controlling authorities (the Environment Agency, Her Majesty's Inspector for Pollution, British Waterways, etc.) by the project-developer *before* the impact assessment is undertaken, therefore those types of standards and possible infringements can be omitted from the study. What water quality impact assessment concentrates on is the composition of the water resulting from the actions of the project *after* the discharges.

After assuming away any impacts (infringements) at source, to look at the effect of pollution on the water we have to distinguish between levels of pollution *at the point of discharge* and the impacts at a distance away (*downstream*, in a flowing freshwater system). The two are related – in a way similar to how air-polluting emissions are related to air pollution downwind – and that relationship is usually simulated with *models* which try to replicate the behaviour of the pollution "plume" in the water. There is a considerable range of water-related models (for a good discussion, see York and Speakman, 1980) and many have been operationalised into software-packages (see York and Speakman, 1980, and also Morris *et al.*, 2001d). The type of "diffusion" models most commonly used addresses combinations of various types of "transport" and "transformation" processes (Atkinson, 1999) in the water, including:

- *Mixing*, as pollutants get diluted more and more as the water flows further from the source, a process dependent on the water flow and on the physical shape of the system; different types of simulations of this effect are used for flowing rivers, lakes, or sea waters, and some models also take *sedimentation* into consideration.
- Changes in the *sedimentation* of solids in the waterbed resulting from the current flow.
- Biological *respiration* (aerobic oxidisation) of the pollutants by the micro-organisms and bacteria in the water, a time-related process

(influenced also by temperature) which consumes dissolved oxygen (DO) in the water.
- *Reaeration* of the water, renovation of the levels of dissolved oxygen in the water from contact with the air at the surface and from biochemical processing by some species at the bottom.

Some models deal with the first effect ("Mixing Zone Models") and there are sedimentation models for rivers and for seawater. One of the most common types (used mainly for rivers) combines all these effects into a standard Streeter-Phelps model (so-called "Dissolved Oxygen Model") for a flowing stream of water (Figure 10.8), with its typical "dissolved oxygen" profile away from the pollution source (see York and Speakman, 1980, or Westman, 1985, for good discussions of all the elements of the model).

This type of simulation will make it possible to *measure* the likely levels of pollution at various distances (and times) away from the point of discharge, so that we can assess their *significance*. One problem related to this is to identify *which* standards apply in each case, which Atkinson (1999) considers part of the expertise in this field, and should be one of the first steps before the impact assessment as such. In addition to expertise, a source for the identification of those standards should be *consultation* with statutory and informed bodies:

- the Environment Agency, the main source of information and advice, in particular HMIP, if the project involves any "prescribed" substances[53];
- British Waterways, if canals are involved;
- the Local Authority.

In addition to the transgression of standards, a criterion to establish significance similar to one used for water quantity can be also applied to water quality (Clarke, 1994): the *comparison* of the levels of concentration

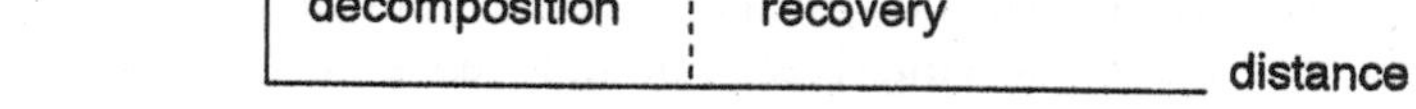

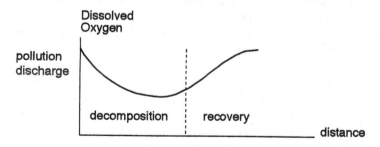

Figure 10.8 Dissolved Oxygen model for a flowing stream.

53 As listed in the Integrated Pollution Control regulations.

of various substances resulting from the project with the past levels as monitored in the baseline study:

- Compare the simulated effects from the project with the monitoring data from the Environment Agency to see if the new effects fall *outside* the range of variations experienced in the past few years.
- If they exceed the normal range of variation, the excess itself will provide a measure of the significance on a continuous scale.
- A more "absolute" measure of significance is to compare the predicted concentrations of pollutants for the water body in question with the requirements of the National Water Council Classification, to see if the new impacts make the *class* of the water system in question change *downwards.*

10.4.3 Water use impacts

In addition to the intrinsic impacts on water quantity and quality – which must meet their own standards and thresholds – variations in quantitative or qualitative aspects can affect *the use* of water:

1. Variations in water *quantity*:
   (a) an increased quantity of water can increase the possibility of *floods*, especially in flood-sensitive areas;
   (b) a reduced quantity of water can affect water leisure activities like boating, water-skiing or fishing.
2. Variations in water *quality*:
   (a) drinking water abstractions downstream can be affected;
   (b) water-contact sports (bathing, wind-surfing, etc.) also.
3. Variations in water *temperature*:
   (a) fisheries are sensitive to changes in temperature;
   (b) bathing can also be affected.

The possible *significance* of any of these impacts can be established in a variety of ways:

1. For quality-derived or temperature-derived impacts, there are usually well-defined standards:
   (a) for drinking water;
   (b) for bathing waters;
   (c) for fisheries.
2. For quantity-derived impacts on some water uses like boating or swimming – which require certain depths to be safe – there will be "thresholds" of significance if the water is likely to go below such critical depths.

3 Also, an essential fail-safe check on the potential significance of any of these impacts can be provided by *consultation* with major (non-statutory) users of water:

   (a) the public, interest groups;
   (b) Water Authorities;
   (c) industries abstracting downstream;
   (d) recreational groups;
   (e) angling clubs;
   (f) residents along the banks;
   (g) riparian farmers;
   (h) the Internal Drainage Board, if flooding can be a problem.

The three types of impacts on water – quantity, quality and use – can be assessed separately but they can also be linked sequentially, as the first two determine the third.

10.4.4 Water impacts mitigation

Impacts on surface water systems are similar to several of the impacts on hydrogeology already discussed, and they are produced by the same project features. Hence it is to be expected that the types of mitigation measures for water systems will quite closely resemble the mitigation measures for hydrogeology. Some water-mitigation measures involve slight modifications – of design or just the materials used – and some involve just improved site management (Figure 10.9):

1 Measures to contain polluting substances:

   (a) treatment units for particular contaminants;
   (b) bunding storage and oil tanks and placing them over impervious surfaces;
   (c) placing diesel/petrol powered fixed plant on impervious drip-trays.

2 Measures to reduce run-off and increase underground recharge:

   (a) remove as little vegetation as possible, and use replanting;
   (b) using gentle gradients in the terrain rather than steep slopes;
   (c) use porous surfaces when paving the ground.

3 Surface-flow retention measures:

   (a) detention/balancing ponds;
   (b) grass swales;
   (c) vegetated channels.

4 Better drainage:

   (a) routing all waste water to sewers or treatment works;

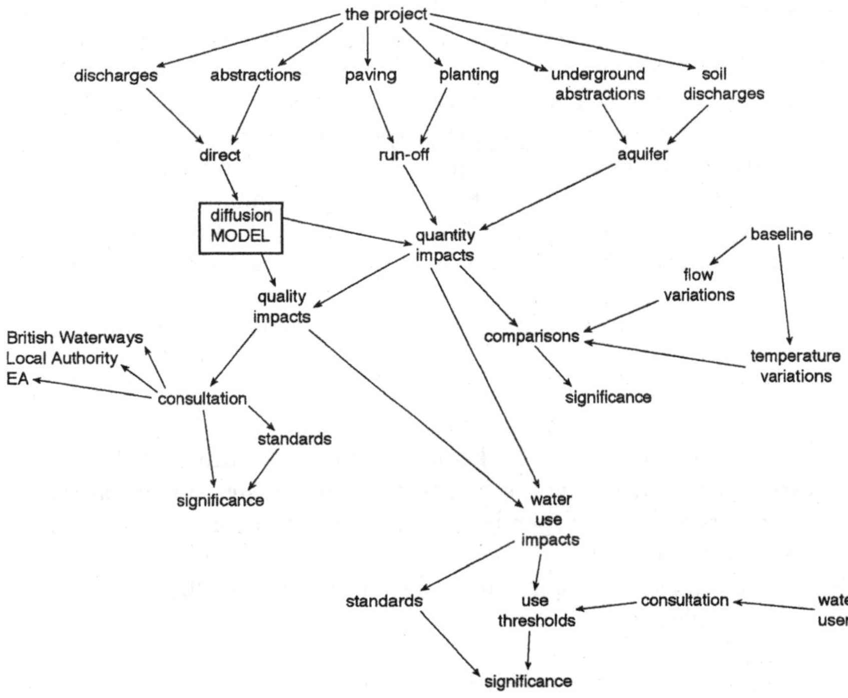

Figure 10.9 Mitigation of water impacts.

(b) site-drainage control to reduce polluted run-off and routing the drainage through siltation lagoons;
(c) using well dimensioned oil-interceptors before the drainage leaves the site.

5 Better site management:

(a) timing construction work to minimise impact on users of the surrounding water environment;
(b) guarding against spillages/leaks or accidental discharges;
(c) metering to minimise water use (if abstractions are a problem).

10.5 FRESHWATER ECOLOGY IMPACT ASSESSMENT: THE BASELINE

Impacts of a project on freshwater ecology are *indirect*: the project has certain impacts on water quantity/quality, and these changes in the water have an impact on the ecology of the surface water system (a river, a lake, a reservoir) and its banks. However, apart from this peculiarity, the logic of freshwater ecology impact assessment is very similar to that of terrestrial ecology (Clarke, 1994), starting from the impacts on the water quantity/quality and assessing the effects they are likely to have on the ecological baseline (Figure 10.10). As with terrestrial ecology, this is again an area of impact assessment characterised by the extent and complexity of the science behind it, but the sequencing of different tasks for the purposes of impact assessment is very similar to the terrestrial case (see Biggs *et al.*, 2001 for a good account).

The first stage is a baseline *desk study* based on published information – like national inventories that exist for certain types of ecosystems – and verbal consultation with well-informed organisations and interest groups:

- the Environment Agency (the most important source for the water ecology as such);
- English Nature;
- the Countryside Agency;
- the Institute of Terrestrial Ecology and the Institute of Freshwater Ecology, for the ecology on the banks of the water system, as well as for insects and birds dependent on the water system;
- Local Authorities;
- local naturalist trusts and interest groups like the RSPB;
- angling clubs.

The *area of study* for the field work is defined, extending in both directions: *downstream* from the point of impact because impacts "travel" with the water; *upstream* also, to see if there are any other similar projects affecting the water in similar ways (Clarke, 1994). In both directions, the study should extend:

- to the closest "monitoring location" of the Environment Agency and, if there are none;
- a minimum of 2 km and a maximum of 5 km depending on the size of the project and the sensitivity of the area.

Next, a *Phase 1 survey* is carried out in the field (a "walkabout") with the same aim as in terrestrial ecology of identifying habitats leading to:

- A *classification* of the habitat type following the Nature Conservancy Council methodology (JNCC, 1993) as discussed in Section 8.2.2 of Chapter 8 (see also Morris and Emberton, 2001a, and Appendix F in Morris and Therivel, 2001b).
- Determining the level of *interest and complexity* of the area to prepare for the next phase of the survey.

Phase 2 aims at establishing the ecological importance of the various habitat areas identified in Phase 1. The class given to that part of the water

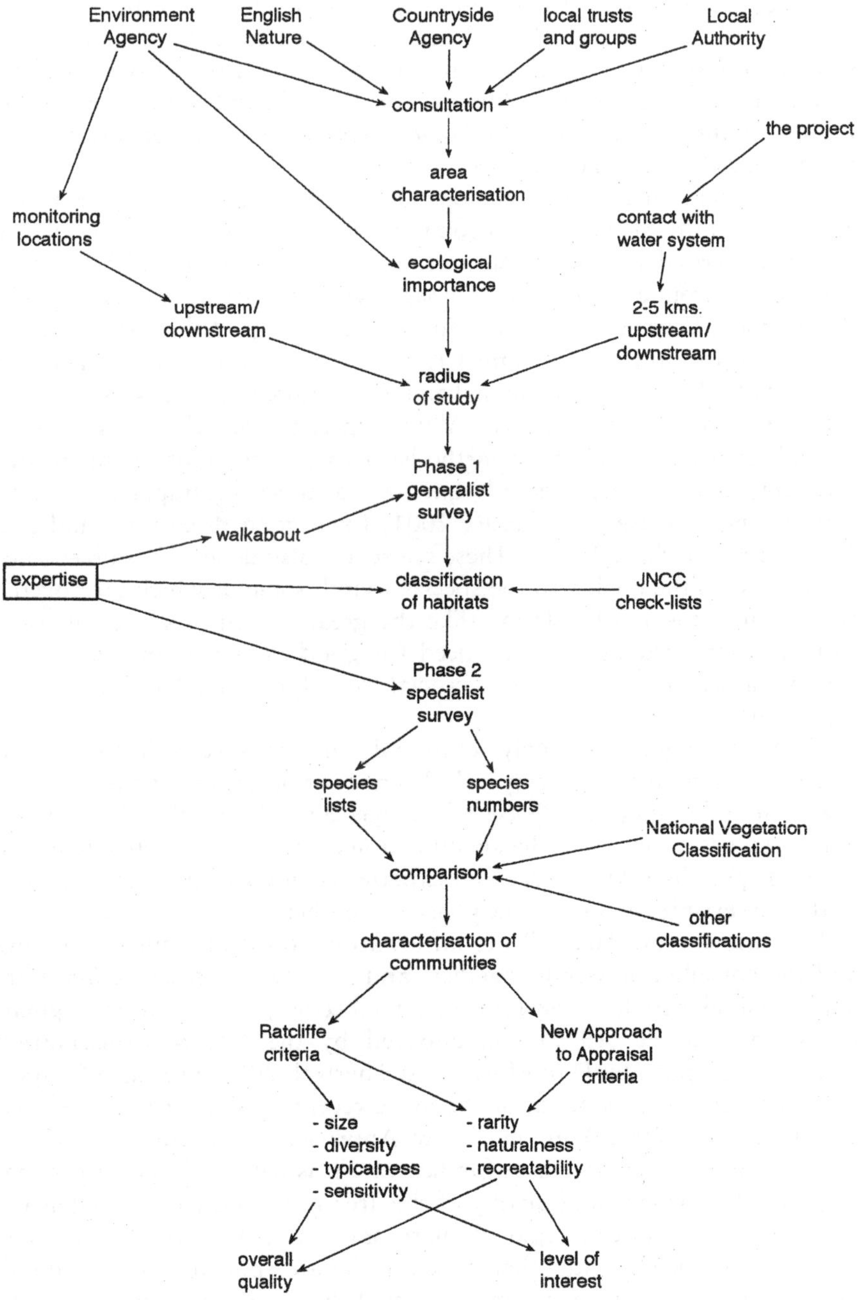

Figure 10.10 The freshwater baseline study.

system by the National Water Council Classification can be used for this purpose, or if the Phase 1 has identified areas of some interest/complexity, then *Phase 2 surveys* are indicated (see Biggs *et al.*, 2001, for a good discussion). These surveys can be done by: (i) using *indicators* of overall water quality like those discussed in Morris *et al.* (2001d); and/or by (ii) *recording the species* directly, including the range of species and numbers present and the presence of high-status species.

As discussed for terrestrial ecology (see Section 8.2.2 in Chapter 8), the aim is to identify the presence of communities and sub-communities, which are characterised by certain proportions of various species. The *vegetal* communities found can then be compared with standard classifications like the National Vegetation Classification (see also Biggs *et al.*, 2001, for a list of other classifications used in the UK), and their "conservation status" can be determined according to national and international standards (Appendices D and E in Morris *et al.*, 2001d). With respect to *animal* species, samples are collected using a range of methods: netting is the most common (for invertebrates, amphibians and fish) but other methods like trapping or fishing can also be used (see Biggs *et al.*, 2001, for a good discussion, and also Morris and Thurling, 2001c). These references also detail the sources to be used to assess the quality of the species found, some of which are covered by special status and legislation (like the great crested newt, or salmonid fish). Authors reiterate also the need for *good timing* when doing these surveys, as the time in the year is crucial (see Morris and Thurling, 2001c, for a yearly chart).

As in terrestrial ecology, only occasionally are rare species (like salmonids or trout, among fish) or a very rich diversity of other species (like invertebrates) found, requiring a *Phase 3* by a specialist. Again, the "bottleneck" at this stage is *expertise*, as the identification of species (more than habitats) requires specialised knowledge and experience, and this problem increases as the surveys progress from one phase to another.

After the survey stage, all the information collected from the surveys and the consultation is put together, and *quality assessment criteria* are applied, similar to those used for terrestrial ecology. The criteria originating from Ratcliffe (1977) and adopted by the Nature Conservation Review (see Appendix D in Morris and Therivel, 2001b; see also Chapter 8 in this book) can be used, or the more recent version (listed in Morris and Emberton, 2001a) in the "New Approach to Appraisal" (DETR, 1998) involving a shorter list of indicators. It is with such indicators that the overall environmental quality of the area is established, as well as the level of interest (local, regional, national, international) it may have. Again it must be reiterated that these criteria are applied in a more qualitative than quantitative way, and the usual three – *rarity*, *naturalness* and *recreatability* – tend to dominate (Clarke, 1994), as they reflect the irreversibility of any impact.

10.5.1 Freshwater ecology impacts and mitigation

Impacts on water ecology can be *direct*, involving physical change to the water environment, or *indirect*, by modifications to the water itself, its quantity, quality or use. The second type derive from considerations such as those covered in our previous discussion (see Section 10.4 above), the first type derive directly from the nature and features of the project, for example:

- construction of a dam/reservoir;
- altering the course of a river (sometimes temporarily during construction);
- building river-bank facilities (like promenades, or leisure facilities);
- building flood defences;
- building in-water facilities supported from the water bed (*not* floating).

Biggs *et al.* (2001) contains an extensive list – and discussion – of the variety of impacts on freshwater ecology and their sources. Whatever the source, the ecological impacts will manifest themselves in similar ways: a piece of construction may reduce the area of a particular habitat, and so can the flooding from an increase in water quantity, just as a decrease in water quantity can expose the banks and the habitats in them. Whether they come from a direct or indirect source, these impacts will take the form of either physical alterations to the habitats or changes in the quality of the water:

1. Physical alterations of the habitats and their species (from construction or changes in water quantity):
   - reduction in habitat area: by paving or building on it, by flooding it;
   - exposure of habitats from reductions in water quantity/level or from modifications in the water course: exposure of banks, desiccation of water areas.
2. Changes in the quality of the water:
   - changes in temperature;
   - pollution.

As with terrestrial ecology, in the case of physical changes affecting a certain extension of an aquatic habitat (all of it, or only part), the *percentage area* affected can be used as an indicator of the importance of the impact (Figure 10.11). If habitats have been mapped (maybe with a GIS), this extension can be measured by superimposition of the project-map on to the habitats map, using simple GIS functions.

To determine the *significance* of such impacts is not easy, as the ecological effects may not be proportional to the extent of the area affected (Biggs

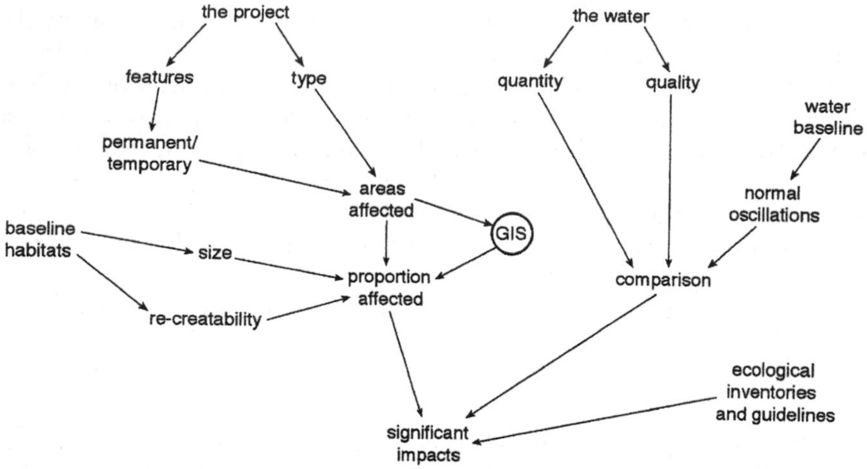

Figure 10.11 Freshwater ecology impacts.

et al., 2001). The area of habitat left may not be sufficient to maintain certain species, which would disappear altogether; or some species, like some fish, may need to migrate between several different habitats, and if one is disturbed, the whole process is damaged. To assess the significance *in general terms*, we can follow the same logic used for terrestrial ecology (Section 8.2.4 in Chapter 8) based on the combination between how "permanent" the impact is and how "recreatable" the lost habitat (or the affected species) is.

With respect to alterations in water quality, the standards of significance will depend on the species affected, and while some species are protected by clear standards (like salmonides) others are not, and the problem is to know what standards to apply (Clarke, 1994). As with habitat loss, to assess the "general" level of impacts we can compare the changes in water conditions with the normal oscillations shown in the monitoring data – the "*natural range of perturbation*" (Biggs *et al.*, 2001) – as already mentioned when discussing the impacts on water quality (Section 10.4.2 above). If the alterations are within the range of normal oscillations, they will be considered non-significant and the reverse will apply if they exceed the natural range.

With respect to *mitigation* measures, we have to distinguish between (i) impacts on ecology that derive from changes in the water; and (ii) impacts that derive directly from the project (an extensive discussion can be found in Biggs *et al.*, 2001). The literature on water-ecology mitigation tends to concentrate on the first type, such as controlling pollution, reducing discharges, maximising aquifer-recharge. Most of this type of mitigation has already been discussed in Sections 10.3 and 10.4 of this chapter when dealing with water impacts (quantity, quality, use). The few mitigations of this type left to mention, overlap with the second type (direct impacts), and they are similar to those already mentioned when dealing with terrestrial ecology in Chapter 8:

- "impact avoidance" by monitoring the environmental situation in areas nearby;
- timing of construction-operations to avoid the breeding season;
- "compensation" by recreating lost habitats somewhere else;
- recreating the communication between parts of habitats which have become fragmented;
- restoration of the habitat-areas left;
- protection of particular species, for instance installing fish-ladders and screens in any abstraction pipes to prevent the fish being sucked in.

10.6 COASTAL WATER ECOLOGY IMPACT ASSESSMENT: THE BASELINE

Impacts on marine water ecology are also *indirect*, derived from the impacts of the project on the *quality* of the water. Changes in water quantity do not apply to seawater in the same way as to freshwater. Certain projects – by their very nature – may involve the desiccation of some areas covered by water (for instance projects for the reclamation of inter-tidal land), but the impacts from those projects will derive less from the change in "quantity" of water, but more from changes in the extension of land covered by it. Apart from this, much of what was said about freshwater ecology impacts can be repeated here, replicating once again the general approach to ecological impact assessment (Ackroyd, 1994),[54] starting from the baseline and the water-quality impacts and assessing their likely effects on the coastal water ecology (Figure 10.12). As with other ecological impacts, this area of impact assessment is also characterised by the complexity of the science that supports it. Thompson and Lee (2001) contain a good discussion of the whole field, and also contain a comprehensive list of legislation, policies and guidance publications in this field, the most important being the Planning and Policy Guidance Note 20 (DoE, 1992) and the National Planning and Policy Guidance Note 13 (SO, 1997), both with the title "Coastal Planning".

The first stage – a baseline *desk study* – is based in the first instance on general information sources like *aerial photographs* and *bathymetric charts*, and also on geological/coastal *maps* that provide good national "inventories"

54 The knowledge acquisition for this part was greatly helped by conversations with Dave Ackroyd, of Environmental Resources Management Ltd (Oxford branch); Andrew Bloore helped with the compilation and structuring of the material for this part. However, only the author should be held responsible for any inaccuracies or misrepresentations of views.

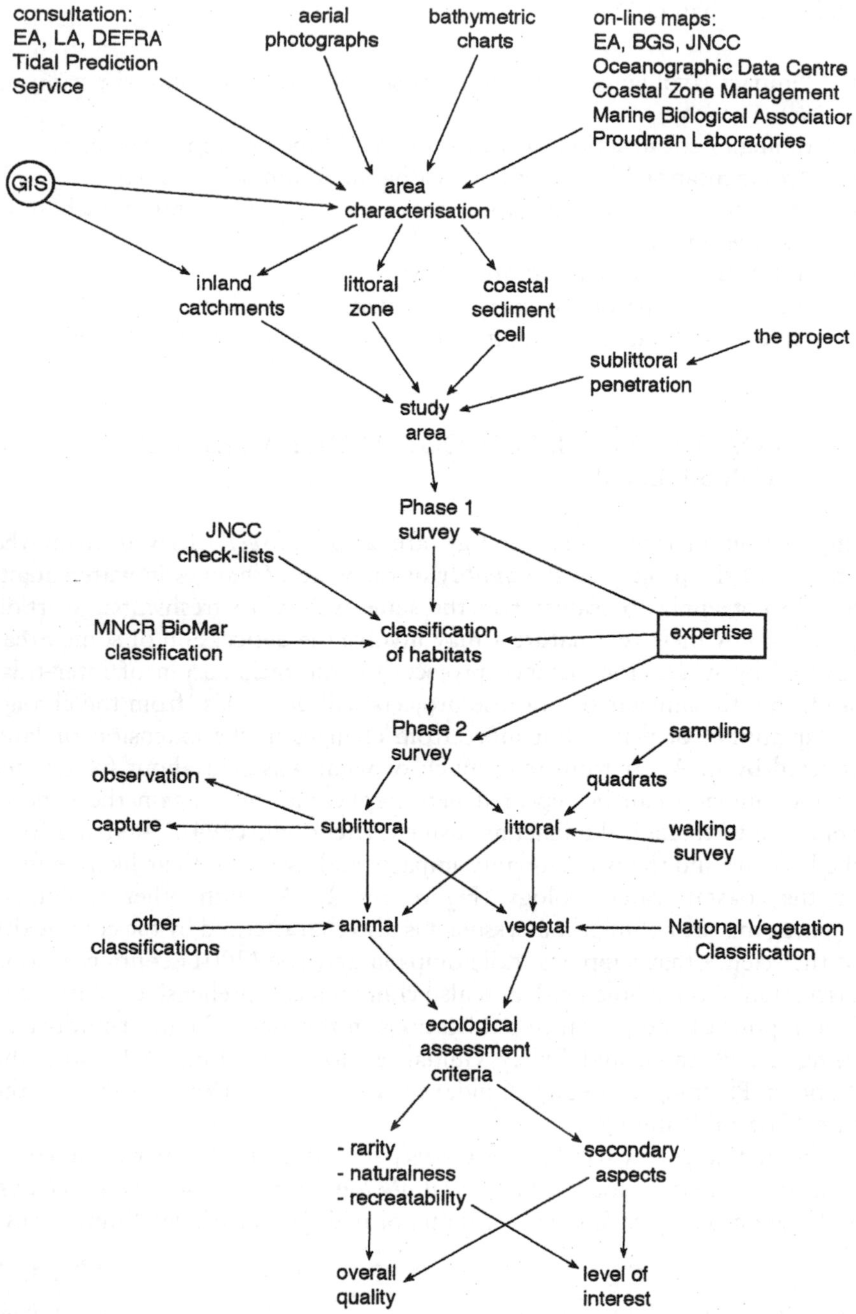

Figure 10.12 Baseline study of coastal water ecology.

of the ecological resources. Thompson and Lee (2001) list a series of such maps available *online* from a variety of sources (* indicates those that we have already encountered when dealing with other impacts), including:

- the Environment Agency (and its counterparts in Wales, Northern Ireland and Scotland) (*);
- British Geological Survey (*);
- Joint Nature Conservancy Committee (*);
- British Oceanographic Data Centre;
- Coastal Zone Management Centre;
- Marine Biological Association;
- Proudman Oceanographic Laboratories (part of the Centre for Coastal and Marine Services).

In conjunction with collecting map-based information, the baseline study is based on *consultation* with key organisations about the resources in the area and their relative ecological importance:

- the Environment Agency (*);
- Local Authorities (*);
- local naturalist trusts and interest groups (*);
- DEFRA (the Department for Environment, Food and Rural Affairs);
- the Tidal Prediction Services.

The *area of study* for coastal water ecology is difficult to define because the "impact area" has quite indeterminate boundaries, both towards the sea and inland (Thompson and Lee, 2001):

- *Inland*, the rainwater *catchment* area – defined by the land sloping towards the coastline – can be used to define the area of study, and GIS can help define the limits of that sloping land (with a Digital Terrain Model).
- *Towards the sea*, the study area should at least include the littoral (inter-tidal) zone – a few hundred metres at the most – as defined in the charts, and how far into the sub-littoral zone it is advisable to extend it will depend on the nature of the project: how far into the sea its structures extend, how deep its effects are likely to get diffused.
- The *lateral extent* of the study area should be determined by "coastal sediment cells" and management plans (under the auspices of DEFRA) which may apply to the area (Thompson and Lee, 2001).

A *Phase 1 survey* is carried out as usual, and the Nature Conservancy Council methodology (JNCC, 1993) discussed in Section 8.2.2 of Chapter 8 (see also Morris and Emberton, 2001a, and Appendix F in Morris and Therivel, 2001a) can be used for the identification and classification of habitat types. But, because this classification does not cover the sublittoral zone, Thompson and Lee (2001) argue that it is best to use the MNCR BioMar classification based on biotypes (Connor *et al.*, 1997a and 1997b; Picton and Costello, 1997) for both littoral and sublittoral surveys.

Phase 2 aims at identifying and quantifying species and communities of various types (see Thompson and Lee, 2001, for a good discussion):

1 For *coastal and littoral* (inter-tidal) species, the approach is the same as for terrestrial ecology (see Section 8.2.2 in Chapter 8), as these areas can be "walked":

   (a) *vegetal* communities found (by "walkabout") can then be compared with standard classifications like the National Vegetation classification;
   (b) with respect to *animal* species, samples are collected using *quadrats*, and *birds* are surveyed from the shore using the instructions from the British Trust for Ornithology;
   (c) the "conservation status" of the species found can be determined according to national and international standards (Appendices D and E in Morris *et al.*, 2001d).

2 For *sublittoral* areas (if needed) the problem is that they cannot be walked:

   (a) for *benthic* species (on the seabed), special equipment and personnel must be used;
   (b) *pelagic* species (free-swimming and floating) present similar problems and can be surveyed by "capture" or simply by observation:

      (i) vegetal *plankton* can be sampled by netting or also by detection from very expensive – hence impractical – aerial and satellite imagery;
      (ii) the presence of *fish* can also be recorded with photography or by fishing (with traps, nets, hook and line);
      (iii) marine *mammals* can be detected with similar methods, but are the most difficult to quantify.

The baseline is assessed with the usual ecological criteria (rarity, naturalness, recreatability, see previous sections and Chapter 8) and its level of interest (local, regional, national, international) is also established. Thompson and Lee (2001) add the importance of considering "secondary" roles that some habitats can play, as *buffers* between terrestrial and marine systems: for example sand-dunes preventing saline intrusion, or saltmarshes acting as oil-traps for spillages.

10.6.1 Coastal water ecology impacts and mitigation

As with freshwater ecology, impacts on coastal water ecology can be more or less *direct*, depending on the extent to which they are caused by the project itself or by its effects on the local hydrogeology or water quality.

There are many different types of impacts that derive from various types of projects (see Thompson and Lee, 2001, for a good discussion) which can be short-listed as:

- loss of habitat;
- changes in water quality: pollution, change in salinity, temperature increases, increased suspended solids, with increased turbidity and light attenuation;
- physical changes to the water: alteration of tidal activity, changes in sedimentation rates and patterns.

Thompson and Lee (2001) also contain a useful list of modelling software used for various areas of *impact prediction*, although it comes with a cautionary note about the use of models on the grounds of how expensive they are and the uncertainty associated with their results.

As before, the *percentage area* affected can be used as an indicator of the importance of the impact (Figure 10.13). If habitats have been mapped (may be with a GIS), this extension can be measured by superimposition of the project map on to the habitats map, using simple GIS functions. The *significance* of such impacts *in general terms* can be determined following the same logic used before (Section 8.2.4 in Chapter 8) combining the "permanence" and "recreatability" of the habitat or species affected. Thompson and Lee (2001) suggest basing the assessment of significance on:

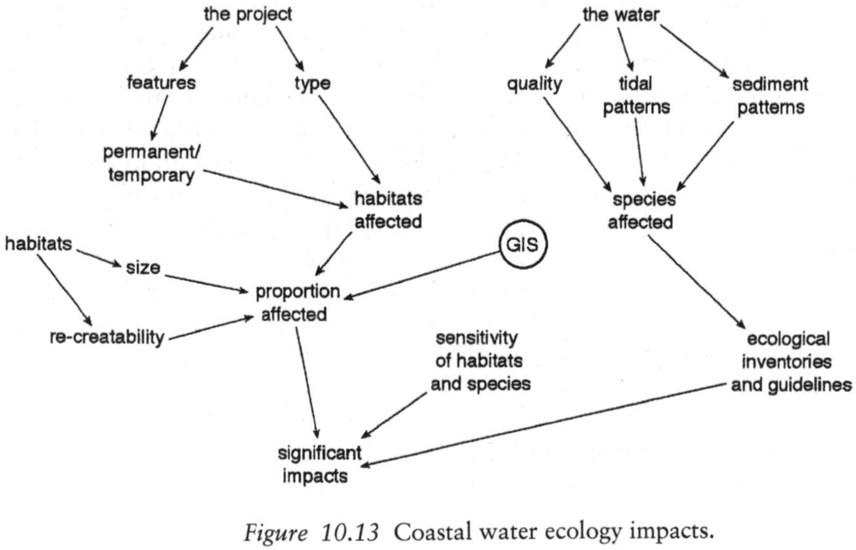

Figure 10.13 Coastal water ecology impacts.

1 sensitivity of the habitat or species affected;
2 how the environment is likely to respond or recover (related to its "recreatability"):

- short-term disturbance;
- long-term disturbance;
- catastrophic disturbance (destruction).

In the UK, *mitigation* measures for coastal-water impacts tend to be based on *engineering* (Ackroyd, 1994), but they are on the whole very similar to other water impact and ecology mitigations we have encountered (Ackroyd, 1994; Thompson and Lee, 2001):

1 At source, *controlling emissions* and treating pollutants before discharge:

- applying nutrient stripping to sewage;
- disinfecting sanitary discharges.

2 Good *management* of the construction/operation of the project:

- using sensible construction methods, like floating platforms;
- dredging only during ebbtide periods;
- excluding vehicles from sensitive areas (like sand dunes);
- protection of coastal areas, for example by re-locating certain activities.

3 Restoration of lost habitats/species by replanting, replacing or re-stocking.

10.7 IMPACT SEQUENCES

We have discussed in this chapter the logic of a number of new areas of impact assessment, which show many of the features already seen in other areas. Hydrogeology is strong on science and can use some modelling (where GIS can help to a limited extent), water is – like noise or air pollution – strong on modelling – and water ecology (fresh or otherwise) resembles closely terrestrial ecology, with a small potential contribution from GIS. Each of these areas have their own "difficulties", some of which reflect potential problems of automating them into expert systems. In *hydrogeology* (Simonson, 1994) the main difficulty in a case can derive from a problematic geology – like fractured bedrock that makes unpredictable which way the water will flow and at what rate – but other typical issues (more related to the "sensitivity" of the case than to its difficulty) usually require the expert's experience:

- previous contamination of the land from previous landuses;
- the presence of a sensitive/vulnerable major aquifer;

- a particularly sensitive client (developer);
- a particularly sensitive Local Authority.

In *water ecology* (Clarke, 1994; Ackroyd, 1994) the greatest difficulty appears in cases where mitigation is difficult (or impossible, when the project is already finalised) or involves measures which are still untested. Other aspects also make cases non-standard:

1 A case can be "big" when:

  - it involves a major physical change of the water environment (a dam, re-routing a river);
  - it affects a sensitive water environment, although "*water is almost always sensitive*" (Clarke, 1994).

2 The difference between expert and "novices" can manifest itself in typical problems of judgement by the latter, including:

  - some of the broad initial questions, like identifying the right standards;
  - accepting the results from a model without thinking;
  - over-emphasising the importance of an impact which is only of temporary significance (water environments are re-colonised quite rapidly);
  - not consulting all the recreational authorities involved when dealing with water use.

Also, what we have encountered repeatedly has been the issue of direct and indirect impacts, as already mentioned in Chapter 9 in relation to traffic impact assessment. A project can generate traffic (a *direct* impact) that, in turn, generates other *indirect* impacts like air pollution or noise (Figure 10.14).

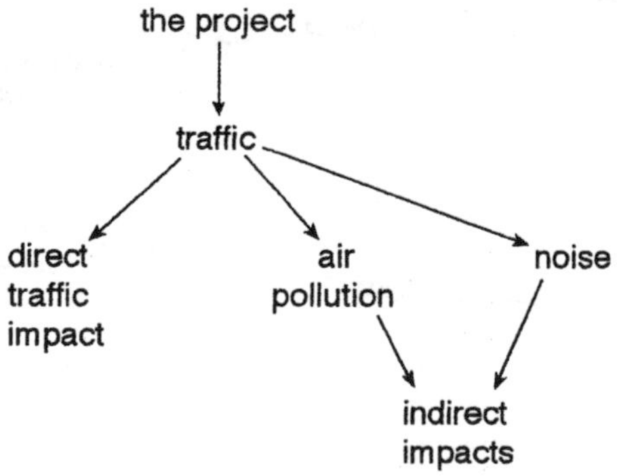

Figure 10.14 Direct and indirect impacts.

In this chapter we have extended this logic into a more complex sequence affecting water (see Section 10.1 in this Chapter), but the logic of the "sequencing" impacts is the same. One of its implications is that the mitigation of an "end-of-chain" impact (like an ecological impact) can focus on any of the stages in the chain: mitigating the project source of the impact or mitigating any of the intermediate impacts. The impacts that are transmitted "down the chain" are the *residual* impacts from the previous link after mitigation (Figure 10.15). For example, for

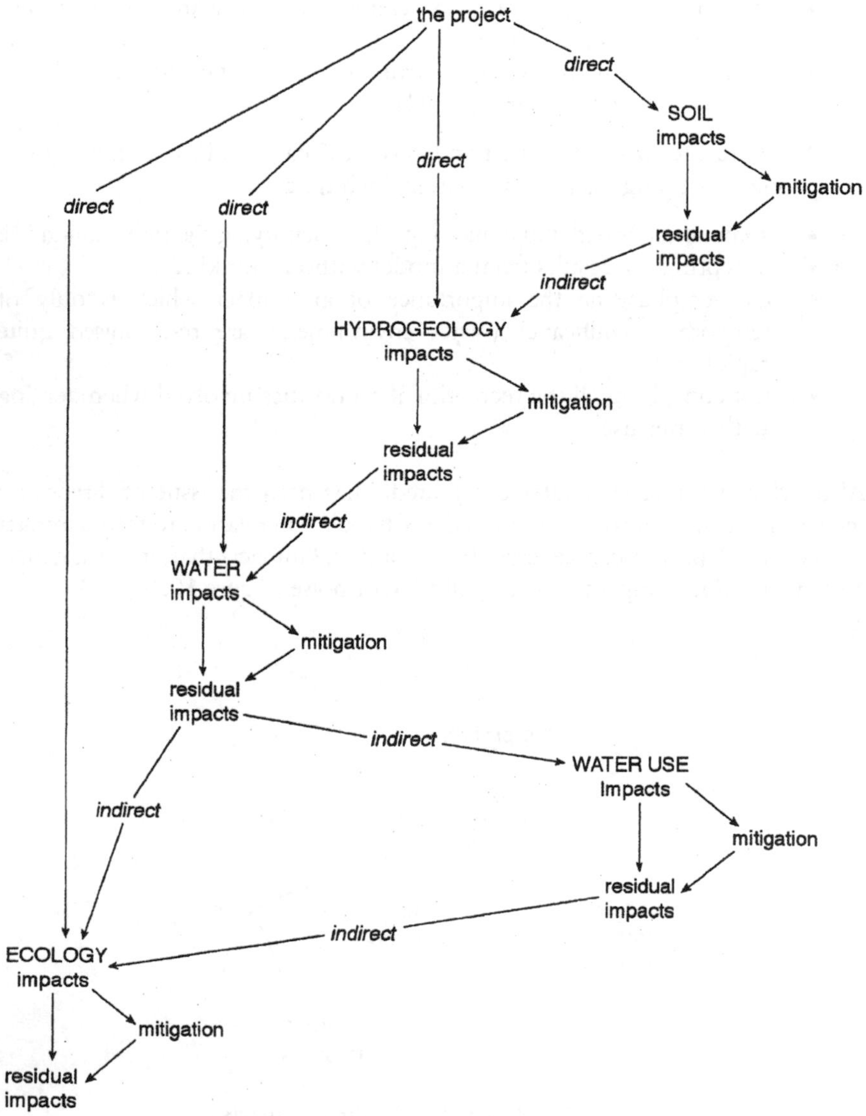

Figure 10.15 Direct and indirect water impacts.

water impacts, mitigation "at source" tends to be preferred in the UK, while mitigation at the "dispersal" end tends to be preferred in Europe (Ackroyd, 1994).

Each area of impact assessment has its own internal logic, but some parts are shared (such as the project description) and some common structures are replicated, like the logic of direct-indirect impacts or the logic of impacts–mitigation–residual impacts. Looking at impact assessment and its possible automation in this light suggests the advantages of using a *modular* approach to impact assessment based on the "parts" of different impact assessment areas which, although different in the information they use and in some of the details, share a common logic and sequential structure. We shall say more about all this in the last chapter.

REFERENCES

Ackroyd, D. (1994) *Personal communication*, Environmental Resources Management Ltd, Oxford.

Atkinson, S. (1999) Water Impact Assessment, J. (ed.) *Handbook of Environmental Impact Assessment*, Blackwell Science Ltd, Oxford (Vol. 1, Ch. 13).

Biggs, J., Fox, G., Nicolet, P., Whifield, M. and Williams, P. (2001) Freshwater ecology, in Morris, P. and Therivel, R. (eds) *Methods of Environmental Impact Assessment*, Spon Press, London, 2nd edition (Ch. 12).

Bourdillon, N. (1995) *Limits & Standards in Environmental Impact Assessment*, Working Paper No. 164, School of Planning, Oxford Brookes University.

CEH (1999a) *Flood Estimation Handbook: Procedures for Flood Frequency Estimation*, Vol. 2: "Rainfall Frequency Estimation"; FEH CD-ROM Version 1, Centre for Hydrology and Ecology, Wallingford.

CEH (1999b) *Flood Estimation Handbook: Procedures for Flood Frequency Estimation*, Vol. 5: "Catchment Descriptors"; FEH CD-ROM Version 1, Centre for Hydrology and Ecology, Wallingford.

Clarke, S. (1994) *Personal Communication*, Environmental Resources Management Ltd, Oxford.

Connor, D.W., Brazier, D.P., Hill, T.O. and Northern, K.O. (1997a) *MNCR Marine Biotope Classification for Britain and Ireland*, Vol. 1: "Littoral Biotypes", Version 97.06, JNCC Research Report No. 229, Joint Nature Conservation Committee, Peterborough.

Connor, D.W., Dalkin, M.J., Hill, T.O., Holt, R.H.F. and Sanderson, W.G. (1997b) *MNCR Marine Biotope Classification for Britain and Ireland,* Vol. 1: "Sublittoral Biotypes", Version 97.06, JNCC Research Report No. 230, Joint Nature Conservation Committee, Peterborough.

DETR (1998) *Guidance on the New Approach to Appraisal*, Department of the Environment, Transport and the Regions.

DoE (1992) *Coastal Planning*, Planning and Policy Guidance Note 20, HMSO, London.

EA (1998) *Policy and Practice for the Protection of Groundwater*, Environment Agency, Bristol.

Eduljee, G.H. (1992) Assessing the risk of landfill activities, in *New Developments in Landfill* (*Proceedings of the 1992 Harwell Waste Management Symposium*), Environmental Safety Centre, AEA Environment & Energy, Harwell, Oxfordshire, UK.

EU (2002) *Water Framework Directive: Guiding Principles on Technical Requirements.*

Hodson, M. (1995) Soils and geology, in Morris, P. and Therivel, R. (eds) *Methods of Environmental Impact Assessment*, UCL Press, London, 1st edition (Ch. 9).

Hodson, M.J., Stapleton, C. and Emberton, R. (2001) Soils, geology and geomorphology, in Morris, P. and Therivel, R. (eds) *Methods of Environmental Impact Assessment*, Spon Press, London, 2nd edition (Ch. 9).

JNCC (1993) *Handbook for Phase 1 Habitat Survey – A Technique for Environmental Audit* (separate *Field Manual* also available), Joint Nature Conservation Committee, Peterborough.

Morris, P. and Emberton, R. (2001a) Ecology – overview and terrestrial systems, in Morris, P. and Therivel, R. (eds) *Methods of Environmental Impact Assessment*, Spon Press, London, 2nd edition (Ch. 11).

Morris, P. and Therivel, R. (eds) (1995) *Methods of Environmental Impact Assessment*, UCL Press, London.

Morris, P. and Therivel, R. (eds) (2001b) *Methods of Environmental Impact Assessment*, UCL Press, London (2nd edition).

Morris, P. and Thurling, D. (2001c) Phase 2–3 ecological sampling methods, in Morris, P. and Therivel, R. (eds) *Methods of Environmental Impact Assessment*, Spon Press, London, 2nd edition (Appendix G).

Morris, P., Biggs, J. and Brookes, A. (2001d) Water, in Morris, P. and Therivel, R. (eds) *Methods of Environmental Impact Assessment*, Spon Press, London, 2nd edition (Ch. 10).

Petts, J. and Eduljee, G. (1994a) "Ground and Surface Water", in *Environmental Impact Assessment for Waste Treatment and Disposal Facilities*, John Wiley & Sons, Chichester (Ch. 10).

Petts, J. and Eduljee, G. (1994b) "Geology and soils", in *Environmental Impact Assessment for Waste Treatment and Disposal Facilities*, John Wiley & Sons, Chichester (Ch. 9).

Picton, B.E. and Costello, M.J. (eds) (1997) *BioMar Biotype Viewer: A Guide to Marine Habitats, Fauna and Flora of Britain and Ireland* (CD-ROM), Environmental Sciences Unit, Trinity College, Dublin.

Ratcliffe, D.A. (ed.) (1977) *A Nature Conservation Review*, 2 Vols, Cambridge University Press, Cambridge.

Simonson, J. (1994) *Personal Communication*, ERM Enviroclean Ltd, Oxford.

S.O. (1997) *Coastal Planning*, National Planning and Policy Guidance Note 13, Scottish Office, Edinburgh.

Thompson, S. and Lee, J. (2001) Coastal ecology and geomorphology, in Morris, P. and Therivel, R. (eds) *Methods of Environmental Impact Assessment*, Spon Press, London, 2nd edition (Ch. 13).

USEPA (1988) *Superfund Exposure Assessment Manual*, United States Environmental Protection Agency, EPA/540/1-88/001, Office of Remedial Response, Washington DC.

Westman, W.E. (1985) *Ecology, Impact Assessment and Environmental Planning*, John Wiley & Sons (Ch. 7: "Air and Water").

York, D. and Speakman, J. (1980) Water Quality Impact Analysis, in Rau, J.G. and Wooten, D.C. (eds) *Environmental Impact Analysis Handbook*, McGraw-Hill (Ch. 6).

11 Reviewing environmental impact statements

11.1 META-ASSESSMENT: REVIEWING ENVIRONMENTAL STATEMENTS

In an impact assessment study, after the various areas of impact are studied, all the "threads" are joined again to arrive at an overall assessment, and the whole discussion must be presented in a report containing the main points of all the areas covered, following specific guidelines. The report as a whole is also the subject of scrutiny as part of the control process, and this is discussed in this chapter. The review of environmental statements is really a completely different stage in the process and requires also a completely different approach, which we can use now to finish our discussion on different types of expert systems for impact assessment.

Reviewing impact-assessment reports can be labelled *meta-IA* as it involves "assessing the assessments" of impacts. Advice on how to prepare Environmental Statements has been forthcoming from the Department of the Environment (DoE, 1995), and the "reverse" task of assessing the quality of such statements has also been progressively explored in increasing depth and detail. Early attempts by Lee and Colley (1990) and by the Commission of the European Communities (CEC, 1993) were followed by the first good-practice guide from the Department of the Environment (DoE, 1994), culminating in the definitive report (DoE, 1996) – which we shall take here as our main source – based on research at Oxford Brookes University on the changing quality of Environmental Statements (Glasson *et al.*, 1997). Weston (2000) has also reviewed this question under the light of the new 1999 legislation, but without developing the methodology any further.

The task of reviewing impact-assessment reports can be summarised as *assessing the completeness and presentation* of such reports. It does not involve assessing the project impacts or the acceptability of the project – that will be for the development control procedures to determine – but assessing the impact report *as a public document* in its suitability for use in that development control process. An obvious implication for our discussion is that GIS is unlikely to play a part in this evaluation, which only involves "comparing" the different parts of the report with some preconceived ideas about what they should contain and how they should be presented.

The assessment must involve many different criteria which have to be assessed on their own merit, and whose partial assessments must be assembled into some form of overall evaluation of the qualities and deficiencies of the report as a whole, following some form of so-called *multi-criteria evaluation* (Voogd, 1983). The questions raised by such an approach can be grouped under several different headings:

1 identifying the aspects to be assessed and how to extract the relevant information about each aspect;
2 determining how to assess each aspect of the report;
3 deciding the nature of any – one or several – overall evaluation(s) of the report to be derived from the partial evaluations of the various aspects;
4 specifying how the partial evaluations are to be combined into the overall evaluation(s).

The details of how these questions are addressed vary depending on the approach. The "Environmental Impact Statement Review Package" – by the Impacts Assessment Unit at Oxford Brookes University (DoE, 1996; Glasson *et al.*, 1999; Weston, 2000) – presents an *expert* assessor with a checklist of criteria grouped under main headings and divided into sections, and the assessor is expected to "score" each of the aspects of the report and, at the end of each section, also the overall worth of that section. Similarly, at the end, the assessor is expected to provide an overall mark for the whole report and to list overall good and bad points for various agents/groups involved: the developer, the local authority, the public, etc. In this approach, the fact that the assessor is an expert has implications for the methodology:

- The assessor can be asked to *judge directly* each of the different aspects of the report and put "scores" to those judgements.
- Similarly, the assessor can be asked to develop in the process an *overall impression in the assessor's mind* of different groups of aspects ("sections").
- Finally, an impression of the *overall quality of the report* is also formed in the assessor's mind as he/she progresses through the detailed evaluation.

If, on the other hand, we are to use an *expert-systems approach* to the same problem, such an approach would be designed for *use by non-experts*, and this has fundamental implications for every step in the process. A non-expert cannot be expected to form opinions to the same extent as an expert about the quality of various aspects of the report, let alone the overall quality of whole sections of the report or of the report as a whole. The aspects of the report for the assessor to consider must be converted into relatively simple *questions* for the expert-system user, and the expert-system designer must "translate" as much as possible any evaluative steps involved into "statements of fact" or descriptive questions which a non-expert can answer. Such questions can sometimes be *yes–no* questions to deal with sharply defined issues ("*Does the Statement contain maps/diagrams describing the project?*"), and the answers to such questions can be converted into "scores" automatically (for example *yes* is "good" and *no* is "bad"). Sometimes it is better for questions to offer a wider range of answers (typically in the form of a *menu*) to which a "scale" of possible scores can easily be attached. For example, the aspect described in the already mentioned ES Review Package by the phrase "*Indicates the nature and status of the decision(s) for which the environmental information has been prepared*" can translate into an expert-system menu-question like:

Does the Statement indicate the nature and status of the decision(s) for which the information has been prepared?

- *yes, in detail*
- *it only refers to a planning application being submitted*
- *it does not specify what type and status of decision it is prepared for.*

This is an example where the question is simple enough and at the same time the three possible answers provide the rudiments of a "scoring scale" from best to worst that is *sufficiently* detailed for our scoring purposes. The approach should avoid starting at the lowest level of the assessment using scales whose excessive detail will get lost once the scores of all the different aspects begin to combine. This aspect of expert-system design requires considerable judgement on the part of the designer, as a balance must be struck between *accuracy* – trying to reflect the true quality of the aspect being assessed – and *simplicity* of the options offered to the non-expert. Extensive lists of options could be offered to deal with every aspect, showing all the possible nuances reflecting the different levels of quality that could be present, but that could make the job of answering those questions excessively onerous for the non-expert. Also, the accumulation of such questions would make the whole evaluation process too long and complicated, and therefore impractical. No simple rule can be suggested to solve this problem,[55] but it is the designer's job to provide sufficient range of possible answers to cover a meaningful scale of "qualities" while at the same time making it easy for the non-expert to understand the answers and the differences between them.

Questions can also take the form of *lists* (multiple-choice menus) from which the user picks out the items, which are relevant: for instance, the item that in the ES Review Package reads "*Indicates the methods by which*

55 This is a variation of the well-known "Law of Requisite Variety" by Ross Ashby (Ashby, 1956), which postulates that an information system should not define its information with a level of detail greater than it is capable of processing.

the quantities of residuals and wastes were estimated, acknowledges any uncertainty, and gives ranges or confidence limits where appropriate" can be broken down into several (one for each project-stage) which in turn can translate into list questions like:

Does the Statement discuss the calculations used to estimate quantities of waste and/or residual materials expected during the construction stage?

- *it indicates the methods used to calculate them*
- *it defines levels of uncertainty associated with the calculations*
- *it gives ranges of confidence limits for the results*
- *none of the above (if this one is chosen, all other choices are excluded)*

The user selects one or several of these options, and the scoring for the purposes of evaluation derives from the combinations of the scores attached to each of the choices.

As can be seen, the transition from a review procedure for experts to an expert-system for non-experts requires a process of translation into simple questions of each and every aspect assessed, and this *simplification* often means that *one* aspect to be assessed translates into *several* questions in the expert system. For example, the item that reads in the ES Review Package "*Describes any additional services (water, electricity, emergency services, etc.) and developments required as a consequence of the project*" can translate (applied to every project stage) into *two* questions for the user:

Does the Statement say if any additional services or development will be required during the construction/preparation stage?

- *no, it does not say*
- *this type of project wouldn't need any additional services/developments*
- *yes, it specifies which additional services/developments it requires.*

(then, if the answer to the previous question is no. 3)

Which additional services or developments will be required during the construction/preparation stage?

- *water*
- *electricity*
- *gas*
- *sewage disposal*
- *waste disposal*
- *additional infrastructure (roads, etc.) to be built*
- *other developments*
- *emergency services*
- *none of the above (if this one is chosen, all other choices are excluded).*

The range of possible variations extends beyond what can be discussed here[56] but the point of this discussion is to show how the translation of aspects to be studied into useful questions for the expert-system user is not trivial and requires careful consideration by the expert-system designer. The following sections now discuss in greater detail the aspects that such an expert system should cover.

11.2 THE BUILDING BLOCKS OF THE ASSESSMENT

The best-practice ES Review Package (DoE, 1996; Glasson *et al.*, 1999; Weston, 2000) groups all the aspects to be reviewed into a structure of numbered sections and headings:

1 Description of the development
- principal features of the project;
- land requirements;
- project inputs;
- residues and emissions.

2 Description of the environment
- description of the area occupied by and surrounding the project;
- baseline conditions.

3 Scoping, consultation and impact identification
- scoping and consultation;
- impact identification.

4 Prediction and evaluation of impacts
- prediction of magnitude of impacts;
- methods and data;
- evaluation of impact significance.

5 Alternatives

6 Mitigation and monitoring
- description of mitigation measure;
- commitment to mitigation and monitoring;
- environmental effects of mitigation.

56 The above examples have been taken from the REVIEW prototype expert system developed at Oxford Brookes University by Agustin Rodriguez-Bachiller for teaching/demonstration purposes, and which is "attached" to the SCREEN and SCOPE prototypes mentioned in Chapter 6.

7 Non-technical summary
8 Organisation and presentation of information

- organisation of the information;
- presentation of information;
- difficulties in compiling the information.

The assessment of each detailed aspect of the report is likely to take the form of one or several of the following:

- stating whether a particular piece of information or elaboration is *present/absent*;
- establishing if certain topics are treated in the *depth/detail* they deserve;
- assessing the *quality of presentation* of the various items.

When translating these aspects into expert-system questions covering *all* possibilities, however, the number of aspects to be covered is multiplied, as some of these areas of investigation must be *repeated* several times:

- some areas of enquiry apply to *each stage* of the project (at least construction and operation);
- some areas of enquiry apply to *each type of impact* covered in the report (noise, air, ecology, etc.).

This type of assessment is based implicitly on a *comparison* between the Statement being reviewed and some form of "ideal" Statement, a kind of "template". However, different parts of this template can originate from very different sources. The need for certain general aspects (maps, summary reports, user-friendly explanations, etc.) can be derived from a *generic* checklist like the one provided by the ES Review Package, and all the expert system needs to do is translate its elements into a series of simplified questions. On the other hand, there are other aspects that are *specific* to each project for which no complete checklist could be prepared in advance; for example some projects may not:

- involve all three stages (the decommissioning stage may not apply, for instance) and applying to them a standard checklist would be wasteful;
- require the investigation of some impacts (in fact, most projects don't), hence their investigation would be superfluous.

The problem is how to let the expert system "know" when some of these parts need exploration and when they do not – an expert user of the system would know, but a non-expert user would not – so that the unnecessary parts of the template are omitted. The REVIEW prototype achieves this by being linked to another two expert systems (SCREEN and SCOPE, mentioned in Chapter 6) that investigate the same project before the review module comes into operation. This means that these other modules (especially SCOPE) will have determined already the major characteristics of the project (and its stages) and will have established in considerable detail (see Chapter 6) which impacts require investigation concerning this project in each of its stages. As such, by the time the expert-system "suite" reaches the Review stage, it has already identified all the impacts that should be explored, and "knows" a lot already about what an impact assessment report *should* contain. This makes the task for the expert system easier, as it guides the system to enquire only about areas of impact that are relevant – and during project stages that are relevant – and will automatically score negatively any noticeable absences in these areas.

The "dialogue" part of the expert system can be organised in a sequence of blocks of questions (some repeated, some not) covering all the aspects above (see Figure 11.1), for example:[57]

1 *General project description*, finding out if the right details and documents are provided in the Statement to give an "administrative" description of the project on the one hand (objectives, identity of the authors, identity of the developer, local planning authority, type of development application and relevant legislation) and, on the other, a general description – including maps – of the site and the area around it.

2 *Description of the project stages* in the Statement (construction, operation, decommissioning if applicable), including two or three blocks of questions covering each stage topic like:

- if drawings are presented describing the stage;
- the methods to be used (method of construction, or operation, etc.);
- description of the materials involved;
- details of any additional services needed;
- wastes and residues produced at each stage;
- details of labour requirements, likely origin, mode of transport;
- consideration of visitors likely at each stage, and their mode of transport.

3 *Identifying the impacts* investigated in the Statement corresponding to:

- each *stage* of the project (construction, operation, decommissioning if applicable);
- whether they are considered *direct* or *indirect* impacts;
- and the *types* of impact (socio-economic, air pollution, noise, landscape, etc.).

57 The sequence presented here is used in the REVIEW prototype already mentioned.

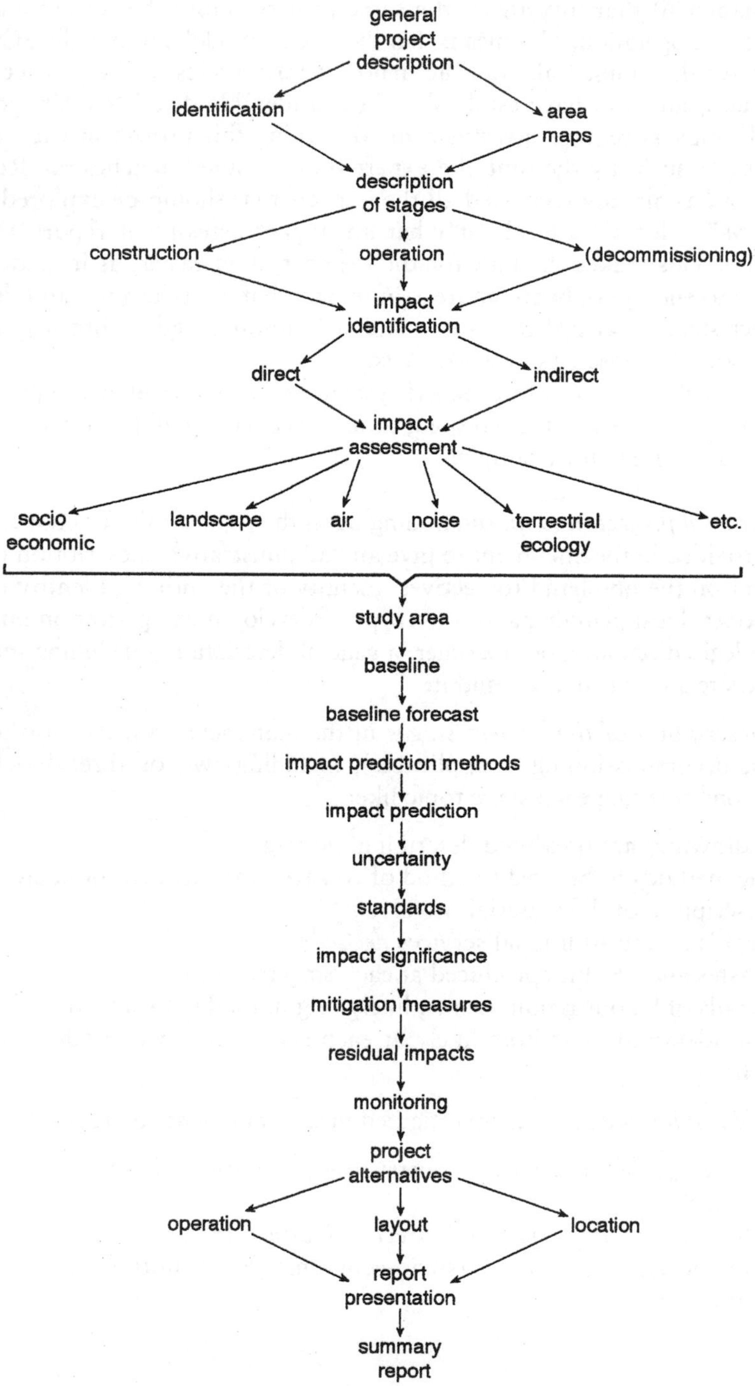

Figure 11.1 The building blocks of Statement review.

This is done by simply asking the user a series of questions (for each project stage and impact type) to acknowledge the presence of the different types of impacts (socio-economic, etc.) and identify on *lists* the specific impacts (housing pressures, income multipliers, etc.) studied in the Statement, for example:

Does the Statement include the forecasting of socio-economic impacts during the construction stage?

(if the answer to the previous question is "yes" then)

What socio-economic impacts does the Statement consider in the construction stage?

– pressures on the housing market
– pressure on social facilities
– cultural/psychological pressures
– employment gains
– none of the above (if this one is chosen, all other choices are excluded).

Note that the list of impacts should not contain *all* possible socio-economic impacts (see Chapter 9), but only those which are present during project construction. This block should include a series of pairs of questions similar to those above, enquiring about direct and indirect impacts for every stage of the project.

4 Detailed *review of each type of impact* present in the Statement, as identified by the previous "block" of questions. This next block will have to be duplicated for every type of impact and, although some variations will exist in the questions asked depending on the specific impact being investigated, it is useful to consider a *typical* series of topics to be investigated:

- the *study area* suitable to each impact;
- study of the *baseline*: information collected relevant to each impact, documentary sources, identification of any consultees used, forecasting of the baseline without the project;
- prediction of the *impact* in question and explanation of the methods used for the prediction, description of the data sources and any gaps identified in them, degree of quantification used for the predictions, form of presentation used, discussion of likelihoods and uncertainties associated with the predictions, discussion of time scales and questions of irreversibility in the predictions;
- reference to existing *standards* relevant to each type of impact;
- discussion/presentation of issues of impact *significance* and detection of any adverse impacts;
- proposition of any *mitigation* measures and justification for choosing them, discussion of the probable effectiveness of the mitigation measures proposed;
- calculation/discussion of any *residual impacts* after mitigation and their significance;

- proposition of any *monitoring* actions;
- general *presentation* of the impact section: is the impact chapter/section overlong, does it have a "summary", is the presentation of the findings appropriate, using a variety of presentational devices (graphics, maps, etc.) and favouring non-technical language.

As will be seen, the aspects covered in this block – with the exception of the last one about presentation – replicate quite closely the points discussed in the previous chapters concerning various impacts. In practice, the questions and possible answers in the expert system corresponding to these blocks would be "taken" from analyses and discussions of specific impacts such as those exemplified in those chapters. For instance, referring to *noise* impacts, this block of questions could look like:

Are there maps of the area surrounding the project?

(if yes)

Do they extend as far as the nearest inhabited buildings or sensitive receptors (hospitals, schools, etc.)?

Does the Statement contain general information about existing noise levels?

– existing noise sources
– information about noise-related complaints from the public
– none of the above (if this one is chosen, all other choices are excluded).

Does the Statement contain information about background noise-levels?

– La90 over 18 hours
– Leq over 24 hours
– La10 over 18 hours
– other (La50, etc.)
– none of the above (if this one is chosen, all other choices are excluded).

(if "none of the above" is not chosen)

Does the Statement indicate what sources have been used for the baseline?

– field surveys
– previous studies or impact statements for the same area
– local authority measurements/forecasts
– none of the above (if this one is chosen, all other choices are excluded).

Does the baseline study of the area include a forecast of noise levels WITHOUT the project?

Does the Statement predict the nature and magnitude of the noise impacts?

– *increases in noise levels*
– *emphasis within 300 m*
– *re-radiated noise*
– *vibration levels*
– *none of the above (if this one is chosen, all other choices are excluded).*

(if "none of the above" is not chosen)

What methods does the Statement use to forecast noise levels resulting from the project?

– *approved noise-simulation models*
– *simulation models developed in-house*
– *borrow noise-level forecasts from similar projects*
– *apply expected vibration levels found in similar projects/situations*
– *comparative study of the likelihood of re-radiated noise*
– *not specified.*

(if "not specified" is not chosen)

How does the Statement present the noise impacts?

– *as quantitative predictions of noise levels*
– *as deviations from the baseline*
– *as yes–no impacts*
– *as qualitative descriptions.*

Does the Statement indicate the availability of accepted standards to assess the significance of noise impacts?

(if yes)

What standards does the Statement use for the assessment of the significance of noise impacts?

– *DoE Advisory Leaflet AL72*
– *British Standard 5228*
– *DoE Circular 10/73*
– *PPG 24*
– *British Standard 4142*
– *Design Manual for Roads and Bridges*
– *etc.*
– *no mention of which standards are used.*

Does the Statement find any significant adverse noise impacts?

– *increased noise levels*
– *re-radiated noise*
– *vibration levels*
– *none of the above (if this one is chosen, all other choices are excluded).*

(if "none of the above" is not chosen)

Does the Statement propose any mitigation measures to deal with the significant noise effects?

- *using quieter plant and equipment*
- *using silencers or noise insulation at source*
- *relocating noisy plant*
- *using enclosures*
- *screening and landscaping*
- *bunding*
- *etc.*
- *none of the above (if this one is chosen, all other choices are excluded).*

(if "none of the above" is not chosen)

Does the Statement mention the reasons for choosing the particular type of mitigation of noise impacts?

Does the Statement explain the extent to which the chosen mitigations of noise impacts will be effective?

Does the Statement make clear any uncertainty about the effectiveness of the chosen mitigations of noise impacts?

Does the Statement consider any adverse environmental effects of the chosen mitigations of noise impacts?

(if yes)

Does the Statement consider the potential for conflict between the positive and adverse effects of the chosen mitigations of noise impacts?

Does the Statement indicate the significance of any residual adverse noise effects after mitigation?

Does the Statement propose any monitoring of noise impacts?

Is the Chapter/Section on noise impacts short (5 pages or less)?

Does the Chapter on noise impacts have a Summary?

Does the Statement present its information and conclusions on noise impacts using a variety of presentational devices (in addition to written text)?

- *tables*
- *maps*
- *diagrams and drawings*
- *none of the above (if this one is chosen, all other choices are excluded).*

Is the information on noise impacts presented so as to be comprehensible to the non-specialist, avoiding technical and obscure language?

– *yes*
– *only in parts*
– *no, it is quite difficult to follow.*

Different impacts require different lists of the options offered concerning many of these aspects – the baseline data would be different, the methods would vary, so would standards and mitigation measures and many more – but the overall format can be kept fairly uniform to simplify the design of the system. The type of "prior knowledge" required does not mean that unless you have designed expert systems to deal with all the impacts you will not be able to design a Review expert system, but the logic followed in one type of system is reflected closely in the other. This is another reason for closing our discussion of "partial" expert systems with one type of system designed to review the product of all the others.

5 Going back to aspects concerning the report as a whole, there is a need to consider *alternatives* to the project, as proposed, and the range of variations:

- if the Statement considers alternatives at all (including the "no-action" alternative);
- if alternatives are considered, the extent of variation they represent (different operating procedures, different layouts, different design and appearance, different locations, etc.);
- the impacts of the alternatives are estimated and compared to those of the project as it stands;
- the Statement discusses the relative advantages/disadvantages of all the alternatives and the reasons for the final choice.

6 Finally, there is one last block dealing with the overall *structure and presentation* of the report: first, whether it contains elements such as a Table of Contents, an Introduction, Technical Appendices and – most importantly – a "Non-Technical" Summary and, second, what are the contents and style of this *Non-Technical Summary*:

- a brief description of the project;
- a brief description of the environment;
- explanation of the overall approach to the assessment;
- main results concerning all the impacts considered;
- the main mitigation measures proposed (if any);
- description of any remaining impacts (if any) after mitigation;
- a mention of the degree of confidence which can be placed on the results.

The previous discussion covers what can be called the "dialogue" part of the expert system, the aspects of the Statement that the reviewer is interrogated about when running the system and some ideas about how to construct such interrogation. The dialogue is all that the user sees of the expert system, but the point of such "advice" or "diagnostic" expert systems is to elaborate the inputs from the dialogue and produce some results, in this case some idea about the worth of the Statement being reviewed. The discussion now turns to those levels of a Statement Review expert system that the user does not get to see and which make up what could be termed the *evaluation* part of the expert system.

11.3 EVALUATION

In the simplest form of pure-inference expert system – like those used in Chapter 2 to introduce this technology – the questions in the dialogue are the "leaf nodes" in the inference tree (the logic tree of the "rules" that make up the knowledge base). The results from using the information obtained in the dialogue are the "conclusions" of certain rules, and they are produced simply by inference from the chain of rules triggered off by the dialogue. More commonly, however, expert systems must do *more* that just derive conclusions from conditions, including:

- Running particular procedures when some extra information is needed (like running a model with dialogue data, or running a GIS to get some geographical information), examples of which have been seen throughout the discussion of specific impact areas in previous chapters.
- Producing some form of *evaluation* of the situation depicted through the dialogue, usually in quantitative form (qualitative evaluations are no different from the inference conclusions mentioned above), and this is what the Statement Review process requires as a conclusion.

Such evaluation of the overall quality of the Statement – based on the responses by the reviewer to many questions about individual aspects of it – is a typical example of *multi-criteria evaluation* (Voogd, 1983), a logic well known in fields such as land use planning, where decisions have to be reached about the suitability of different options (possible locations for instance) for various requirements (different land uses for example). Although the complete logic of the evaluation process involves several steps as listed in Section 11.2 there are two basic *types of tasks* involved which present completely different problems to the system designer:

- evaluating the individual aspects being looked at;
- combining those partial evaluations into an overall assessment.

11.3.1 Scoring individual aspects

When the review is undertaken directly by an expert (as with the ES Review Package) the evaluation takes place *in the expert's mind* and he/she only registers the result, using a pre-arranged code (DoE, 1996):

A – the work has generally been well performed with no important omissions;
B – generally satisfactory and complete with only minor omissions and inadequacies;
C – just satisfactory despite some omissions or inadequacies;
D – parts are well attempted but on the whole it is just unsatisfactory because of omissions or inadequacies;
E – not satisfactory, revealing significant omissions or inadequacies;
F – very unsatisfactory with important tasks poorly done or not attempted.

In an expert system to be used by a non-expert, the evaluation has to be done in the computer *automatically*, and achieving the richness of the scale shown above can be quite cumbersome. It can be deemed sufficient to have a much coarser evaluation of each aspect, as long as it satisfies the overall objective of the exercise. Such evaluation with an expert system means translating the answer to every question related to a particular aspect into a *score* for that aspect, and the way it can be done will vary depending on the different forms that questions take. With direct *yes–no* questions, scoring is easy, as the score can only take one of the two values: the maximum and the minimum, whatever they may be (such as 1 and 0). For example, the question "*Does the Statement contain maps/diagrams describing the project?*" will generate an obviously good "yes" answer and a bad "no" answer. With *menu questions* that offer a range of possibilities, the conceptual difficulty of this translation becomes apparent: unless the answers to the question are quantitative, only a *ranking* scale can be matched to the different answers, with no indication of the interval between successive answers. For example, if we have the following menu question:

> *Does the Statement describe the existing landuses on the project's site(s) and its surrounding areas?*
> *– no, it does not describe present landuses*
> *– it describes the existing landuses in the site area*
> *– it also describes landuses in the surrounding area.*

We can attach a "ranked" score to each of the three answers (worst–better–best) even if we are not able to attach any specific value to the intermediate options. When the answers to the menu questions are simply *lists* of approximately equivalent aspects, the problem is even worse, as we can only treat such answers as *categories* without even being able to establish a ranking between them. For example, in a factual question like the following:

> *Does the Statement indicate what sources have been used for the baseline study of existing socio-economic conditions?*
>
> *– the Census of Population as published*
> *– the Census Small Area Statistics (or the Local Base Statistics)*

– *the NOMIS database (National Online Manpower Information System)*
– *other Environmental Statements or research reports for the same area*
– *none of the above (if this one is chosen, all other choices are excluded).*

Apart from the last option (obviously bad) all other answers would have approximately equivalent value, and the best we can do is *count* them, unless we can attach specific score values to each based on additional information.

These difficulties highlight the problems of trying to derive a quantifiable evaluation from a set of qualitative answers to questions. One option is *not* to try to "quantify the unquantifiable", which can mean only considering best/worst cases and putting all others into an "intermediate" category. This is what a relatively elementary system like the Oxford Brookes REVIEW prototype does:

- In *yes–no* questions, one answer is taken to be "best" and the other "worst".
- In best–worst menu questions, one extreme answer is taken as "best" and the opposite as "worst".
- In list-questions, one extreme answer (usually the "none of the above") is taken as "worst" and any others as intermediate.

The scales of scores obtained have a "span" of two or three values (*bad–intermediate–good*) and do not have the richness of the A–F scale shown at the beginning of this section, but they suffice for the exercise. Having decided – by whatever means – how to score the answers to the individual questions in the review expert system, the problem remains how to derive from those an overall evaluation of the Statement being reviewed.

11.3.2 Overall evaluation

If individual factors are scored on a quantitative basis, combining them into an overall index involves well-known steps: (i) transforming all factors into a common scale (referred to as "standardising"); (ii) working out the relative importance of every factor (their "weight"); (iii) multiplying each standardised score by its weight; (iv) combining arithmetically (usually adding together) the weighted scores involved in the overall index. But when the scores are themselves not quantitative, it cannot be done so systematically. The Statement Review Package simply expects the assessor to add an overall assessment of each "section" of the review – using the same A–F scale mentioned at the beginning of Section 11.3.1 – and another overall assessment for the whole report at the end. If an expert system is used, the difficulty is compounded by having to use a procedure that can be applied automatically by the computer. The REVIEW prototype simplifies the scales used for "scoring" all the aspects of the review. Similarly, an overall evaluation of the Statement and its different sections follows a rather simplified approach.

First, review aspects are classified into two groups: those that are *essential* for the quality of the Statement (like the presence of the relevant impact predictions, or the use of maps/diagrams) and those which are not. Some aspects can be scored as "bad", "intermediate", etc. Combining these aspect scores with their classification, the system determines that:

- There is something drastically wrong in one of the review sections (including presentation) when an "essential" aspect is missing (there is a *gap*) or is scored as "bad" (of low *quality*).
- There are problems of "intermediate" level if all the problems found concerning any of the aspects are no worse than "intermediate".

Second, a final three-point scale for the overall assessment is devised:

- The Statement is good in every respect.
- It has major problems that put it below the Government's relevant standards of acceptability and should be resubmitted.
- Although of a low overall quality, it meets the minimum standards of acceptability, but it has gaps and problems which could be corrected now to avoid future problems.

The Statement will be in the first category if *all* its sections are complete and no essential aspects are missing. It will be in the second category and will be unacceptable if *any one* of its sections is drastically wrong in terms of gaps in it *and* in terms of quality. The Statement will be judged to be in the third category (barely acceptable but "improvable") if the quality of all of its sections is satisfactory but there are important gaps. At the end of the review, a *feedback message* indicating the overall assessment is produced for the user. In addition, if the diagnosis is not perfect (meaning that there are still gaps and problems), a *list* will be added at the end of all the problems and gaps that should be put right, and this is detailed further in an Appendix. Seen below is a hypothetical example of such final feedback:

Environmental Statement Review Report

SUMMARY/RECOMMENDATIONS

The Statement seems to have major problems that put it below what are considered by the Government to be minimum acceptable standards of quality (see the list below):

– some areas of impact are not discussed at all
– the assessment of impact prediction is unsatisfactory.

The Statement should be re-submitted when all these problems have been resolved and it meets the required standards.

Also, there are smaller gaps and problems in the Statement, and the need to resubmit it should be seen as an opportunity to put these right (see the Appendix for further details):

- *general organisation of the Statement*
- *gaps in the baseline for some of the impacts*
- *the consideration/discussion of alternatives*
- *problems with impact mitigation/monitoring.*

APPENDIX: DETAILED COMMENTS

Minor gaps and problems encountered in the Statement:

GENERAL ORGANISATION OF THE STATEMENT
- *there is no Table of Contents*
- *etc.*

DIRECT IMPACTS
- *the Statement does not study pressures on the housing market during the construction stage*
- *etc.*

LANDSCAPE IMPACTS IN THE STATEMENT
- *the area described around the project for the baseline study does not include all the areas of likely visibility of the project*
- *etc.*

CONSIDERATION OF ALTERNATIVES
- *the developer did not consider the "no-action" alternative*
- *etc.*

For this feedback report it is common practice in expert systems to use *canned text*, pre-prepared strings of text representing each individual message:

- whole paragraphs conveying the overall message;
- lists of individual strings of text, each string linked to a particular deficiency of the Statement;
- link-phrases/paragraphs in between, linking the different parts to make the whole readable.

The first part of the feedback report – with the overall impression and the major reasons for it – is designed for the development controllers, to help them make a decision about the Statement. The second part is really designed as feedback to the *authors* of the report, with all the details of what they should modify and improve.

11.4 CONCLUSIONS: EXPERT SYSTEMS FOR HIGHLY STRUCTURED QUANTITY–QUALITY CONVERSIONS

At the end of our journey through various areas of impact assessment, we find ourselves again discussing the application of the expert-systems approach to *highly structured problems* – problems "dissected" and discussed in the literature in considerable detail – raising issues similar to those encountered at the beginning of our journey, when discussing project screening in Chapter 6. Such well-defined problems present in both cases very simple logics to the expert-system designer, mainly based on sequences of questions using *checklists* to establish the compliance of the case being studied with certain criteria:

- In the case of project screening, the aim of the "checks" is to *classify* the project in its different parts to determine which category it belongs to among those determined by the regulations as requiring (or not) impact assessment.
- In the case of statement review, the aim is to *qualify* the different aspects of the project to establish their level of quality, so that their compliance with accepted standards can be established and an overall evaluation of the document can be reached.

Besides this relatively "trivial" sequential questioning of the user, these two types of systems are left with problems that are beyond expert-systems technology as such, in particular, the problem of *qualitative–quantitative conversion*, working in opposite directions:

- In the case of project screening, the determination of *significance* for cases in the "Schedule II grey-area" (see Chapter 6) – a qualitative determination – using information (mostly quantitative) about the project to determine the likelihood of "significant" effects.
- In the case of statement review, the conversion is in the other direction: the transformation of different "qualities" into quantitative values, in order to combine them into higher-order quality-indices (which may in turn have to be recombined further) following the logic of multi-criteria evaluation.

Expert systems are ideally suited to deal with the first set of problems but, when it comes to the second set, all expert systems can do is "present" them in a logical and transparent way to the user to extract the right information from him/her, but the conversion as such must be resolved using other methods, maybe statistical, maybe based on "fuzzy" logic, as already mentioned in Chapter 8 when discussing landscape assessment.

REFERENCES

Ashby, W.R. (1956) *An Introduction to Cybernetics*, Chapman & Hall, London (Ch. 11).

CEC (1993) *Checklist for the Review of Environmental Information Submitted under EIA Procedures*, DG XI, Commission of the European Communities, Brussels.

DoE (1994) *Evaluation of Environmental Information for Planing Projects: A Good Practice Guide* (prepared by Land Use Consultants), Department of the Environment, HMSO, London.

DoE (1995) *Preparation of Environmental Statements for Planning Projects that Require Environmental Assessment: A Good Practice Guide*, Department of the Environment, HMSO, London.

DoE (1996) *Changes in the Quality of Environmental Statements for Planning Projects: Research Report*, Department of the Environment, HMSO, London.

Glasson, J., Therivel, R., Weston, J., Wilson, E. and Frost, R. (1997) EIA – Learning from Experience: Changes in the quality of Environmental Impact Statements for UK Planning Projects, *Journal of Environmental Planning and Management*, Vol. 40, No. 4, pp. 451–64.

Glasson, J., Therivel, R. and Chadwick, A. (1999) *Introduction to Environmental Impact Assessment*, UCL Press, London (2nd edition, 1st edition in 1994), Appendix 4.

Lee, N. and Colley, R. (1990) *Reviewing the Quality of Environmental Statements*, Occasional Paper No. 24, EIA Centre, University of Manchester.

Voogd, H. (1983) *Multi-criteria Evaluation Methods*, Pion, London.

Weston, J. (2000) *Screening, Scoping and ES Review Under the 1999 EIA Regulations*, Working Paper No. 184, School of Planning, Oxford Brookes University.

12 Conclusions

The limits of GIS and expert systems for impact assessment

12.1 EXPERT SYSTEMS FOR IMPACT ASSESSMENT

We have discussed in some detail a wide range of types of impacts, reducing them to relatively simple logical processes with a potential for automation as expert systems. Although not all the standard areas of impact assessment have been covered, there has been enough variety to illustrate most of the problems and issues involved when "translating" expert behaviour and judgement into a simple logical process that a non-expert can follow. The logic followed in the discussion so far can be summed up in Figure 12.1, showing the structure of what could be some kind of "super expert system" to deal with the whole process of impact assessment. After the initial stages focussed on the need for impact assessment and the areas of impact to be studied (discussed in Chapter 6), the logic breaks out into many different lines of enquiry for the different areas of impact, as discussed in subsequent chapters. Finally, all the "threads" are joined again to arrive at some form of overall assessment, and the whole discussion is presented in a report containing the main points of all the areas discussed before (covered in Chapter 11), and the report itself is also the subject of scrutiny as part of the control process (Figure 12.1).

The first two stages (Screening and Scoping) can be programmed into reasonably straightforward expert systems, examples of which were discussed in Chapter 6. Although either of these two systems can be self-contained, they overlap considerably in terms of the information they require (details about the project), and the most efficient arrangement is to have both systems *linked* into one, so that the information used to screen the project can then contribute to help with the scoping. Beyond these initial stages, when it comes to the impact assessment as such, there is a basic choice of strategy, to design an expert system:

- for each type of impact (each *column* in the matrix in Figure 12.1) to deal with the different stages of the assessment itself (defining the study area, studying the baseline, etc.); or

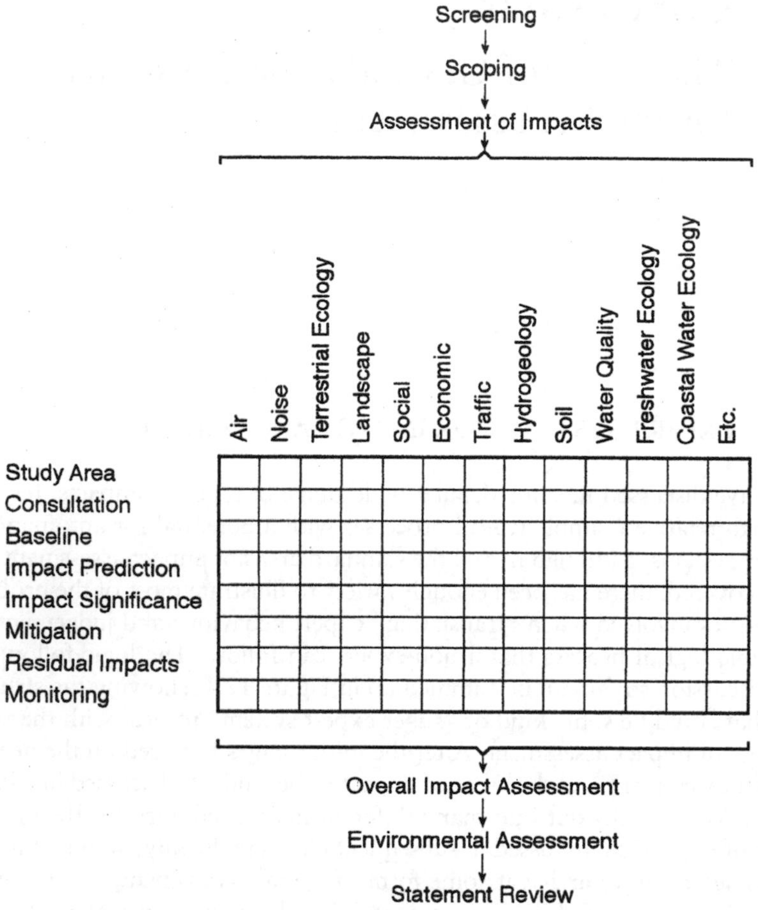

Figure 12.1 The overall impact assessment process.

- for each "stage" of impact assessment (each *row* in the matrix in Figure 12.1) including the different variations, to deal with the different types of impact.

The discussion in the previous chapters (by impact types) has implicitly adopted the first approach, but the possibility of adopting the "row" approach – programming each stage of the impact assessment for all types of impacts – should be considered, if only for completeness, looking at what the different approaches have in common that could be handled by the same type of system.

Starting with the *definition of the study area*, there is a basic commonality of many types of impacts, using the identification of "sensitive receptors" to define the study area: human receptors in the case of noise, traffic or landscape impacts, animal and vegetal receptors in the case of the various ecological impacts, or even physical receptors with the different water-related or geology impacts. But also, the existence of data for an area can be a major factor, as with impacts that rely on existing monitoring data, from air pollution to water quality or traffic. Apart from such common aspects, the approach and scale can be quite different, from a few hundred metres for noise to several kilometres for air pollution or for landscape. And the approach can also vary drastically: from the "fixed" area approach of many impacts (noise, landscape, etc.) to the "flexible" area-of-study approach typical of traffic and socio-economic impacts, where the final scale (where to stop extending the study area) will depend on the findings.

The *consultation* stage also has a few commonalities for many impacts, as typical organisations are always expected to be consulted – like Local Authorities, Ordnance Survey, local interest groups and newspapers – for most types of impacts. Beyond these, the diversity of impacts starts to reflect in the bodies expected to be consulted – some of them by statutory obligation – particularly the government agencies and organisations responsible for the resources being affected by the impacts (such as the different sections of the Environment Agency, the Countryside Agency, English Heritage, etc.). And finally, many different bodies are to be contacted as the holders of important information needed for the study, like the various Institutes (for Ecology, for Landscape, etc.).

The diversity of approaches found in the *baseline study* is even greater, as the study is directly linked to the type of information needed for each impact, and the commonality between impact types virtually disappears. Only impact types which share common methodologies also share similar approaches to the baseline study, such as all the ecology impacts (sharing similar "Phase 1 – Phase 2 – evaluation" approaches) or all water-related impacts, where they are collecting the same type of data (habitats and species for ecology, bio-chemical composition for water). Beyond these, the baseline studies are quite specific to each type of impact in terms of the data collected and even in the overall approach, some requiring field visits and/or data collection and some not. Even the relative "weight" that the baseline carries as part of the impact study can vary: while in impacts like noise or air pollution the baseline study provides just the starting point for the impact predictions, in ecological studies the baseline *is* virtually the impact assessment itself, as it is the quality of that baseline that determines the magnitude of the impact.

Moving on, the discussion in the previous chapters has illustrated the extent to which the logic and the mechanics of *impact prediction* are specific to each impact-type, maybe with the exception of the various ecological impacts. Some parallelisms may be drawn between some areas of impact – maybe between heritage/archaeology and landscape, or between air and river pollution – but such similarities are rare. Impacts are even expressed in totally different units and forms – from decibels to square metres of land, from milligrams per cubic metre to multiplier values. Some impacts are predicted using models (of very different kinds) and some are not, some use subjective judgement and some do not. The list of "dissimilarities" could be endless, and it must be concluded that it would be practically impossible to design a computer framework of the expert systems type that would meet all such requirements.

The assessment of *impact significance* is often undertaken using a common logic, by comparing the predicted impacts to certain standards, even if each standard is specific to particular impacts and comes from different sources (for example, different pieces of UK legislation, or the World Health Organisation). On the other hand, some impacts derive their significance in other ways: from the importance of the receptors affected (such as ecological impacts), from public opinion (such as social impacts), or even from subjective judgement (as with landscape).

When it comes to *mitigation measures*, their degree of diversity varies with the level of mitigation. At the most general level, mitigation can involve project changes (from changes in the design or in the layout to relocation) which affect many impacts in a similar way and can be decided out of a "joint" consideration of impacts that would benefit from integrated programming. At an intermediate level, some mitigation measures can have effects on more than one type of impact (for example, "bunding" can help with noise and also with run-off water) and can be discussed jointly. At the most specific extreme, each impact carries its own set of possible mitigation measures which are specific to that impact alone and cannot be decided and "shared" with any other impact.

Finally, *monitoring* is also quite specific to different impacts, and even its role in the whole process can be quite different. In most cases, monitoring is simply a "check" on the performance of the project. But in some cases it can have in itself a "mitigating" effect, just by being in place, reducing public anxieties concerning some aspects of the project and the dangers it poses to local communities, and increasing developer awareness of obligations.

It can be seen that there would be advantages in automating across the board some impact assessment stages more than others – consultation and mitigation seem to be the best candidates. However, an overall approach based on designing expert systems "by rows" to deal with the central part of the assessment (baseline-impacts significance) seems out of the question, suggesting it is more sensible to keep the "columns" approach followed in the structure of the discussion, at least for that central part of the assessment. The advantage of such an approach is that each impact type considered worth encapsulating in an expert system is programmed separately and all the stages in the process are tailored to that impact – its sources, data and procedures – instead of trying to design expert system structures that are applicable to *all* the possible variations for all the impacts in each stage.

Once we move past the impact assessment as such in the Figure 12.1 above, a more synthetic approach is again possible, as *all* the impact assessments are put together into a report that is submitted to the relevant control authorities for review. This hints at another difference between the stages in the above diagram, the *clientele* of the various expert systems which can be designed also varies:

- Potential Screening and Scoping expert systems would be directed mainly to development controllers – to help them decide on projects – but could also be used by the developer's advisers to "try out" different project options and decide before submitting the final design to the scrutiny of the development controllers.
- On the other hand, expert systems to predict potential impacts (the "matrix" in the diagram above) will be of real use to the technicians undertaking such studies at the developer's request.[58]
- Finally, Review expert systems also share the same clientele with the first group, as they could be used by controllers or by the developer's consultants when deciding how to present the impact report.

12.2 CONCLUSION: THE LIMITS OF EXPERT SYSTEMS AND GIS

The potential "on paper" of new technologies such as expert systems and GIS for a fast-expanding field of professional work like impact assessment seemed quite strong at the start of our discussion. Expert systems could make a significant contribution to the ongoing *diffusion* of the best-practice methods and techniques needed for IA, also adding an element of "political correctness" to this diffusion by the top-down *technology transfer* (within and/or between organisations) implicit in expert systems, seen in this respect as ideal "enabling" tools.

And GIS are built to deal quickly and accurately with geographical information, central to all areas of IA, making them obvious candidates for incorporation into the mechanics of IA. In order to explore these hypotheses – after reviewing the use of these technologies in IA by others in Part I – this text has sought to synthesise the best-practice approaches to a variety of aspects of IA into what could be seen as the rudiments of "paper" expert systems. This was done with the dual purpose of taking the first step towards that synthesis on the one hand, and at the same time trying to establish the true practical feasibility of such an approach.

58 The introduction of Strategic Environmental Assessment (for example, following the 2001 EU Directive in Europe) would pass on much of the burden of producing impact-assessment studies to the planners and government technicians in charge of preparing such documents, and this would make these groups also potential clients for this type of expert system.

12.2.1 GIS and IA in retrospect

With respect to GIS, its suitability for IA has always carried a "question mark" – as cost considerations always dominate in the debates about the practical use of GIS – and the exploration seems to have broadly confirmed those reservations. First, GIS can be used in purely "presentational" roles for *map production*, generating maps showing some of the results of impacts for instance (like contours of air pollution). In such contexts, GIS can be useful by improving the appearance of the results, but the real quality of the results will be dictated by their accuracy (in the case of air, by the accuracy of the model), and GIS is not really crucial. When it comes to *analytical* roles, the list of GIS functions with potential use for IA is relatively short:

- *Map-overlay*, to identify if parts of the project "touch" or overlap with relevant areas: environmentally sensitive areas (when screening a project), ecological or agricultural areas (when assessing impacts).
- *Buffering*, to find out if environmentally sensitive areas or potential "receptors" are within a certain distance from some parts of the project (like roads, or noise sources).
- *Clipping* – a logical extension of buffering or polygon overlay, often used in combination with them – to measure or count the features inside (or outside) buffered or overlaid areas: for example, the number of residential properties entitled to compensation for noise pollution.
- *Measuring areas*, for example the areas overlaid, clipped or buffered using some of the other functions: for example the area of good agricultural land lost by impingement of the project.
- *Measuring distances* – a one-to-one version of buffering – between the project and relevant points or features (like water systems).
- *Visibility analysis* – based on 3D terrain modelling – between specific points or defining visibility areas.
- *Determination of slopes* in a terrain or in underground geological layers – also based on 3D modelling – to identify potential run-off directions.
- *Map algebra* can also be used in IA – although it has not been considered in the discussion of individual impacts – if the impact study is interested in combining *in space* all the impacts, working out some kind of "overall" index of impact to be calculated for every part of the territory. The reason it has not been considered is that it only makes sense if all impacts are expressed *quantitatively*, and the impracticality of that for some of the impacts has been discussed.

Even with respect to these functions, the question must be asked about the precise contribution of GIS to them. In the specific discussions of areas of IA in Part II, GIS was introduced at various points by indicating that certain tasks could be performed automatically with GIS. But it was more a question of pointing out that certain jobs "could be done with GIS", rather than GIS being able to perform tasks too difficult by other means. The comparison between using and not-using GIS was never made, and it is seldom made in the literature, often too busy trying to demonstrate the qualities of GIS. It could be said that the discussion there was almost a question of justifying the *feasibility* of using GIS rather than the *convenience* of using it. For example, identifying potential receptors within a certain distance is a job that can be done visually with little or no training, in fact it can be done almost "at a glance" just by looking at a map with a ruler in your hand, taking virtually no time or resources to do it. The question that should be asked is whether there are some IA tasks that *only* GIS can perform (or that GIS can perform *best*), and the list above should be qualified in this new light. With the exception of specialised tasks like *buffering and 3D-based analysis*, most other tasks on the list can be performed by non-experts without difficulty, and even such specialised tasks would not all be impossible "by hand", but they would present varying degrees of difficulty: (i) at the lower level of difficulty, *buffering* does not present theoretical difficulties if done by hand, but only the practical problem of "sliding" the buffer-distance along complex or lengthy lines; (ii) at an intermediate level of difficulty, *slope analysis* using topographic or geological maps is probably the easiest and, even if GIS can do it more accurately and quickly, a human with relatively little experience (or with very little training) in reading topography maps can also do it visually with sufficient accuracy; (iii) at the top of the difficulty-scale, *visibility analysis* is probably one task for which it can be said that GIS is ideally suited – even if GIS sometimes do not do visibility analysis with the detail that is assumed[59] – as it is a form of calculation that would prove too difficult to do by hand. It can be argued that these tasks where GIS can make irreplaceable contributions occupy a relatively small part in the whole impact assessment, and some (like geology) are quite infrequent.

Considering these different degrees of contribution that GIS can make to IA, the final question which must be asked to reach some kind of assessment of the worth of this technology is about the *costs* of making those GIS contributions to IA. As mentioned in Chapter 3, the costs of the technology itself are quite high – even if they are coming down – but the *data costs* of maintaining the map bases necessary for IA can be prohibitive. Also, for GIS *linked to an expert system*, another type of cost appears, which is the cost in *running time* (see Chapter 6): one of the problems of linking GIS to the expert system is that to run the system you must first pass on to it the necessary information about the structure and contents of the GIS

59 As pointed out by Hankinson (1999) and as any GIS user with experience in visibility areas knows, GIS-generated areas are good enough as starting points for the analysis, but often require adaptation to specific local circumstances (vegetation, etc.)

map base: the maps available, their names and contents, the items of information in each map and their names, etc. This can take some considerable dialogue time, and can be a crucial drawback in a type of system (expert systems) that has precisely as one of its objectives to speed up the problem-solving process for the non-expert user.

The moment the emphasis is shifted *from technical feasibility to cost-effectiveness* of using GIS, the whole assessment of its worth starts changing towards the negative, and this is probably what is behind the trend in the bibliography detected in the discussion in Chapter 3 with the interest in GIS for EIA increasing fast in the early 1990s and then levelling off. Only in a professional environment where GIS costs – especially data costs – can be shared, is it possible to anticipate its use in IA growing, maybe by transferring the responsibility for impact studies to the public sector (as Strategic Impact Assessment would probably do), or maybe by subsidising from the public sector the availability of GIS data in the public domain.

12.2.2 Expert systems and IA in retrospect

Turning now to expert systems, the implications of the discussion so far are less straightforward. That discussion has tried to show how much of IA can be expressed in a relatively simple sequential logic of successive problems to resolve. That sequence can be translated into an interactive computerised system to guide non-experts – maybe as an expert system, maybe as a succession of expert systems – and the discussion has been presented as the *first step* in that translation process, leading to the production of what can be called *paper expert systems*. The discussion has used a form of presentation for these translations that departs from the tree-like structures introduced in Chapter 2 as typical of expert systems. That chapter showed inference trees that start from a top goal (a conclusion) and branch down into pre-conditions which, in turn, are taken as sub-goals and branch down further into more pre-conditions, etc. On the other hand, the "sequences of problems" into which we translated the different parts of each impact assessment were presented as flow charts in virtually the opposite order: starting from the data collection, reaching partial goals (definition of the study area, baseline), and building up into more ambitious results (impact prediction, significance, etc.) to reach the final "goal" of determining the impacts remaining after mitigation. These two approaches can look superficially as opposites, but they are mutually equivalent and the conversion from one to the other is quite straightforward. For example, from all the discussions of different types of impacts there is an emerging overall approach that can be simplified into a flow chart like the one in Figure 12.2.

This diagram expresses the process visually in the same order in which it progresses in reality (from data collection to calculations and conclusions) but the corresponding (virtually symmetrical) inference-tree can be easily constructed showing the process in a reversed order (Figure 12.3), not as it

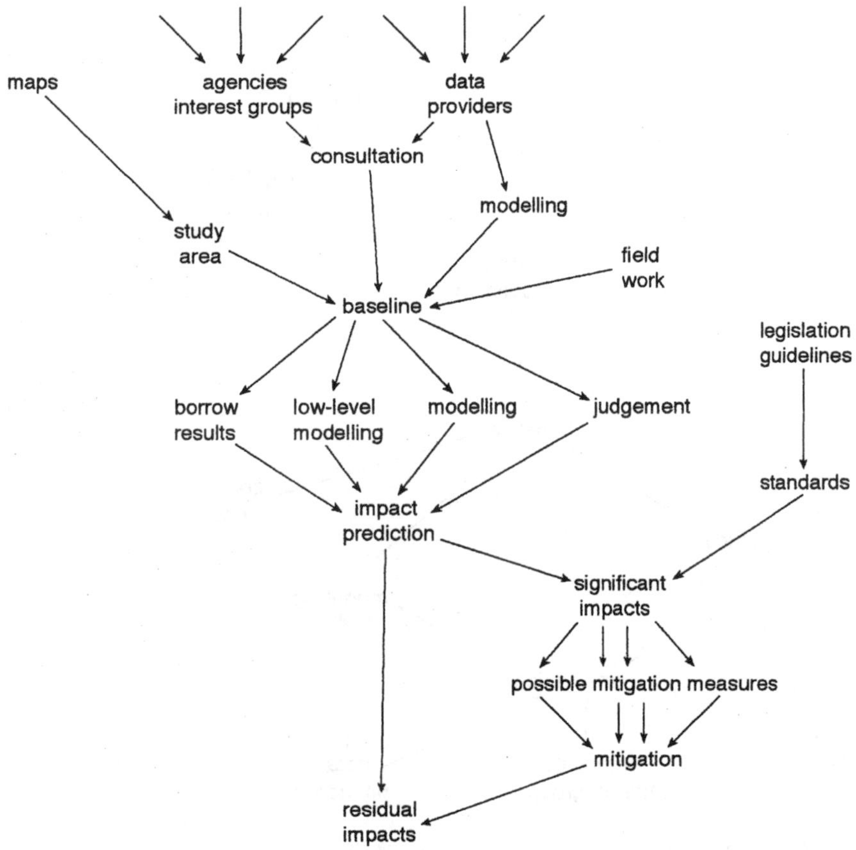

Figure 12.2 The overall progress of an impact assessment study.

progresses in reality, but how its logic is constructed, deriving the particular from the general.

In terms of representation, there is a direct correspondence between the sequential diagrams used and possible inference trees we might want to construct. In terms of *content*, however, the flow charts used contain more than logical steps in a deduction process, and in this respect there began to appear more important differences from the simple inference trees introduced in Chapter 2 and often used to exemplify the very essence of expert systems. The "shape" of such trees is determined by the logical steps in a deductive process used for problem-solving, and the search for information (the "dialogue" in an interactive system) is determined by that shape. One important implication of this is that only the *minimum* information necessary is used and, as soon as enough has been obtained to complete the inference, the dialogue stops and the conclusion is reached. Elementary

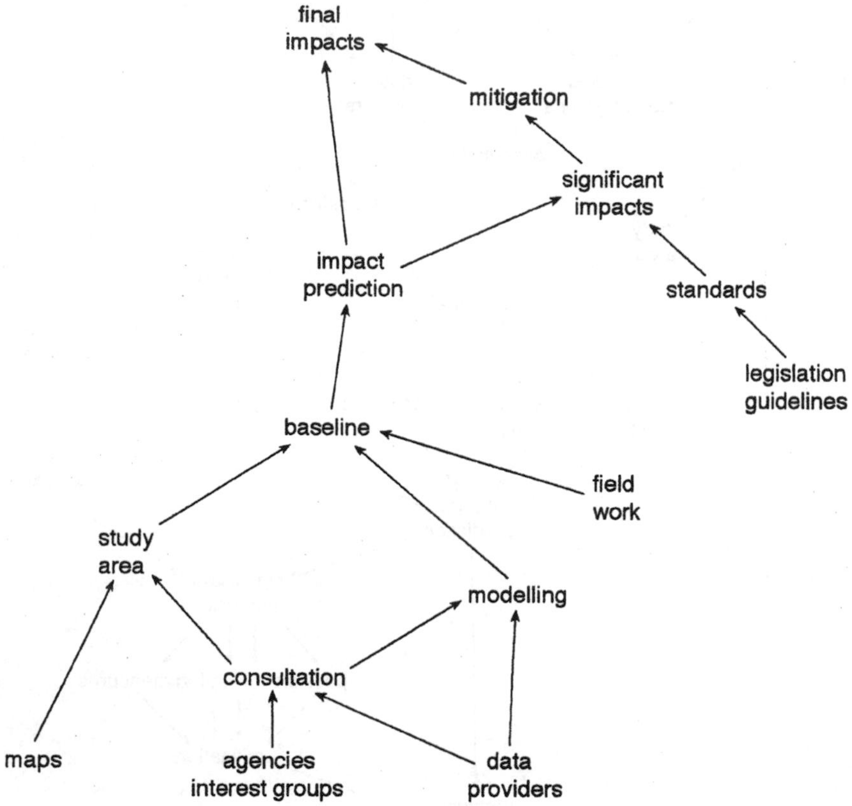

Figure 12.3 The backward-chaining logic of an impact assessment study.

"diagnostic" systems following this *minimalist* approach can be appropriate for the simplest problems, like deciding *should I take my umbrella when I go out this afternoon?*. However, when dealing with real problems like those discussed in this book, we find that very soon their complexity exceeds the capacity of such an approach. Stopping the investigation as soon as an answer to the main question has been reached may not even be appropriate for relatively simple but realistic diagnostic cases. Taking the *screening* question for example, to know if a project will require an Environmental Statement, any one of the possible answers to that question is going to require a rather exhaustive exploration of the project:

- To determine that the project does *not* require a Statement, *all* its aspects will have to be investigated and cleared, and the satisfactory conclusion will only be reached after checking that the project does not fail any of the criteria;

- and the opposite conclusion (that the project *does* require a Statement) also requires an exhaustive investigation, as finding only *one* – the first – reason for failure is not sufficient. There may be multiple reasons for the project to fail, and giving only one could give the erroneous impression that correcting it would be enough for the project to be acceptable. Even when the project "fails" the screening, *all* the reasons for failure must be detected to help the developer re-submit a new version of the project; even if the decision to fail only required *one* such reason.[60]

Such need for an exhaustive search suggests that the highly focussed inference tree is likely to be insufficient, and that other structures common in mainstream computer programming – maybe less elegant – will be needed to complement it. Typical examples of these can be:

- *Checklist* structures to guide series of enquiries. For example, to review an Environmental Statement a series of aspects must *all* be covered: description of the project, description of the environment, scoping, consultation, etc.
- *Classification* structures to "categorise" the case being examined so that the enquiry can follow the right direction. For example, when screening a project, all its elements (roads, infrastructure, buildings, incinerators, etc.) must be identified so that each can be investigated in turn.
- *Evaluation* structures where different elements are given relative weights in order to achieve some form of collective assessment of groups of elements (as in the Review of impact reports).
- *Cyclical* structures, repetitions of sequences of operations changing some of the variables. For example, widening the area of traffic impact prediction after evaluating the significance at a lower scale and repeating the whole process all over again until significant impacts are no longer present.

In practice, all these structures are usually needed *in combination*, and it is relatively common to find the need to put them in standard sequences, for example:

1 A project whose Statement is being reviewed is identified and "classified" according to its type.
2 For the type identified, a "checklist" of aspects to investigate is followed exhaustively.

60 This comment can easily be generalised to other diagnostic expert systems: for example, one cannot imagine a medical expert system stopping as soon as *one* problem is diagnosed without having explored all possibilities of other illnesses being also present.

3 Each aspect on that checklist is diagnosed using a standard "pure-inference" tree.
4 Weights are attached to each of the diagnoses of all the aspects and an overall "evaluation" is reached of the project as a whole.

The effects that these structures achieve can be replicated using the syntax of pure-inference trees, but it can complicate the programming to such an extent that it can be more productive to use more traditional programming syntaxes for the overall framework within which the different elements are combined. Inference structures can be part of such combinations, but not necessarily the part that controls the overall performance of the system.[61]

Another typical structure that has been encountered that does not conform to the inference logic is *modelling* in one form or another, sometimes using off-the-shelf models, sometimes using homemade ones (with spreadsheets to do demographic analysis for instance), or just using some form of simple calculation like the income multiplier. Even expert-systems "shells" have had to accommodate the possibility of attaching models, routines and "procedures" of any kind at any point in the inference, when the logical process is *suspended* while certain calculations or procedures are applied. Modelling can be one high-level example of such procedures; extracting information from a GIS can be a low-level example. As seen in previous chapters, modelling is not always used but, when it is, it can "shape" the structure of the whole approach (as with air pollution or noise) so that the main objective becomes selecting the right model and finding the data for it. But even in such cases modelling is only one of several possibilities, and the modelling option could be seen embedded in some logical mini-structure like that in Figure 12.4.

Finally, a major problem encountered in some cases vis-à-vis the possible automation of the process has been simply the virtual impossibility of computerising certain operations that need to be performed, highlighting the fact that sometimes *experts are irreplaceable*. This appeared to be for several reasons:

- *Theoretical*: the theoretical complexity of the problem in hand – as in the case of ecology or geology – that makes it too difficult to reduce it to a simple-enough set of rules and procedures that are universally accepted and can be automated.
- *Perceptual*: the necessity to observe "first hand" certain phenomena during fieldwork (as in ecology) for which the expert is irreplaceable.
- *Judgmental*: some problems (like landscape assessment) need to be addressed involving subjective judgement (by experts and also by others),

61 One of the implications of this is that so-called expert-system "shells" are very rarely suitable for complex problems like those discussed here, as they tend to be organised around a central logic of standard inference and, although other functions can be attached to them, the central control mechanism is usually an inference tree.

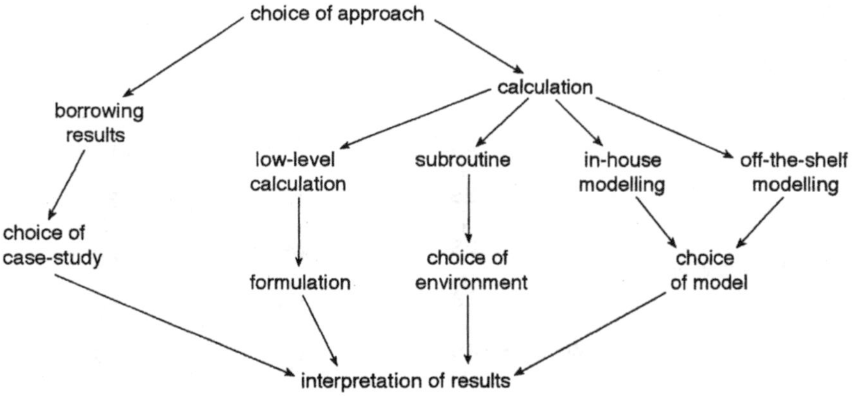

Figure 12.4 Modelling and its alternatives.

even if in the case of experts it can be based on professional experience – "novices" were considered by some of the experts consulted to be lacking in judgement – which brings this issue also back to the first point above.

In practice, this means that *gaps* appear in the structure and any computerised process used needs to be stopped for these tasks to be performed by humans, and then their results are fed back into the automated process, which can then proceed. In a way it represents an interruption similar to that of modelling, but in the case of modelling the "diversion" can still be automated and "seamless", while in the case of these difficulties it is probably better not to try to automatise them, as it could lead to "black-box" approaches of questionable credibility.

12.2.3 Conclusions

It can therefore be concluded that expert systems have a definite potential for problem solving in IA, but we must once and for all "divorce" the idea of expert systems from specific forms of computer programming like the syntax of inference trees, which they have traditionally been associated with. Expert systems should be just seen as *interactive*[62] *computer systems that encapsulate the problem-solving procedures of experts for the use of non-experts*, without identifying them with any particular form of logic or computer structures, leaving them open to any type of approach for their implementation. The discussion clearly points out in the direction of a flexible framework within which *chains* of expert systems can be used to

62 Interaction with a human user in the case of diagnostic systems like those discussed here, or with sensors and control mechanisms in the case of real-time control systems.

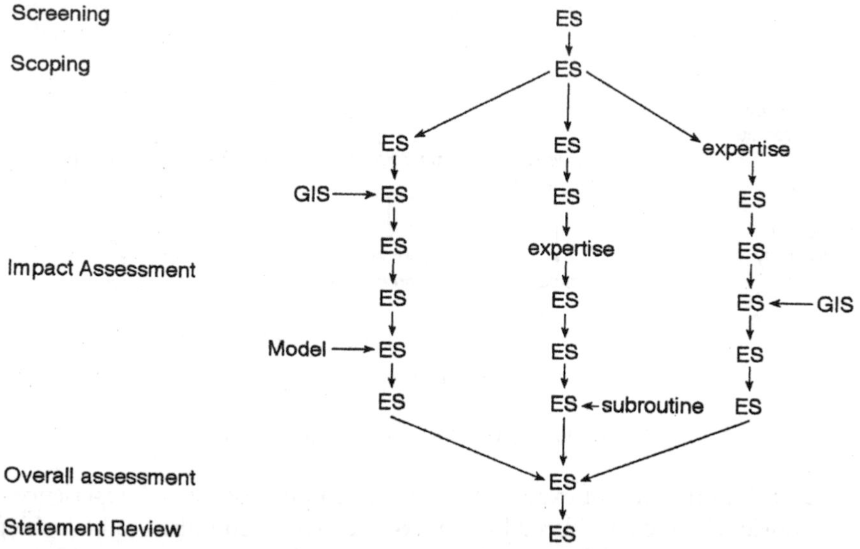

Figure 12.5 Chains of expert systems, models, GIS and expertise inputs.

"think through" IA problems (Figure 12.5): some in combination with models and fully automated (like maybe noise or air pollution), some with GIS routines or other procedures (like landscape), some even leaving gaps for some stages where purely human input – expertise – is required (like ecology).

As we saw in Chapter 5, such situations have been in the past the fertile ground on which decision support systems (DSS) have flourished. But DSS were originally designed to *support experts* with complex management decisions involving forecasting, evaluation, optimisation, etc., using a range of techniques and data sources to "try out" different approaches and identify the most robust results – which these systems could "learn" and remember. Such systems are not supposed to guide the user but be guided by one – because the user is an expert – and a crucial difference with the type of system envisaged here is that in these networks of expert systems and procedures there is still a need for the user (a non-expert) to be guided by the system – this is the point of the whole approach. Maybe a more appropriate denomination for such systems could be a more modest decision support systems "with lower case" or maybe simply Decision Guidance Systems.

REFERENCE

Hankinson, M. (1999) Landscape and Visual Impact Assessment, in Petts, J. (ed.) *Handbook of Environmental Impact Assessment*, Blackwell Science Ltd, Oxford (Vol. 1, Ch. 16).

Index